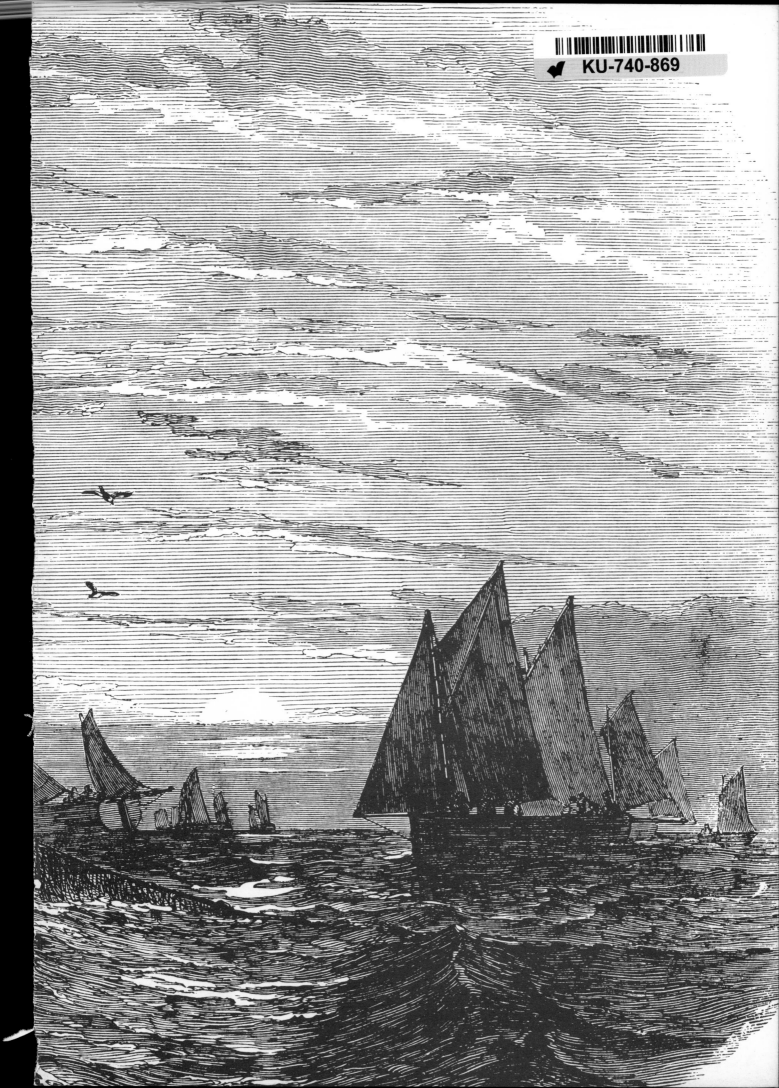

SECRETS OF THE
OCEANS

Above: this diver, at work
outside a lockout craft, faces
many hazards in spite of great
technological advances in deep
diving equipment.

Endpapers: an engraving of
about 1847 showing a fleet of
boats fishing for mackerel off the
south coast of England.

Aldus Books London

SECRETS OF THE OCEANS

Robert Barton

Editorial Coordinator: John Mason
Art Editor: John Fitzmaurice
Designer: Gill Mouqué
Editor: Mitzi Bales
Researcher: Marian Pullen

ISBN 0490 00461X
© 1980 Aldus Books Limited London
First published in the United Kingdom
in 1980 by Aldus Books Limited,
17 Conway Street, London, W1P 6BS.

Printed and bound in Hong Kong
by Leefung-Asco

Title page: an underwater
picture of a shoal of grunts in the
Caribbean Sea.

Introduction

Most people are so much bound to the land as their environment that they seldom realize what a large part of our planet consists of water. In fact, water covers more than two thirds of the globe. This sheer immensity has helped the world's oceans to keep many of their secrets from human curiosity. Nonetheless, people began to explore the seas long before recorded history, and the great spurt of scientific interest and activity that started only 100 years ago has made remarkable progress. This book examines the mysterious seas from many angles. There are chapters explaining what is known about ocean size, waves and currents, tides, and marine life. Other chapters describe the latest advances in fishing, oil and gas exploitation, and power generation from the oceans. Finally there is a look into the future to see how ocean resources could be developed further, and what international problems might arise from such development.

Contents

Chapter 1

Enigma of the Waters

Earth is a water planet, and nearly all that water is in the oceans. All life came from the oceans, and people are dependent on the oceans. They are the most important part of a system that keeps planet Earth in ecological balance. The way the oceans move – their currents, tides, and waves – plays a vital part in regulating the climate. The ocean's resources are huge. There are vast, as-yet untapped resources of protein throughout the world ocean, for example. Seawater itself contains vital minerals, and the seabed is littered with essential minerals. Beneath the seabed are plentiful supplies of oil and gas.

Yet we know so little about the oceans. We can describe their size, physical features, currents, and tides. We have many hints of their potential riches. But we cannot explain the why and how of it all.

The oceans are therefore a mystery, even today. Yet a proper understanding – an unlocking of those mysteries – holds out the greatest promise for the survival of mankind.

Opposite: there is a majestic beauty in an expanse of water lit by the sun. Such scenes are found the world over on a planet that is more than two thirds ocean.

Our Watery Planet

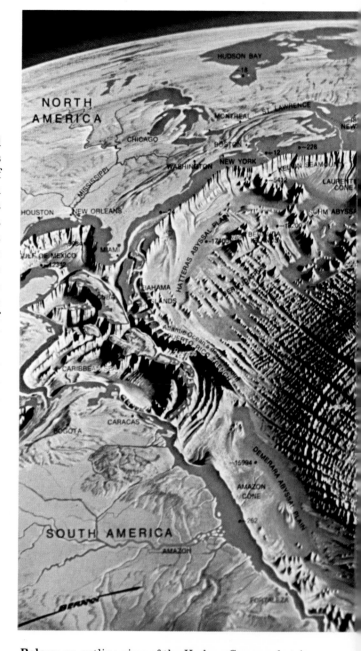

Land takes up less than one third of the planet, and not all of that is habitable by humans. Yet thousands of years of conditioning have made people think of land as predominant. They also think of land and sea as separate: the land ends, the sea begins and that's that. It might be surprising, therefore, to learn that some basic, seemingly irrefutable facts taught about Earth are just not so. For example, try to answer this: "What's the tallest mountain in the world?" If the answer was, "Mount Everest," it was wrong. It is Mauna Kea (White Mountain) on Hawaii, which rises from the sea floor of the Pacific to a total height of 33,476 feet. More than half of this mountain is below sea level.

That is just one example of how people tend to view Earth not as a single planet, but two. However, if the oceans could be drained, people's viewpoint of planet Earth would be dramatically changed. The skimpiness of land as compared to water would be startling, and the physical features that would open up would be spectacular: huge mountain ranges, gaping chasms, vast deserts, all far greater in size and proportion than anything to be found on land. On this drained planet it could quickly be seen that the land often does not end and the sea begin at the shoreline. The beach slopes gently away at an average inclination of one yard in every 500 yards, and this gentle slope forms the continental shelf whose width varies throughout the world. The drained planet would show that the British Isles are in fact a part of the European continent below the mere puddle of water that separates the two; that the Falkland Islands are joined to Argentina by a vast shelf; that off northern Russia the shelf, which extends to 800 miles, is the widest; and that off the west of South America there is little or no shelf, leaving the Andes to slide straight into the Pacific.

At its deepest the continental shelf reaches a depth of 1200 feet. On average, however, it has a depth of around 460 feet. In international law the shelf is defined as the part of the seabed that adjoins a country and extends to a depth of 600 feet.

Geological characteristics mark the continental shelf as being part of the land rather than of the ocean floor. Generally, the shelf is at its widest off the mouths of large rivers and glaciated coasts. This is because it is the depository of sediment washed from the land into the sea. On average the shelf is 42 miles wide. Its seabed and subseabed resources belong to the nation it adjoins. Where there is a continuous shelf between two countries, it is divided down the middle and half allowed to each.

Below: an outline view of the Hudson Canyon showing its contour and depth. This deep gorge that starts near the mouth of the Hudson River in New York is one of the best-known of the world's submarine canyons. Like most of the others, this chasm has short sharp side slopes and a long concave profile. It is about 2 miles wide and about 1200 feet deep, although the true bottom is estimated to reach nearly 11,000 feet where it extends into the Atlantic.

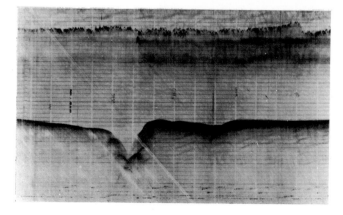

Left: the Atlantic ocean floor, pictured as if drained of its water. The clear S-shaped feature is the Mid-Atlantic Ridge. The areas abutting the coasts of the various continents are part of the continental shelves.

abyssal plain – the biggest single area in the world, covering some 46 percent of the Earth's surface.

These figures for widths, depths, and gradients of the continental shelf are averages, and descriptions of "gentle slopes" and "sharp drops" are general. Sometimes, however, the continental slope does not have a sharp drop but is simply an extension of the continental shelf, even though it carries sediments washed from the land. For years oceanographers

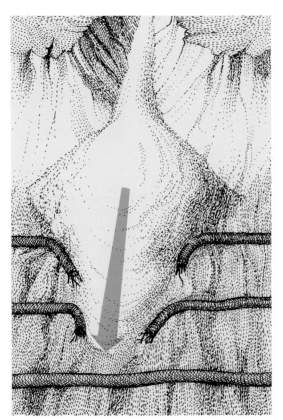

Above: diagram illustrating how turbidity currents snapped a series of transatlantic telegraph cables after an earthquake in 1929. The cables were located on the continental slope south of Newfoundland. Oceanographers worked out that the currents causing this underwater landslide were traveling at about 50 miles an hour.

The continental shelves take up only a small part of the total area of the world: just about five percent of the Earth's surface and seven percent of the underwater region. Nonetheless, because they are relatively accessible and because they support so much marine life, they are vastly important in terms of the resources they yield.

On going down the gentle gradient that is the continental shelf, the angle suddenly sharpens until it becomes the continental edge. Over the edge there is a sharply angled drop called the continental slope. Continental slopes take up about 17 percent of the ocean and mark the true boundary between the ocean floors and the continents: on our drained planet it can now be seen that the land starts at the ocean floor to rise like huge bluffs on a desert plain.

The continental slope begins to even out at from 4700 feet to 6500 feet below the surface, and this milder incline is called the continental rise. The continental rise flattens out at about 12,300 feet in depth into the

were puzzled by one of the features of this type of continental slope. It should, in theory, have been fairly featureless, being built up of smooth sediments as it was. However, echo sounding and photographs often showed it to be split and gouged by huge canyons and gullies like a violent landscape on land. Overseas telegraph operators also used to be puzzled when a sudden break in transmission was later discovered to be the result of a broken cable on the ocean floor. How could a cable that had been laid in deep water on smooth terrain get broken? What mysterious force had caused the damage?

The mystery has not been solved by direct observation, but by logical deduction based on simulated

11

Right: a coral reef atoll in the Pacific, known as the island of Aldabra. Formed by myriads of skeletons of coral polyps, the atoll is the tip of a sunken volcanic island. The slopes of an atoll take a sudden deep plunge to the depths of the ocean, unlike those of continental shelves that have a gradual gradient.

scale tests in laboratories. From these a theory has been developed to explain the unexpected deep canyons and snapped cables. This theory holds that sediment mixes with water at the top of the continental slope to form a dense mixture which, simply by the effects of gravity, flows down the slope. As it flows it picks up more sediment and gets bigger and bigger in the same way that a snowball set rolling from the top of a mountain grows into a mighty avalanche. This sediment mixture is called a turbidity current, and it would be frightening to see one. Oceanographers have calculated that these torrents crash down over great areas at speeds of up to 30 miles an hour, ripping gullies and canyons out of the soft sediments and parting subsea cables as if they were cobwebs. These currents travel downhill for

hundreds of miles before gradually slowing up on the flat abyssal plain where they deposit the material they have carried from the heights. Their activity explains another oceanic riddle: that of the occurrence on the seabed floor of material characteristic of the continental shelf or the land above. It is now thought that such foreign material is largely carried down by turbidity currents, even though the much less dramatic effects of gravity – a slow creeping motion of material down the slopes – may be responsible for some of the unexpected deposits.

Great tracts of the abyssal plain are flat featureless deserts and represent the most tedious part of a journey across our drained planet. There is not much to see except the tracks and burrows of animals in the soft sediments, or the occasional wreck of a ship or

Left: undersea volcanic activity along the northern part of the Mid-Atlantic Ridge threw up the island of Surtsey in 1963. The huge columns of vapor are caused by the flow of hot lava into the cooler sea. Surtsey is off the coast of Iceland, itself a part of the Mid-Atlantic Ridge that has broken through the surface of the ocean.

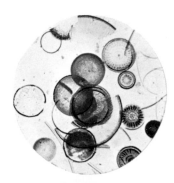

plane, looking as sad and lonely as the skeletons of cattle on a land desert.

The journey across the ocean's abyssal plains would not be easy from the first. Leaving the continental slope travelers would frequently be confronted by steep-walled chasms called trenches – some of them the deepest canyons in the world. Off the Philippines, for example, the Mindanao Trench reaches a depth of 36,000 feet below sea level. This and other trenches plunge to depths of over 30,000 feet in seas whose average depth is only 12,000. As surprising as their depth is their relative nearness to land. Given the overall slope of the ocean floor away from the land, the logical place for them would be in the middle of the oceans. But far from it. Having skirted some of the deep trenches and made their way to the center of the ocean, travelers would come not to trenches

but to mountains – huge towering peaks that form the most striking and certainly the biggest single feature on planet Earth. This is the Mid-Oceanic Ridge. It is the biggest mountain range in the world, 35,000 miles long. It goes through the Atlantic, Indian, Southern, and Pacific Oceans with elevations of up to 10,000 feet from the floor of the abyssal plain. In places it is three miles wide. Through its center runs a giant rift valley. Occasionally it breaks the surface of the sea to form a mid-ocean island.

The Mid-Oceanic Ridge was found to be a continuous feature only earlier this century. Its existence, together with that of deep ocean trenches fringing the continents, started scientists on an exciting detective hunt that ended in the theory of continental drift. This has revolutionized thinking about the formation and continuing development of the Earth itself.

13

The World Ocean

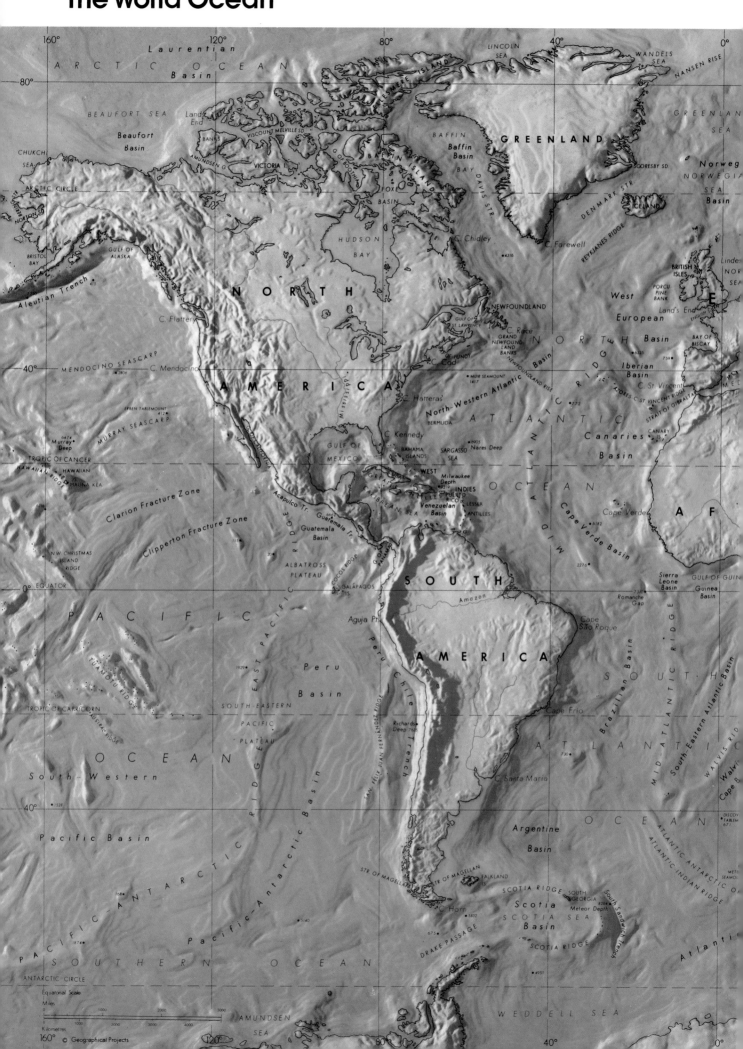

ARCTIC OCEAN
Laurentian Basin

BEAUFORT SEA
Beaufort Basin
Land End
VISCOUNT MELVILLE SD
BANKS
AMUNDSEN G
VICTORIA I
G. OF BOOTHIA
BAFFIN ISLAND
FOXE BASIN

CHUKCH SEA
ARCTIC CIRCLE
NORTON SD
BRISTOL BAY
GULF OF ALASKA

ELLESMERE ISLAND
LINCOLN SEA
WANDELS SEA
NANSEN RISE
GREENLAND
GREENLAND SEA
Norweg
NORWEGIA SEA Basin

BAFFIN Baffin Basin
BAY
DAVIS STR
C. Chidley
•4318
C. Farewell
SCORESBY SD
ICELAND
DENMARK STR
REYKJANES RIDGE
BRITISH ISLES
Linde
NOR
SEA

HUDSON BAY

NORTH
AMERICA

C. Flattery

Aleutian Trench

MENDOCINO SEASCARP
C. Mendocino
•2806

ERBEN TABLEMOUNT 412
MURRAY SEASCARP
6474
Murray Deep
TROPIC OF CANCER
HAWAIIAN RIDGE
75 HAWAII
MAUNA KEA
6188

Clarion Fracture Zone

Clipperton Fracture Zone

N.W. CHRISTMAS ISLAND RIDGE

EQUATOR
0°

PACIFIC

TUAMOTU RIDGE

TROPIC OF CAPRICORN
AUSTRAL RIDGE

South-Western

OCEAN

Pacific Basin
•1529
40°

GULF OF ST. LAWRENCE
NEWFOUNDLAND
C. Race
GRAND NEWFOUNDLAND BANKS
C. Cod
C. Fundy
BERMUDA

MUIR SEAMOUNT 1417
North-Western Atlantic
C. Hatteras

C. Kennedy
GULF OF MEXICO
BAHAMA ISLANDS
GREAT
WEST
Milwaukee Depth
6212
PUERTO RICO
Nares Deep
6905
SARGASSO SEA
INDIES
Venezuelan Basin
LESSER ANTILLES

Acapulco Tr.
Guatemala Tr.
Guatemala Basin
115
20
EAST PACIFIC RIDGE
ALBATROSS PLATEAU
COCOS RIDGE
GALÁPAGOS
15

PANAMA

SOUTH
AMERICA

Amazon

Aguja Pt
Peru
Basin
1929

SOUTH-EASTERN
PACIFIC
PLATEAU

Richards Deep 765
PERU-CHILE TRENCH
SAN FÉLIX JUAN FERNÁNDEZ RIDGE

NEWFOUNDLAND RISE
NORTH
ATLANTIC RIDGE
6125
759
West European Basin
Iberian Basin
C. St. Vincent
AZORES C. ST. VINCENT RIDGE
STRAIT OF GIBRALTAR
772
Canaries
CANARY IS.
Basin
MID
ATLANTIC
OCEAN
RIDGE

PORCU PINE BANK
Land's End
ENGLISH
BAY OF BISCAY
MED

AF

Cape Verde
6182
Cape Verde Basin
2276

Sierra Leone Basin
GULF OF GUIN
Guinea Basin
730
Romanche Gap

Cape São Roque

SOUTH

Brazilian Basin
MID-ATLANTIC RIDGE
South-Eastern Atlantic Basin
WALVIS RID
Walvi
Cape

Cape Frio
730
ATLANTIC
Argentine
Basin

STR OF MAGELLAN
STR OF MAGELLAN
Falkland
C. Santa María

C. Horn
5832
DRAKE PASSAGE
675
5140
1874

PACIFIC—ANTARCTIC RIDGE

SOUTHERN OCEAN

ANTARCTIC CIRCLE

PACIFIC—Antarctic Basin

Scotia Basin
SCOTIA RIDGE
SCOTIA SEA
Scotia
SOUTH GEORGIA
Meteor Depth
SCOTIA RIDGE
SOUTH SANDWICH TRENCH

ATLANTIC ANTARCTIC RIDGE
ATLANTIC INDIAN RIDGE
DISCOV TABLEM 671
6284
METE SEAMOU
Atlanti
4951

Equatorial Scale
Miles
0 1000 2000 3000
Kilometres
0 1000 2000 3000 4000

160° © Geographical Projects 120° 80° 40° 0°
AMUNDSEN SEA WEDDELL SEA

Rich Resources of the Deep

Pick up an attractive shell from the seashore and you have exploited the oceans. Humans have always exploited the sea from the time they first picked up a shellfish from the beach and found it good to eat. But though exploitation of a vast range of food and materials has gone on for thousands upon thousands of years, people have stayed close to shore trapping, trawling, dredging, and drilling for the resources of the continental shelf. Only in recent years has anyone ventured over the continental edge to look for deep ocean resources.

As far as living resources such as fish are concerned, there is good reason for staying on the continental shelf. The great bulk of the world's food fish – some 90 percent of the world catch in fact – is limited to a few species, most of which live on the continental shelf. Deep ocean fish are far more sparse and spread over such immense distances and depths that it would be uneconomical to exploit them. In any case,

most of these species taste foul. Tuna is one of the few exceptions, and consequently it is widely exploited.

The world fish catch comes to about 70 million tons a year. This could be vastly increased because there is an estimated 100 million tons a year that could be caught without damaging the total resources. But people would have to change their taste from traditional species like cod and haddock to more exotic

Above: equipment used in the search for diamonds by the De Beers Mining Company off the coast of Africa. Although mining for this ocean resource was suspended by De Beers in 1971, prospecting still continues.

Left: this relief showing a man fishing is from the palace of Sennacherib, built by that Assyrian ruler in the 7th century BC. Fishing was one of the earliest forms of exploitation of the waters of the world.

fish, and a new technology would have to be found to harvest them efficiently. Certainly there is enough protein food in the oceans to alleviate many of the dietary deficiency problems of the hungry nations of the world. The irony is that there are huge unexploited stocks of fish close to the shores of the very nations that have the least to eat.

There is an irony connected with other resources as well, in that humans equip expensive ships to get back from the seabed material that was washed there from the land in the first place. That is what marine dredging is all about. Mineral deposits are washed down from the land. They are worked on by currents, tides, and gravity, and become concentrated in areas across the continental shelf. Such resources include a wide range of useful raw materials from sand and gravel to tin, gold, and even diamonds. Just a few miles beneath the water's surface lies a self-renewable resource of copper, cobalt, manganese, and nickel, all in handy-sized lumps. They are simply waiting to be picked up. Yet, although space vehicles today can record data from distant planets, and space explorers have gathered rock samples from the moon, only in the past few years have methods been found to pick up a vast resource only two miles under the sea. At last, methods used to recover continental shelf deposits are now being extended to the great depths of the abyssal plain, where large areas of this marine desert are covered with potato-sized lumps of material that are crammed with the minerals of which the world finds itself in increasingly short supply.

Fortunately, there are marine resources right on our doorsteps, contained in seawater itself. In every ton of seawater there may exist all the elements we know on Earth. So far this is only supposition because most of the elements occur in such minute quantities that incredibly sensitive instruments will have to be invented merely to detect them. However, salt and bromine are already being extracted from seawater successfully, as is the biggest single resource of all – fresh water.

The oceans represent the last great deposit of resources for humankind. Careful exploitation of its wealth could solve most material needs of the present and future.

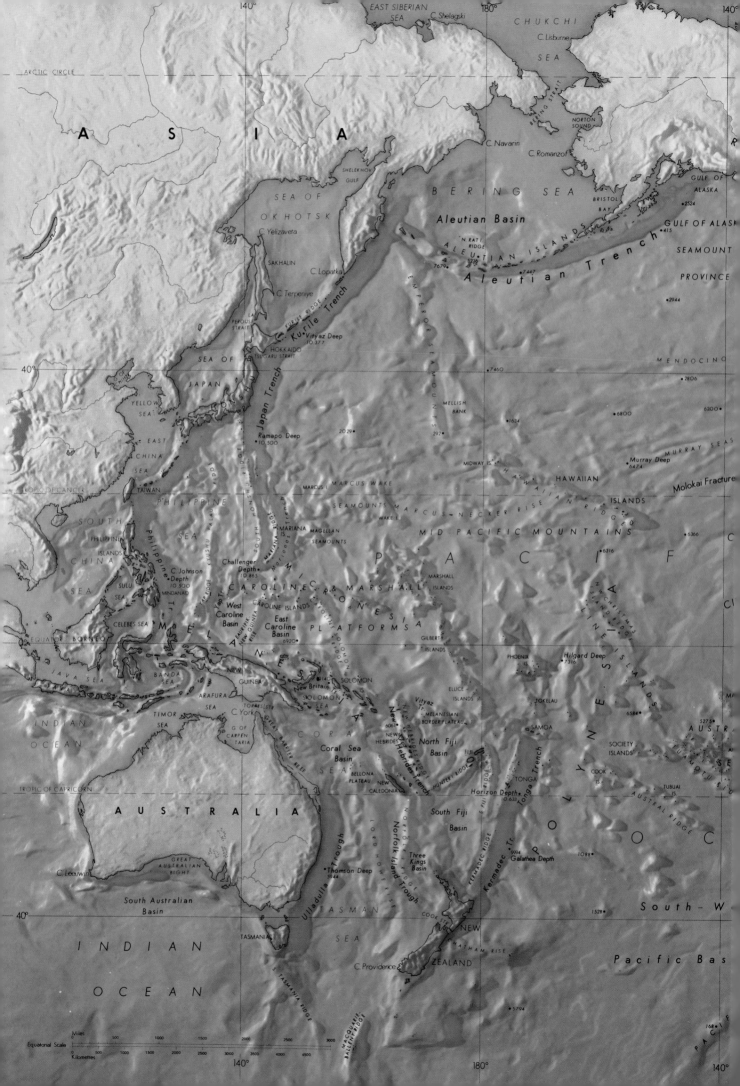

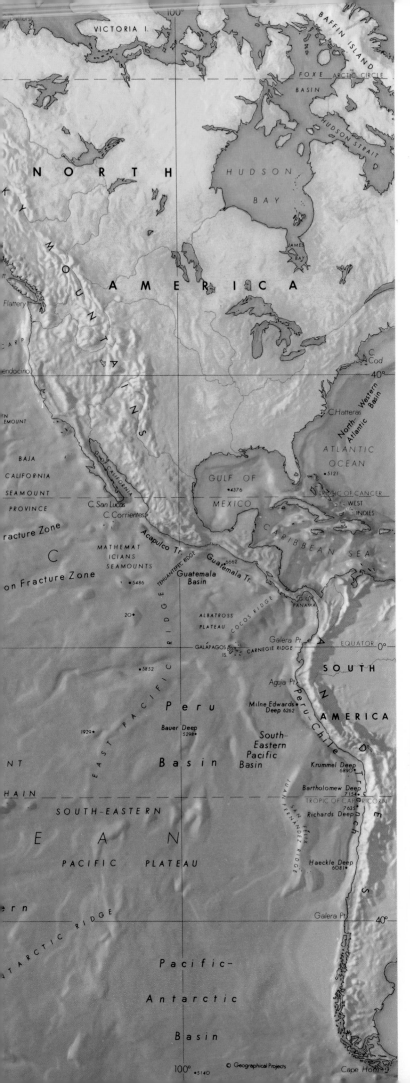

The Pacific Ocean

Below: three stages in the development of a
coral atoll. In the first, coral grows on the sides
of an extinct volcano. In the second, the volcano
is sinking back into the Earth's crust, but the
coral keeps growing. In the last, the volcano has
sunk, but the coral atoll remains.

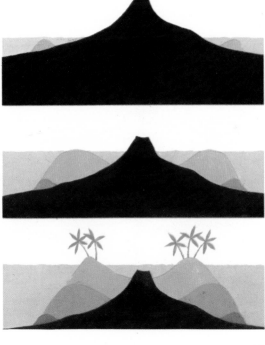

Chapter 2

The Powerful Sea

The world ocean is the heat regulator for planet Earth: it stops Earth from getting too hot or too cold. This is because of the remarkable properties of water in bringing about the transfer of heat by absorbing heat from the sun and sending it back into the atmosphere. The oceans are the biggest single part of a land/sea/air system that maintains life – 98.5 percent of all the water on Earth is in the oceans.

The oceans are far from being benign, however. They will batter, bury, crush, or corrode anything that is put into them by design or accident. There are two aspects of the oceans in seeming paradox: they are both responsible for all life and they are a powerful force that seeks to destroy it.

Opposite: the heavy rush of huge waves against the shore – as seen on an island of the South Pacific – can inspire a justified fear of the sea's great power.

21

How the World's Sea Was Formed

How did all the water get where it is in the first place? To answer that it is necessary to go back still further and ask how the Earth came to be. The answer, of course, is that nobody knows, but theories galore wax and wane in popularity. For years most experts believed that a lump of gaseous material broke away from the sun – perhaps sucked away by a star that drifted a little close – and that this gas cooled down to become the Earth. Then there was a shift in opinion to the theory that great clouds of dust particles gradually stuck to each other and then contracted to form the massive planet we know as Earth.

Above: the sea rolling against the shore and lapping the edge of the land conjures up many pictures and emotions: whitecaps in powerful play, awe at the depth and extent, thoughts of infinity. In all this, probably few think of the ocean as the cradle of all life – which it is.

Left: a lagoon nebula in the constellation Sagittarius, caused by dust clouds. The study of star formation, which is constantly in progress, has contributed to the discussion of theories about how the Earth was formed.

The theory goes on to say that as the cloud contracted it produced heat, and so did the decay of radioactive elements. Volatile constituents of such elements, such as water, ammonia, carbon dioxide, and others, were produced. These reacted with molten rock to form an atmosphere. As the Earth cooled down the atmosphere condensed and it started to rain. This rain was a deluge lasting for centuries – enough to fill the ocean basins.

Another theory says that the whole process was much more gradual, and that in fact it still goes on today. According to this theory, water poured out of the Earth's interior to form an atmosphere which condensed into small oceans. Water continued to ooze through cracks in the crust, making the oceans larger. Some scientists believe that they probably reached their present level 500 million years ago.

Human beings appeared on land no less than a million years ago, but their ancestors came from the sea. The sea is the cradle of all life. The elements that make up 99 percent of all living matter – carbon, hydrogen, nitrogen, and oxygen – were all present in the early ocean. They combined and reacted in a molecular way, encouraged and affected by heat, radioactivity, light, movement, and probably a great flash of lightning. Compounds with protein and nucleic acids – the bases of life – evolved to produce simple organisms capable of feeding themselves by absorbing organic matter. Along with this, the oceans provided the other right conditions for the evolution of life: an equable temperature, an even distribution of organic matter, and a consistent chemical make-up.

Certain organisms were more efficient in absorbing organic matter than others. They grew faster, divided more speedily, and passed on the same characteristics to their offspring. To gain energy, increasingly efficient cells consumed more and more organic matter. The waste products of the energy-making process –

carbon dioxide and acids, for example – also consumed organic matter. So there soon developed an early food shortage, and competition for necessary sustenance grew. The shortage could have led to the end of the only recently formed life on Earth, except that there began the fight for survival that Charles Darwin dubbed "the survival of the fittest."

The more clever of the marine organisms found a way to make their own food from carbon dioxide and inorganic materials in the ocean. They used the energy of the sun and the action of chlorophyll in the process called photosynthesis. In this way plants were born – microscopic organisms that were the ancestors of forests and grasslands.

There was no free oxygen in the early atmosphere, but one by-product of photosynthesis is oxygen. The atmosphere therefore became oxygen enriched as the result of the photosynthesis of plants. Today, all the oxygen people breathe is completely renewed by plants every 2000 years.

The next stage in evolution was for plant cells to use oxygen as part of the process of obtaining energy from organic substances. Such action led to respiration and, combined with fermentation, it gave organisms an energy surplus over that needed merely to get food.

The next set of organisms to make better adaptation were those that found the day-long process of photosynthesis unsatisfactory. They began to feed on other organisms that had already done the job of photosynthesis. These were the first animals. They developed ways of moving from one plant to another without relying on the vagaries of oceanic movement. They developed ways of short-cutting the food chain by eating each other. Then they evolved means of protecting themselves from each other by encasing themselves in shells. The line of evolution went from a crablike creature that ruled the oceans some 600 million years ago to fish that crawled onto the land and led an amphibious life. The amphibians evolved into reptiles. Reptiles become birds. These became land-dwelling mammals. Finally came human beings.

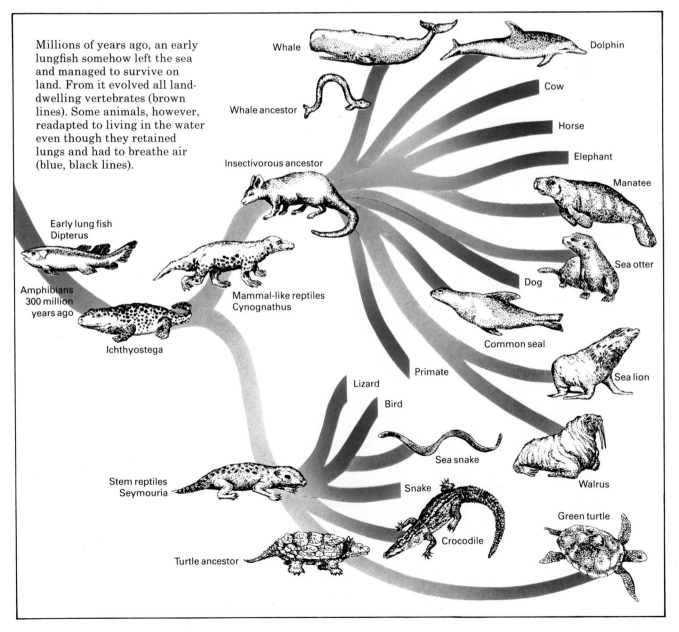

Millions of years ago, an early lungfish somehow left the sea and managed to survive on land. From it evolved all land-dwelling vertebrates (brown lines). Some animals, however, readapted to living in the water even though they retained lungs and had to breathe air (blue, black lines).

Whale

Dolphin

Whale ancestor

Cow

Horse

Elephant

Manatee

Insectivorous ancestor

Sea otter

Early lung fish
Dipterus

Dog

Amphibians
300 million
years ago

Mammal-like reptiles
Cynognathus

Common seal

Ichthyostega

Primate

Sea lion

Lizard

Bird

Sea snake

Stem reptiles
Seymouria

Snake

Walrus

Green turtle

Crocodile

Turtle ancestor

Interaction of Air, Land, and Sea

If people go out into the tropical sun after months in a chilly North European winter, they burn. Ultraviolet radiation from the sun will blister them. But people are shielded from the worst effects of ultraviolet radiation by a layer of ozone that encircles the Earth in the upper atmosphere. This layer was created when the process of photosynthesis flooded the atmosphere with oxygen. The gas reacted with the ultraviolet rays and formed ozone. This filtered out much of the ultraviolet radiation that was one of the key influences in the formation of organic material. The

Below: a vast cumulonimbus cloud sheds some of the water it contains as rain. There is a continuous exchange of water between the Earth and the atmosphere, regulated and assisted by the interrelation of land, air, and sea.

Above: a satellite photograph in which the convective pattern over the southeastern Pacific is interpreted as heated air rising from the sea. The ocean and the atmosphere are linked in determining weather.

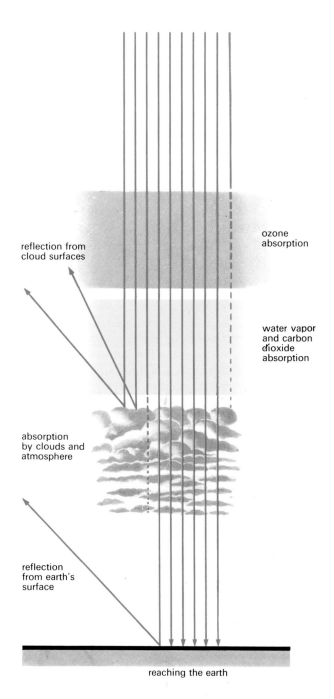

reflection from
cloud surfaces

ozone
absorption

water vapor
and carbon
dioxide
absorption

absorption
by clouds and
atmosphere

reflection
from earth's
surface

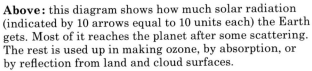

reaching the earth

Above: this diagram shows how much solar radiation (indicated by 10 arrows equal to 10 units each) the Earth gets. Most of it reaches the planet after some scattering. The rest is used up in making ozone, by absorption, or by reflection from land and cloud surfaces.

Above: waves shimmering in the sunlight. The way in which water and sun exchange heat is vital to the maintenance of the Earth's moderate temperature. Evaporation of the water's surface by the sun's rays also plays a role in providing precipitation from the clouds.

oceans became far less of a bubbling cauldron and settled, biologically, into the reasonably quiescent state in which we find them today. In this newer state of mildness, the oceans were able to support far more complex organisms climbing the evolutionary ladder toward advanced mammals.

Eventually a natural balance, which applies today, was struck. The oceans, the land, and the atmosphere form an inseparable, mutually dependent system in which the oceans are by far the most important factor. There are many reasons for this, but one of the main ones is the remarkable property of seawater in having such a high specific heat. Water is capable of absorb-

ing more heat for a lesser rise in temperature than any other substance except ammonia. Conversely, it can lose heat in huge amounts without its temperature falling appreciably. Put most simply, this means that the oceans heat up and cool down slowly.

There are clear examples of this heat neutrality of seawater. Plunge into the sea on a summer's day for instance, and it will feel cold. Listen to those hardy souls who insist on swimming in the northern hemisphere sea on January 1 and they will tell you that it is not nearly as cold as one might expect from the atmospheric temperature.

These heat-neutrality properties of water mean

Above: a cumulus cloud build-up. Massed cumulus that are well developed are usually a sign of rain. They build up during the morning and make the sky heavy with cloud by late afternoon. They then often release a light showery rain. Sunlit parts of the cumulus show up as a brilliant white in color.

that the oceans have a big effect on the weather. In fact they make the weather. To do so they need energy, and they get their energy from the sun. Some 99 percent of the Earth's heat energy comes from the sun and the rest from volcanoes and other splits in the crust. Because of their size the oceans absorb most of the heat energy that reaches Earth from the sun, but they absorb it inefficiently because it reaches them from the top. Think of boiling a pan of water. Heat is applied from underneath. Apply it from the top and the water will take an extremely long time to

heat up. That fact, coupled with the high specific heat of water, means that the oceans act to keep world temperature within the limits that support life as we know it.

In fact the oceans act as a great radiator: much of the heat absorbed from the sun is sent up into the atmosphere and space. Frequently too the air above the oceans is colder than the water below and another form of heat transfer takes place. But the greatest use made of the sun's heat energy is in the evaporation of water from the sea surface. Expressed as percentages of the total heat input to the oceans from the sun, evaporation dissipates over 53 percent of the energy; radiated heat accounts for over 40 percent; and heat transfer to cold air above the oceans just over six percent.

In these ways the oceans keep the Earth's heat balance in an acceptably neutral state. Evaporation, as can be seen, is the greatest single factor in removing water from the oceans. As the water evaporates it leaves its salts behind and forms fresh water clouds which are driven by winds, eventually to move over the land. The clouds become so heavy with water that they condense and rain or snow falls. Most of it falls back into the oceans simply by a law of averages because the oceans cover 71 percent of the Earth's surface. Of the water falling on land, some evaporates on the way down. The fraction that reaches the land then tries to get back to the sea: it evaporates from vegetation, soil, lakes, rivers, and reservoirs to reenter the sea/cloud/land/sea system. It runs off the land into rivers, which in turn flow to the sea, and the cycle begins again.

The amount of fresh water that flows from the land into the oceans is tiny compared with the total amount of water in the oceans, which contain 98.5 percent of all the Earth's water. Only a tiny amount of water exists in the fresh state. Even the 1.5 percent of water outside the oceans, which is fresh, is mostly tied up in Arctic and Antarctic ice.

Waters that flow into rivers and seas have an added ingredient: the soil they carry. Even the mere trickle

Right: a diagram illustrating the hydrological cycle, which is the constant exchange of water between the Earth and the atmosphere. Precipitation (solid lines) is the result of evaporation (dotted lines) from water and land surfaces. Transpiration from plants is also a source of water exchange.

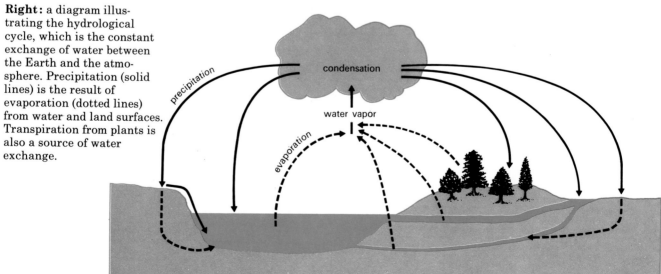

Left: a satellite view of the Nile Delta. Deltas are an example of how flowing waters change the face of the Earth, tearing down in one place and building up in another. Rivers form deltas by depositing the silt they carry where they flow into a shallow sea or lake with weak currents, so allowing silt to collect.

Below: poisoned fish lie dead in a river, victims of toxic wastes dumped into the water by factories. It is not yet fully known what effect polluted waters have on the all-important evaporation/precipitation/run-off cycle.

of water represented by the flow-off of such great rivers as the Nile, Amazon, Mississippi, and Ganges sweeps vast deposits of land into the sea. Around 4000 million tons of land a year goes into the sea in a dissolved state, and 13,000 million tons in suspension. Currents, tides, and waves also scrape at the land to pull material into the sea. Water has the remarkable property of being able to dissolve more substances in greater quantities than any other liquid.

The minerals in the soil carried to the ocean is what makes seawater salty. With the steady annual addition of millions of tons of minerals, the oceans could be expected to become steadily saltier. But because of the sheer size of the oceans, the yearly addition of soil is nothing in terms of the age and the volume of the oceans.

The oceans already contain such a huge amount of dissolved salts that it cannot be expressed in a figure with less than 17 zeroes in it. Next to that, a mere 17,000 million tons a year may be compared to the effect one extra grain of salt might have if added to a spoonful of salt already in a pot of soup.

There is a danger in the quality of the grains of salt people are adding to the oceanic soup. Some particularly nasty compounds are being dumped into the oceans, and their full effect on the evaporation/precipitation/run-off cycle is not yet clear. A plutonium rain cloud is an unhappy prospect, for example.

When the Waters Show their Power

Land, sea, air, and sun are parts of a system in which there is a multitude of checks and balances to keep the Earth physically, chemically, and environmentally stable. The oceans absorb and reflect heat; water is evaporated to maintain life on land and flows back into the sea, carrying with it the minerals and other materials that help to maintain oceanic life. The process has gone on for millions of years and should continue for millions more. It is a refined system, with the oceans working quietly to sustain life on the planet.

Sometimes, however, the oceans behave like an enemy. Anything that humans put into the sea, it will seek to destroy: by battering, burying, crushing, or corroding. It is almost as if any human intervention in the working of the ocean is an affront to it and must be removed as quickly as possible. Here is one story of a battle against the sea.

In the early 1970s a pipelaying barge was working in the North Sea in 500 feet of water. The pipe was to bring North Sea oil to Great Britain. The weather was gray and moody, and drizzle misted the bridge windows. The barge captain kept checking the weather forecasts. Gales were due in a few hours, but for the moment the pipe was snaking gently to the seabed. The giant barge was winching itself along steadily on its anchors. Tugs scurried around like pilot fish in attendance on a whale. All was going well, but the captain kept an anxious eye on the horizon because the wind was freshening and the sea was beginning to whip around the bows of the tugs. He knew he would have to allow time to break out the

barge's huge anchors and to stop the pipelaying operation before the time the gale was expected to strike. Only 10 minutes later the wind was whistling ominously through the radio aerials on the top of the bridge and the barge was beginning to roll. That was completely unexpected.

Twenty minutes later all was confusion. Only occasionally could the captain see the mast light of a tug as it reared up from the trough of a wave. The barge was rearing and bucking. The captain knew that two of the anchors had not been broken out, but the wind and sea were hurling his vessel around and the anchors – each as tall as three men – were dragging awkwardly through the seabed.

However, the squall passed as quickly as it had come and the barge crew prepared to start work again. The captain reflected that an unforecasted squall had wrecked his work program. The pipeline, the end of which had been capped and unceremoniously dumped on the seabed, would have to be picked up and the anchors set again before work could resume. A submarine was sent to locate the end of the line and its pilot came up with grim news. The seabed looked like a battlefield. Great trenches had been scoured by the dragging anchors. Worse, their flukes had ripped into the line so that parts of it looked like spaghetti. Days – perhaps weeks if the weather was bad – would be lost while the damaged sections were cut away and laying could begin again. A fortune would be lost both in terms of the costs of repairs and the time the expensive barge stood idle. All for a storm that had blown up in 30 minutes.

In some ways this event was a trivial incident in spite of the heavy financial loss. Nobody was hurt, no major items of equipment were lost. Just a bit of

Below: an engraving of 1665 by W. Hollar, entitled *Four Warships in a Storm*. To be lashed by the fury of the sea in a small craft is probably one of the most terrifying experiences anyone can go through. High rough waves, strong sweeping winds, and dark lowering clouds are not only frightening, but also are as dangerous as they look.

mangled pipe. But it shows how the sea had again done its best to remove a foreign body.

Off the Philippines a telephone cable-laying company had been having similar problems. As fast as the cables were laid they were broken by some powerful unknown force on the seabed some 10,000 feet below. The company could not understand why and decided to find out by measuring the ocean floor currents. A team of oceanographers was brought in and the experts devised arrays of equipment, each of which consisted of an expendable weight that would carry the array to the sea floor. Above the weight were acoustic releases. An instrument to measure and record current speed and direction on a magnetic tape was fitted above the releases. Above this current meter were floats. Each array was launched without a cable connecting it to the ship and was allowed to make its own way down to the sea floor where it could settle to record data for days at a time if necessary, while the ship moved on to launch or recover one of the other arrays. When the ship returned it sounded a coded acoustic signal through the water. This triggered one of the acoustic releases which parted the array from the ballast weight. The equipment became buoyant and floated gently to the surface.

All went well during the trials with the equipment

Left: seaside strollers hasten away as rough waves smash against a promenade barrier at Hastings on the south coast of England. A photograph taken in 1880.

Below: even with nothing in their path, thrashing waves in a stormy sea evoke a threat of destructiveness.

Above: huge piles in the harbor of Rio de Janeiro are buffeted by Atlantic ocean waves, their rustiness giving evidence to the corrosiveness of seawater. It is a constant battle to preserve anything that stands in water.

Below: a sacrificial anode on a cross bracing after several years of immersion, showing a high degree of corrosion. It is called sacrificial because, while it takes the full brunt of corrosion, it protects what it covers.

and on the first deployments. But at the end of one recovery operation the oceanographers spotted the floats breaking the surface, hooked them in – and found nothing. No current meter, no acoustic releases. Connected to the floats there was only the mangled tie-bar of the current meter, a one-inch thick straight stainless steel rod. The scientists could not imagine what had caused it to become so twisted and bent. What had happened? What enormous force had ripped away the current meter and mutilated the bar? Other data showed that it could not have been caused by a turbidity current. A fish? Animals have been known to attack underwater equipment, but the extent of the damage hinted at monsters. There are many theories, but nobody knows for certain what happened down there on the deep ocean floor. The distorted tie-bar, mounted and displayed on one oceanographer's desk, pays mute and dramatic tribute to the unknown forces of the ocean.

One day in the early 1970s a large chunk of a hotel in Bahrain collapsed. The hotel had been built on land reclaimed from the sea, and the steel piles supporting it had been carefully driven deep into the sand. But the sand had not been desalinated, and the sea salt had steadily chewed its way into the piles until they

crumbled. The sea had made its mark even on land to destroy human works.

It can be seen how the millions of tons of salts washed from the land each year to play their part in the ocean/atmosphere system are a mixed blessing. For, while contributing to ocean fertility and providing convenient concentrations of valuable minerals, they also make seawater one of the most corrosive substances known. Seawater eats metal quickly. In

Above: a glass sphere protected by a cage while being prepared for pressure chamber tests. Although glass has the advantage of being noncorrosive as well as extremely strong, it is very difficult to make a sphere free of flaws.

Left: a diver cutting away the bolts from a clamp on a leg of an oil platform. The smallest fault in equipment or structures beneath the water must be taken care of as soon as possible.

one case, two incompatible metals in contact with each other in seawater for only five weeks were eaten into by a millimeter.

In order to exploit ocean resources fully, enormous amounts would have to be spent to develop materials and equipment that can resist the sea's inexorable corrosion. Although there are totally corrosion-resistant substances such as glass, ceramics, and carbon fiber, it is necessary to adapt them for use in making equipment, which takes time and money. Other materials, mostly alloys, must be developed and ways found to protect existing materials in use. At present protection is achieved by two main methods, one by coating metals with a corrosion-resistant material such as a highly expensive alloy, a polymer, or precious metal. The other is cathodic protection by which a block of easily corroded material such as zinc is clamped to the material needing protection. The sea then attacks the outer metal, which has to be replaced when it has been eaten away. Look at the picture on page 236 of an oil production platform about to be launched. The hundreds of little white flecks are cathodic anodes. Their replacement is a major task for the underwater engineer.

Bury, batter, corrode – and crush. At the sea surface pressure is 14 pounds per square inch, a measurement known as one atmosphere. That pressure increases by one atmosphere for every 33 feet until at the bottom of the ocean trenches it can be 14,000 pounds per square inch or 1000 atmosphere. At that level of pressure faulty equipment can implode with terrifying force or water can rush through the most microscopic crack. Even at depths of as little as 180 feet humans cannot have ordinary air pumped down because it poisons them.

Imagine, then, what care must be taken for the seemingly simple task of lowering a piece of equipment to the ocean floor. The ship has to be designed to take the very worst that the weather can hurl at it. Special launching gear has to be devised for the equipment so that it is not slammed against the side of the ship. The equipment itself must be designed to resist accidental slams; it must be corrosion-proof; and it must be able to withstand pressures far in excess of those inside the combustion chamber of a rocket motor. Take all these precautions and even then the sea may have another unexpected trick to play. Recall the unexplained twisted tie-bar for frightening testimony to its mysterious powers.

31

The Year of the Great Storm

Every day the sea works to combat and reject human efforts to understand and exploit it. In this process the sea can sometimes be stirred to unprecedented and terrifying fury. All we can do then is hide.

On the stormy evening of January 31, 1953 a young couple left their home in a small seaside resort on the coast of Lincolnshire, England. They struggled against the ferocious gale that had been blowing for over 24 hours and made their way to take shelter elsewhere. Floods had been forecast for the next high tide and the theater was the gathering point for those in greatest danger. The two had been most pleased with their newly acquired house, which had fine views of the sweeping beach and the North Sea. They had been reluctant to leave it, as if merely by being there they could in some way protect it. They slept little that night as the gale heightened and news of flooding began to come in. Their house would be flooded, they knew, but they hoped that the furniture and carpets they had carefully stored upstairs would be safe. The next day they were able to venture out to look. There was no house. It had gone, along with all its contents, leaving only the foundations behind. Such had been the sea's fury.

That young couple was lucky as it turned out. Farther south in Essex, sea walls failed and water

Above: the town of Stavenisse, the Netherlands, after the great flood of January 1953. Only a few houses remain. **Below:** flooded houses in Tilbury, England, during this disaster. No one was prepared for this North Sea assault.

raced through the streets, trapping people in their homes. Over 300 people were killed in England in that one night.

In the Netherlands, which is particularly vulnerable to the anger of the North Sea, it was far worse. Some 38 percent of the land area of the Netherlands lies below high water level, and some 20 percent lies actually below mean water level. Seven million people – about half the population – live below high water level; airports, ports, and major industries are so placed as well. For centuries the Dutch have been masters at increasing their small land area by reclaiming it from the sea. It is a slow, laborious task and there has to be constant vigilance to protect the new-won territory. No other nation knows better what safeguards have to be taken. But even the Dutch were not prepared for January 31, 1953.

Even before high tide the sea was level with the tops of many of the dikes. As the tide heightened, water began to flow over the tops so that a two-way erosion on the insides and the outsides of the walls began to take place. Before midnight the sea surged. Over 50 dikes just could not stand the strain. Most of them burst and walls of water swept into the country,

Below: a contemporary engraving of a flood that took place on the east coast of England in 1607. A report of the day said that the overflow killed many thousands of men, women, and children, destroyed whole towns and villages, and drowned numerous sheep and cattle.

killing 1800 people and covering over 625 square miles of land.

Why that particular night? Why should the sea have struck then? It had to do with tides, the shape of the North Sea, and the wind. First the wind. It had been blowing hard and steadily from the north for days, pushing great quantities of water from the Atlantic into the North Sea and whipping up huge waves. The North Sea is shaped like a funnel with only a tiny opening – the Strait of Dover – at the southern end. Little water can escape through this narrow end. So, when water cannot escape through the strait, it flows down the west side of the North Sea (the east coast of Scotland and England) and turns to flow back past the Netherlands. All the time it is being "squeezed" by the surrounding land, getting higher and higher. Even more height is added because the North Sea shelf slopes from north to south from depths as great as 1000 feet to an average of 60 feet in the south. This again makes the water pile up. On that fateful night a high or spring tide was due in any event. This would normally have gone unnoticed by most people but on that occasion it combined with the storm surge to reach heights some 10 feet greater than had been predicted. Waves up to 16 feet high on the coast completed the combination – and the result was devastation on a scale that northern Europe had not seen for centuries.

Could it happen again? Yes. There is nothing we can do to stop it. All we can do is to be better prepared. The coasts of England and Holland are today far more strongly protected. Better weather prediction methods are also available to tell us when the combination of wind and tide is building up to produce a deadly surge.

That is all we can ever do when the sea is at its most furious: simply hide behind a wall and wait it out until the fury abates.

Top: the beginning of construction work on a dike across a Netherlands estuary in 1956, part of the country's defense system against floods.

Above: the completion of the dike, watched by a cheering crowd from shore.

Left: one of the Netherland's new protective dikes. It runs for about 19 miles in Friesland, a northern province directly on the North Sea.

33

Tsunami: Highest Waves of the Sea

The fateful, crucial incident in Joseph Conrad's novel *Lord Jim* occurs when Jim is on watch as chief mate in a bedraggled steamship taking pilgrims from the Far East toward the Red Sea. It is a hot night and there is a flat calm sea. Jim and the skipper are on the bridge listening to the drunken ramblings of the second engineer. Suddenly the engineer falls. Jim and the skipper also stagger. When they recover they look, amazed, at the still sea and the stars.

"What had happened? The wheezy thump of the engine went on. Had the earth been checked in her course? . . . . A faint noise as of thunder, of thunder infinitely remote, less than a second, hardly more than a vibration, passed slowly, and the ship quivered in response, as if the thunder had growled deep down in the water. . . . The sharp hull driving on its way seemed to rise a few inches in succession through its

Above: an engraving of the British mail ship *La Plata* encountering a tsunami off the Virgin Islands in the West Indies in 1867. Although the popular name for tsunami is "tidal wave," it is a misnomer because the phenomenon has nothing to do with solar or lunar tides. Tsunami are caused by undersea earthquakes.

Below: a woodblock print by the 19th-century Japanese artist Utagawa Kuniyoshi of a boat being overwhelmed by a giant wave. This kind of event would be a natural subject for Japanese paintings and writings because Japan has suffered many onslaughts of the towering walls of water known as tsunami. The word itself is Japanese.

whole length, as though it had become pliable, and settled down again rigidly to its work of cleaving the smooth surface of the sea. Its quivering stopped, and the faint noise of thunder ceased all at once, as though the ship had steamed across a narrow belt of vibrating

water and of humming air."

Conrad described with accuracy the passing under the ship of a tiny wave – only a few inches high but moving at a speed of perhaps 500 miles per hour.

Such a wave is called a tsunami. It is unlikely that Conrad had ever heard of it, and he probably had no knowledge of the reason for a slow-moving ship on a calm sea to be suddenly subjected to a momentary jolt. But his description of that jolt is surprisingly, if unwittingly, close to the mark: "as if the thunder had growled deep down in the water." For, similar to thunder, these potentially devastating little waves are formed as shock waves during an earthquake in the deep ocean. The shock waves reverberate through the rock strata and the ocean floor moves.

On land the effects of an earthquake are immediate and often catastrophic. Buildings tumble, water and gas mains burst, fires break out, people are killed.

In the deep ocean, however, the effects of an earthquake are not so immediately apparent. The thundery waves growl deep down in the water and the energy is translated into a water column to send a ripple racing across the ocean.

At sea the passing of this ripple goes unnoticed by big ships. Conrad's character describes his steamer as going over "whatever it was as easy as a snake crawling over a stick." Sailors have often thought that their ship has touched a rock, with the result that some nonexistent reefs have appeared on charts.

Often these ripples fade away, their energy burning out across thousands of miles of ocean. But if a land mass impedes their progress, trouble starts. The liquid shock wave approaches the shore across an often steeply shelving seabed. All that energy is compressed into a smaller and smaller space and the water has nowhere to go but upward. As the wave rips in toward the shore it creates what is in effect a vacuum, perhaps sucking the water out of a harbor to leave the seabed completely exposed for a few moments. That is the last warning. The sea returns a few minutes later and balances out the instant low tide with a wave as high again as the normal depth of the harbor – and moving frighteningly fast.

Tsunamis occur mostly in the Pacific and reach colossal heights when they touch the shore. One knocked a radio mast off the top of a 100-foot cliff in Alaska in 1946. That same subsea earthquake sent waves speeding in the other direction to cross 2300 miles of the Pacific in four hours. The captain of a ship moored off Hawaii felt nothing as the deadly ripples passed under his vessel, but on turning around he saw waves of over 50 feet high crashing into the harbor, destroying much in its wake.

Nothing can be done to stop tsunamis. All that is possible is to warn coastal dwellers that one or a series of them is on the way. The shock waves from earthquakes travel through rock faster than tsunamis rip across the oceans. So by pinpointing the source of an earthquake and using ocean monitoring systems to detect the length of the waves as well as their speed and direction, it is possible to give a few hours warning. There is an international tsunami warning system of this type in operation in the Pacific, and it has saved thousands of lives.

Above: an information check at the Pacific Tsunami Warning Center, part of the international warning system.
Left: Hilo, Hawaii lies devastated after a tsunami on May 24, 1960. The wave bore the large boulder inland.

35

An Ocean Mystery

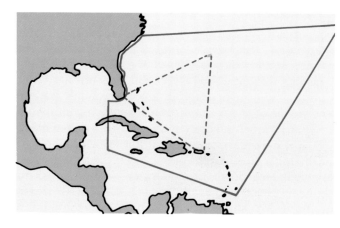

Some areas of the world ocean have for many years been regarded with a kind of superstitious dread, earning themselves such names as the "Devil's Sea," the "Graveyard of Lost Ships," or the "Sea of Fear." Comparatively recently, notoriety has attached itself to a roughly triangular part of the Atlantic Ocean that encloses the slow-moving Sargasso Sea. Known most commonly as The Bermuda Triangle, this stretch of the ocean has other more chilling designations such as the "Triangle of Death" and the "Graveyard of the Atlantic."

Records have been kept only since the mid-19th century, but in that time at least 50 ships and 20 planes have disappeared in The Bermuda Triangle without leaving any wreckage. It is the absence of bodies and floating debris that compounds the mystery of the region, giving rise to theories and speculations about supernatural forces. In fact, there have not been satisfactory explanations for many of the recorded disappearances.

The Bermuda Triangle burst into public consciousness in the last month of 1945 when a bomber mission on a routine training flight vanished without trace

Above: The Bermuda Triangle. The dotted line traces the traditional area; the solid line shows modern extensions.

between a United States Air Force base in Florida and Bermuda. The flight consisted of five planes and 14 crew members. The tragedy and mystery deepened when a flying boat was sent out to rescue the bomber crews and in turn disappeared as if into thin air, claiming 13 more lives. Some 300 planes criss-crossing the region in an air reconnaissance failed to unearth a single clue to the double disappearance. Nor did the massive surface rescue party that moved across the Sargasso Sea and its environs in as thorough a search as was humanly possible. The eight Coast Guard boats, four navy destroyers, several submarines, and hundreds of volunteer private yachts and boats found nothing at all.

It was in a post-war book on sea mysteries that Vincent Gaddis drew attention to the many unsolved

Right: Avenger torpedo bombers of the same type that vanished on a routine training flight off the east coast of Florida in 1945 – and started a resurgence of interest in The Bermuda Triangle mysteries.

disappearances in the area and coined the phrase "The Bermuda Triangle" to describe this relatively small but apparently perilous area of the Atlantic. Charles Berlitz, now the writer most closely associated with The Bermuda Triangle, intensified the mystery by listing 61 disappearances before 1945 and 80 in the 31 years after that.

Exactly where is The Bermuda Triangle? Put scientifically, its latitude is from 25° to 40°N and its longitude from 55° to 85°W. Put in everyday terms, it can be drawn as a triangle from Florida to Bermuda on one side, from Bermuda to Puerto Rico on another, and from Puerto Rico back to Florida through the Bahamas on the third. It covers 1,500,000 square miles.

The Sargasso Sea, which it encompasses, is a sea

strange disappearances over and in The Bermuda Triangle? One is that, within this area, a magnetic compass points true north rather than toward the magnetic north, and these two points can vary as much as 20° in the course of circumnavigating the Earth. If this compass variation is not compensated for, a plane or ship could find itself far far off course. When instruments go awry, as they often do in the danger area, there is no way of making this vital compensation. Serious trouble is sure to follow.

Other natural conditions make the region a menace. There is the unpredictable and wild Caribbean-Atlantic weather pattern that calls up sudden thunderstorms and waterspouts. Another factor is that the submarine topography of the region is ex-

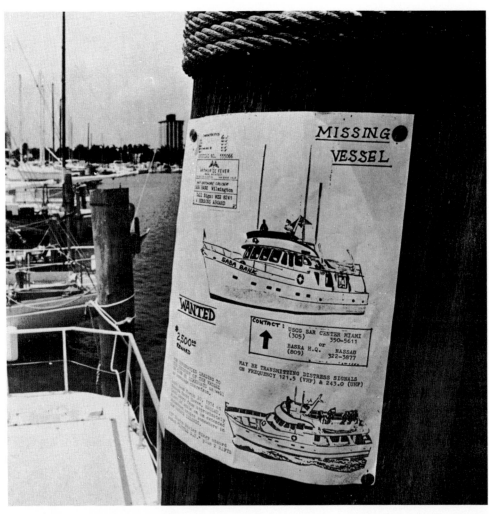

Left: this poster offers a reward for information about the missing yacht *Saba Bank*, one of a number of boats that disappeared under mysterious circumstances in the water and skies of The Bermuda Triangle since 1945.

within the Atlantic Ocean, the quiet center of the massive circular current system in the Atlantic north of the equator. Because of its lack of currents, it was the scourge of sailing ships which often found themselves becalmed there. It is named after the seaweed *Sargassum*, which floats in and on it in enormous quantities. This nearly motionless mass and the lack of fast-moving currents gave rise to legends of sailing vessels being trapped forever in the sea's grasp, floating eternally in a midwater limbo when they sank.

What are some of the logical explanations for the

ceptionally complex, with extensive underwater cave systems and deep trenches as well as long shoals. Another possible factor is the sharp contrast between the sluggish Sargasso Sea and the turbulent currents around the Bahamas.

None of these known factors provides a complete explanation for a specific disappearance within The Bermuda Triangle – neither of the 1945 bomber flight nor of the other planes and ships lost since then. That leaves the many mysteries of The Bermuda Triangle unsolved, emphasizing once again the perilous and unpredictable nature of the world's ocean.

Chapter 3

Understanding the Ocean

The ocean is always moving. Mighty currents flow across its surface and some, like the Gulf Stream, have profound effects on the climate of the land. There is a constant but slower flow at all depths: fingers of near-freezing Antarctic water, for instance, creep across the ocean floors to reach as far north as Europe.

The tides provide more motion as, under the gravitational influence of the moon and sun, the bodies of water in the ocean basins rock to and fro twice a day. And the winds whip up waves in the most dramatic-looking ocean motion of all.

All this movement means that the salts flowing from the land into the oceans are well and truly mixed so that the chemical composition of seawater is remarkably constant all over the world. Currents, tides, waves, and ocean chemistry represent one of the most exciting and rewarding areas of study left to us today.

Opposite: the jagged coastline of southwest England, though dramatic to see, is clear evidence of the remarkable power of water to eat away and reshape even huge rocks.

The Five Parts of the World Ocean

There is only one ocean. Throw a cork into the sea off Singapore and there is nothing, theoretically, to stop you picking it up on the beach of Long Island, New York, after it has drifted around the world ocean. It shows that Earth is a water planet on which small areas of land float. These vast watery areas are divided for convenience of reference into five main oceans: the Pacific, Atlantic, Indian, Southern, and Arctic.

On its surface the world ocean has some parts that are warmer than others. Some are perpetually wracked by huge waves, some stay relatively calm for most of the year. However, these physical boundaries often bear little relation to the geographical boundaries that have been imposed upon them.

Looking at conventional maps gives no real indication of the sheer size of the world ocean. Maps are made for the convenience of humans who are interested primarily in land, so they show the land on an exaggerated scale. The map on this page is an equal area projection of land and sea, and this starts to give an idea of how puny the largest continents really are. It shows that the Pacific alone covers a third of the Earth's surface.

The simplest way to appreciate the extensiveness of the oceans is to sail across them, but even just flying across a major ocean can convey something of

Above: an ancient bas-relief depicting an expedition of warships. It was among the early peoples of the eastern Mediterranean region that exploration of the sea began – the Phoenicians, for example, reached the Atlantic Ocean.

its magnitude. Travel from Singapore to San Francisco and there is nothing but dots of islands at which the plane touches down, such as Guam and Hawaii. Cross the Pacific from Panama to the Gulf of Thailand and the journey has been halfway around the world. Set off from Alaska for a trip to the icebergs of Antarctica and only a few islands are to be seen all along the way.

The Pacific is the deepest of the oceans, the distance from the surface to the bottom of the Marianas Trench, off Guam, being seven miles. It is also the widest and the longest as well as having by far the greatest share of the sea's resources.

This equal area projection map shows clearly how much less land Earth has than water.

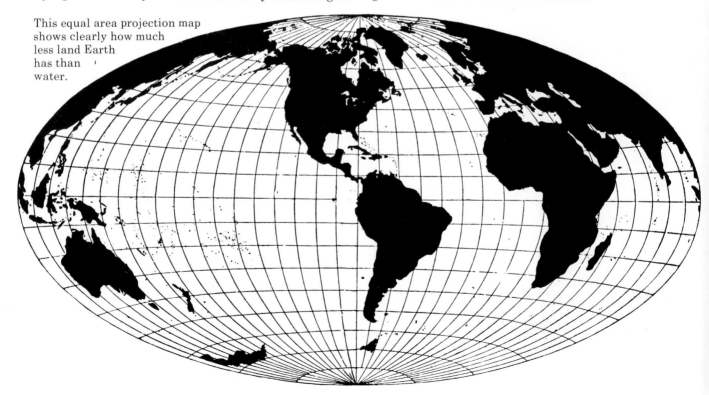

Measurement figures are so huge as to defy comprehension, but just for the record, the Pacific Ocean occupies an area of nearly 64 million square miles, and has a volume of 250 billion cubic feet. Its average depth is a much more comprehendable number: 14,049 feet.

The boundaries of the Pacific are arbitrary and authorities differ on how they should be defined. Most agree, however, that the eastern limit is the turning point of Cape Horn. There the Atlantic Ocean begins.

The Atlantic is about half the size of the Pacific. It has an area of nearly 32 million square miles, a volume of over 100 billion cubic feet, and an average depth of 12,880 feet. It is the second biggest of the five oceans, the Indian, Southern, and Arctic following in that order.

There is more information on the Atlantic Ocean than any of the other five. That is because it has been known longer and because it has played such a vital part in the history of human trade and war. The Atlantic was the first ocean to be recorded – discovered by the Phoenicians who sailed bravely through the Pillars of Hercules, now called the Strait of Gibraltar, to marvel at its immensity. The Atlantic was the link between the Old World and the New after the discovery of the Americas, and it became necessary to learn more about it to help send traffic speedily and safely between the two worlds.

First, knowledge had to be gained about prevailing winds and currents, particularly the mighty Gulf Stream, so that sailing ships could travel by the fastest and most efficient routes. Later came a requirement to know more about the features of the ocean floor so that telephone cables could be laid, and about its depths and the distribution of marine life so that fishing could be made more efficient. More recently the need arose to determine subsurface temperature gradients so that deadly war games could be played.

Out of all these pursuits has emerged a picture of the features of the Atlantic that is considerably clearer than that of the other oceans. Its depths are well charted, and maps of them show a distinct S-shaped ridge running down its center. This follows the shape of the adjacent continents and represents part of the greatest mountain range in the world – the Mid-Oceanic Ridge. The discovery of this ridge in the Atlantic, and the later realization that it extended throughout the world ocean, played a major part in the formulation of the theory of continental drift as the answer to how the continents were formed.

The Indian Ocean is the smallest of the three major oceans, but nevertheless it is 4000 miles wide and has an area eight times that of the United States. There is much less known about it than about the Pacific and Atlantic. The first important oceanographic expedition to the area was by HMS *Challenger* during its famous voyage of discovery from 1872-76. Almost 100 years later a big international effort was made to glean more knowledge of the physical, chemical, and biological characteristics of this ocean by the International Indian Ocean Expedition of 1960-65 when research ships from 12 nations explored the area.

The Indian Ocean has an average depth of 12,700 feet, reaching a maximum of just over 24,400 feet. The mighty Mid-Oceanic Ridge sweeps around Africa and takes an inverted Y shape. But not all the ridges in the Indian Ocean are a part of the Mid-Oceanic Ridge. During the International Indian Ocean Expedition it was discovered that one of these ridges was extraordinarily long and straight. It runs 1500 miles south from the Bay of Bengal and is named the Ninety East Ridge after its longitudinal position. It is the longest rectilinear geotectonic structure in

41

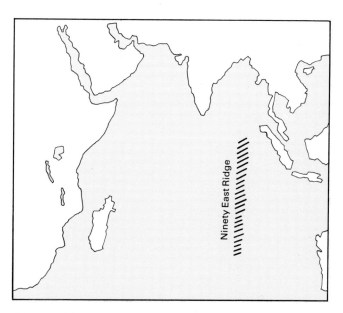

Above: this map shows the location of Ninety East Ridge in the Indian ocean, discovered for the first time during the International Indian Ocean expedition from 1960–1965. Of basalt composition, it is about 125 miles wide.

The differences between an ocean and a sea are not at all simple to define, although the North Sea can easily be seen to be a simple extension of the Atlantic Ocean. It is a good example of a marginal or adjacent sea which is connected individually to the larger body of water. The mediterranean type of sea consists of a group of seas linked to each other but collectively separated from the ocean. The Mediterranean complex, made up of the Mediterranean, Black, Adriatic, and other seas and linked to the Atlantic only by the narrow Strait of Gibraltar, is a good example.

In some areas it is difficult to understand why there should be such a large number of individually named seas in what seems to be a continuous stretch of water. The west side of the Pacific is a good example of this. Broadly speaking, the division of one sea from another is usually decided by the shape of the adjacent land. However, other factors also have to be taken into account. These include the features of the seabed, the degree of independence of currents, tidal and atmospheric circulation, and the distribution of water temperature and salinity gradients.

the world. Put simply this means that it is the world's longest straight physical feature.

Not very much is known about the Arctic Ocean. In fact it was only discovered to be a true ocean – in that it has continental shelves and an abyssal plain – nearly 90 years ago. This fact emerged as a result of the epic voyage made between 1893 and 1896 by the Norwegian explorer Fridtjof Nansen in the wooden ship *Fram*. The very inaccessibility of the Arctic has seemed to prove a magnet to explorers.

The Arctic is practically landlocked, and one third of it consists of continental shelf. It is hardly surprising that little else is known about the Arctic. Not only does it have a permanent cover of ice, but also it is in darkness for half the year. It is isolated and has an extremely harsh climate. Approaching the area from the south, exposed flesh freezes, ships accumulate so much ice on their superstructures that they can become top heavy and capsize, and icebergs are a constant threat to safe passage.

The huge continental shelf of the Arctic Ocean has the geological structures that elsewhere have proved to contain oil. So at the top of the world people work to harvest natural resources – but at a high price.

The Southern Ocean encircles the continent of Antarctica. On the north it has no apparent physical boundaries, but there is an invisible line which quickly distinguishes it from its northern neighbors. Because of the movement of currents and winds, cold water drifting north meets less dense water flowing south at a latitude of about 50° S, and sinks beneath it in a massive, slow, subsurface waterfall called the Antarctive Convergence. Crossing it going south there is a quick change in both air and water temperatures, as well as a sharp difference in animal population. Many birds, fish, and plants found on one side of the convergence are rarely seen on the other.

Left: the Norwegian explorer Fridtjof Nansen looks out on the summer landscape in the Arctic in 1894. His epic voyage in the *Fram* proved that the frozen Arctic is really ocean.

Below: the golden glow of the summer sky over the pack ice of northern Greenland. Both the Arctic Ocean and the land areas immediately around it are covered with ice the year round.

The World's Oceanic Currents

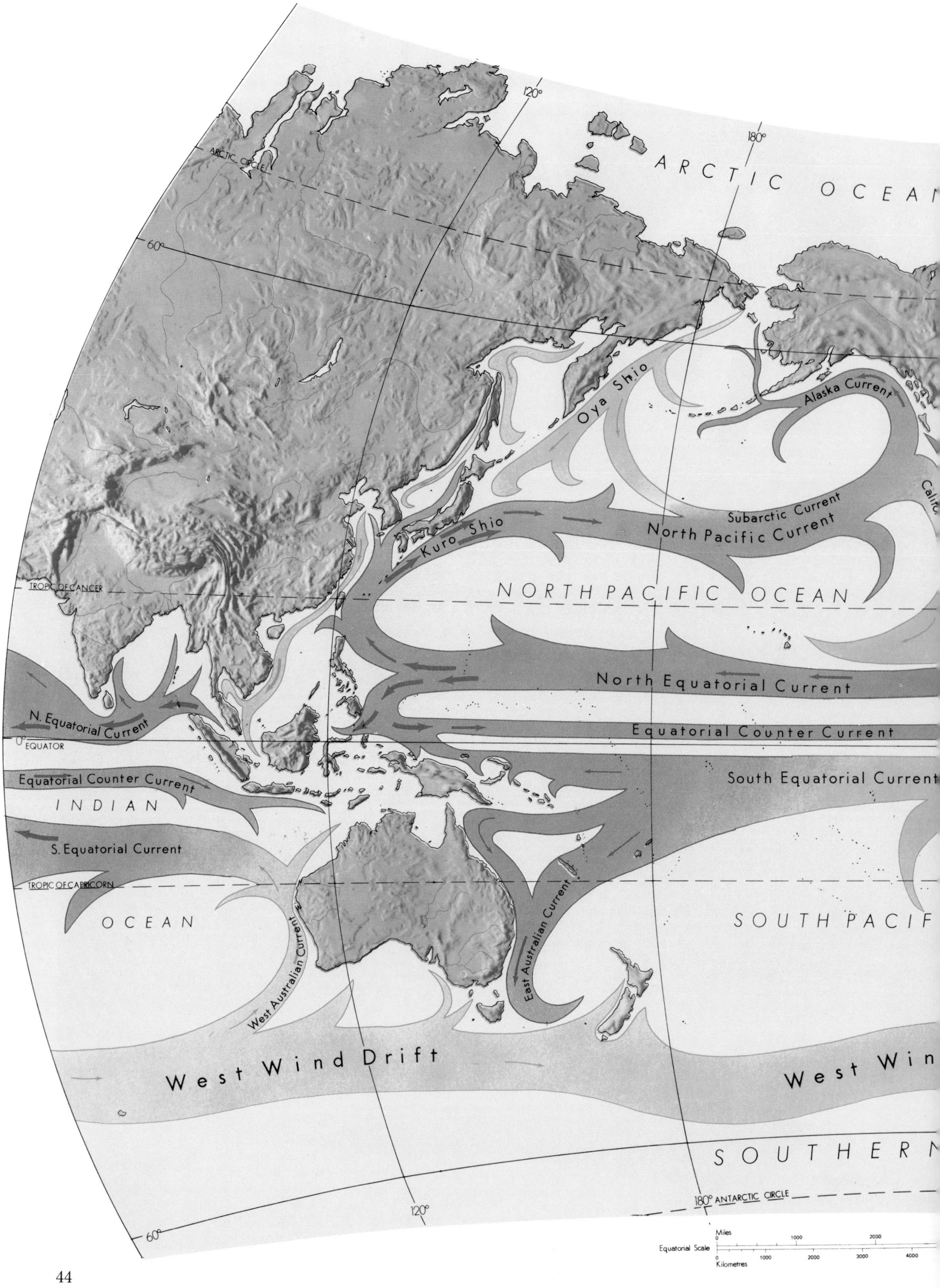

ARCTIC CIRCLE
ARCTIC OCEAN
120°
180°
60°
Oya Shio
Alaska Current
Calif
Subarctic Current
North Pacific Current
Kuro Shio
TROPIC OF CANCER
NORTH PACIFIC OCEAN
North Equatorial Current
N. Equatorial Current
Equatorial Counter Current
0° EQUATOR
South Equatorial Current
Equatorial Counter Current
INDIAN
S. Equatorial Current
TROPIC OF CAPRICORN
OCEAN
SOUTH PACIF
East Australian Current
West Australian Current
West Wind Drift
West Win
SOUTHERN
180° ANTARCTIC CIRCLE
120°
60°
Miles
0
1000
2000
Equatorial Scale
0
1000
2000
3000
4000
Kilometres

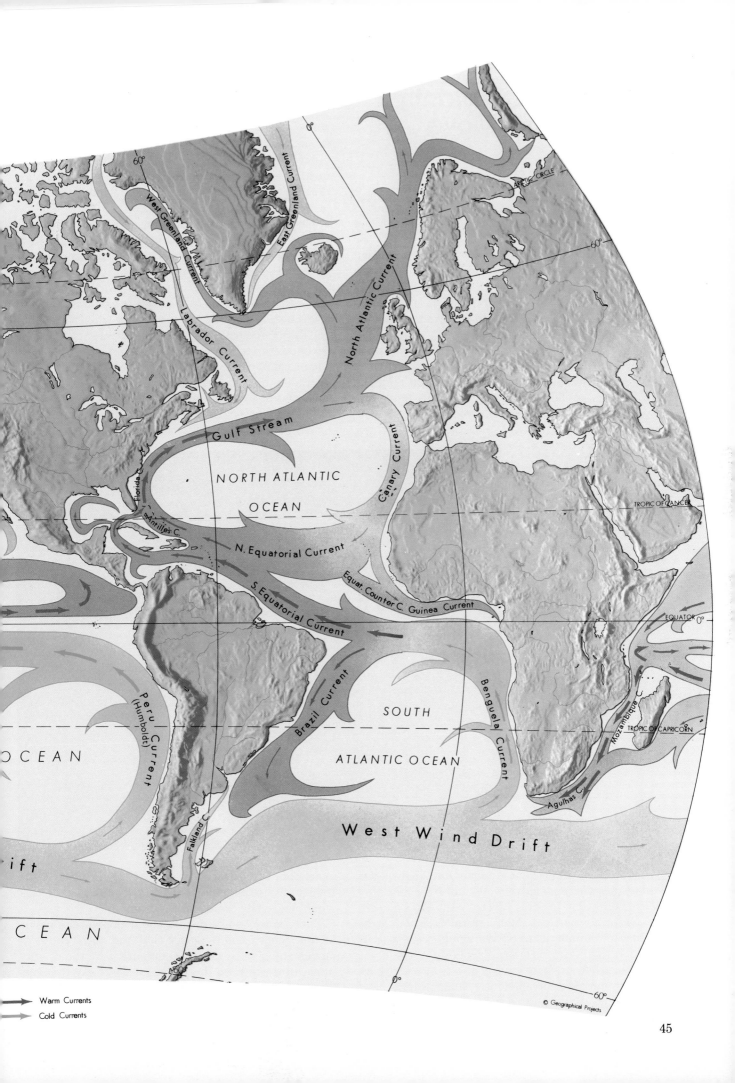

Warm Currents
Cold Currents

© Geographical Projects

45

Waters Constantly on the Move

The world ocean is constantly on the move – and in every direction. Anyone who stands on a beach can see this. Tides ebb and flow. The surface ripples. Waves rush in and break. All is in constant motion. However, these readily observable movements are mere ripples on the surface of the world's oceans compared to the underlying total flow. It is all on the move, steadily and inexorably, in the great worldwide flow of oceanic currents.

These currents have been known for a long time. Sailors have used them for centuries to speed their progress across the oceans. Some of these, like the Gulf Stream, are fairly well charted. But whatever is known about them, the mystery of why they flow is not yet known.

Right: a satellite picture of a typhoon vortex that occurred in the south central Pacific Ocean southeast of New Zealand. All scientists agree that winds affect currents, but some put a greater importance on them.

long-range weather forecasts. It is easy to see the great advantages of reliable weather prediction. Farmers throughout the world would be able to know exactly when to sow crops. They would know when to water them, when to fertilize them, when to reap them. They would know when to move their animals to different pastures, and how much winter food to store up for them. Human life could also be saved by having the ability to predict accurately the flooding of a river or the arrival of a storm surge.

Why don't we have more information about oceanic currents? One reason is that only in recent years have scientists become fully aware that oceanography, the study of the oceans, and meteorology, the study of the weather, should not be treated as different branches of the earth sciences. It might seem amazing that this awareness has developed so recently when the oceans and their currents play such a major part in the system that maintains life on this planet. But although fairly comprehensive charts of currents and winds have been available to mariners since the mid-19th century, each major current was considered in isolation and not much connection between the currents and winds had been established. It is only within the past 70

This piece of knowledge is important. If we knew what makes currents move we would be a long way toward being able to chart precisely the total pattern of oceanic circulation. Such information would be invaluable on many counts. We would have a basis on which to predict with some accuracy those areas where large fish stocks occur; on how to move cargoes around the world more efficiently; and, perhaps most important of all, on how to produce accurate global

years that scientists have been studying directly the theory and principles of oceanic circulation to ask what makes the oceans move around.

There is still some debate about the water's movements. Some scientists say that the main cause of movement of most oceanic currents is by the drag of prevailing winds on the surface. Others say that it is due to differences in water density, uneven heating, rainfall, and evaporation, combined with the effects of wind. The first theory is more widely accepted at the moment. It explains currents like this:

The sun heats the atmosphere unevenly, being hotter at the equator than at the poles. If the Earth – and hence the atmosphere and the oceans – was still, hot air would rise at the equator to be replaced by cold air flowing from north and south. There would be a simple north-south flow. In its wake, however, there would be a heat surplus at the equator and a deficit at the poles; the tropics would get steadily warmer, the poles steadily colder. In fact, of course, the Earth rotates, so the flow of the air streams is twisted from its north-south path in a phenomenon known as the Coriolis effect. This effect is at its maximum at the poles and at its lowest at the equator. In the northern hemisphere the Coriolis effect deflects winds to blow from the northeast to the southwest. In the southern hemisphere they are deflected to blow from southeast to northwest.

The Coriolis effect also works on the water flow. In this case, instead of flowing downwind the surface oceanic currents are deflected to the right in the northern hemisphere and to the left in the southern hemisphere. North and south of the equator, therefore,

Above: a drogue, which is an instrument for tracking currents. It is fitted with a pole and flag to make it easier to follow on its course, and it is tracked either from a boat or a shore site.

Left: a team of researchers tracking currents. Oceanographers use colored dye to trace the movement of currents in the layers of water. These layers are sometimes only a few inches deep.

47

Right: the Coriolis effect on the major currents of the Atlantic Ocean. It gives currents north of the equator a clockwise turn and those south a counterclockwise turn.

Below: a diagram showing how the South Equatorial Current, blown across the Atlantic by westward trade winds, is deflected on nearing the continental land mass. Part of it combines with the North Equatorial Current in a counterflow.

two east-west flowing currents are set up. These are the north equatorial and south equatorial currents, and they would flow completely around the world except for the interruption of land.

Take a bowl of water and blow steadily across it. The water will pile up only slightly against the far edge before finding its own level by deflecting and setting up a countercurrent. This is what happens in the oceans. The currents sweep into the eastern side of a continent and are deflected, some to the north and some to the south. In the northern hemisphere the southern deflected component of the equatorial current meets its counterpart flowing north from the southern equatorial current. The two combine to set up a counterflow running parallel to, but in the opposite direction to, the east-west flow of the equatorial currents. This weaker current is known as the equatorial countercurrent and it flows from west to east in the area between the two great trade wind airflows known as the doldrums.

In the northern hemisphere, this is what happens to the northern-flowing component of the equatorial current:

After flowing north, parallel to a coastline, the current reaches an area where westerly winds predominate. These westerly winds are caused by warm subtropical air moving north to meet cold polar air coming south, and at their meeting point they form complicated patterns in an area between 30° and 60°. Their effect is to drag the current away from the coast, sending it looping and spiralling first in a northeastern direction, then spinning in a general clockwise direction around the entire ocean basin in a

flow system known as a gyre.

This description gives a very basic notion of the great ocean currents. In the northern hemisphere there are two clockwise gyres, one in the Pacific and one in the Atlantic. In the southern hemisphere there are four counterclockwise gyres.

The most famous of these gyres is in the northern hemisphere and is made up of the northern equatorial current and the Gulf Stream. The Gulf Stream is the best charted of the major oceanic currents because the North Atlantic was the first of the great oceans to be explored, to be traversed for extensive colonization, and to be used for intercontinental trade. Explorers from Columbus onward were aware that the Atlantic's circulatory pattern could speed or impede their progress east or west, and theories about its overall make-up were being developed as early as the 16th century.

The first chart of the Gulf Stream was made in 1770 by Benjamin Franklin who was then Postmaster-General of the American colonies. He had noticed that mail packets were taking two weeks longer to cross the Atlantic westward than were merchant ships. He talked to mariners, especially fishermen, and calculated that the mail ships were sailing against the current. His charts showed the mail packet captains how to sail between the flow of the adverse current and the treacherous shoals off Newfoundland and New York in order to save time.

The Gulf Stream has its measurable beginnings in the 50-mile gap between Cuba and the United States – the Straits of Florida. Northward-flowing water from the southern equatorial current has met the westward

48

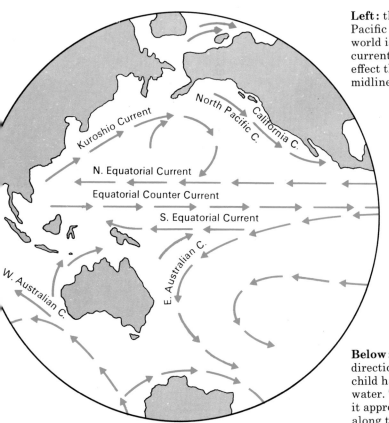

N. Equatorial Current

Equatorial Counter Current

S. Equatorial Current

Kuroshio Current

North Pacific C.

California C.

E. Australian C.

W. Australian C.

Left: the Coriolis effect on the major currents of the Pacific Ocean. Because the equator in this part of the world is in an area of mostly open ocean, the ocean currents move in a nearly straight line. The Coriolis effect therefore dwindles near the Earth's imaginary midline.

Below: an easy illustration of how currents change direction on deflection. In making the boat move, the child has created a current across the top of the tub of water. This current turns to the right and to the left when it approaches the opposite side of the tub, and moves back along the outside edges toward the place it started.

flow of the northern equatorial current off the Lesser Antilles chain of islands. The combined flow sweeps into the Caribbean Sea, moves northward, skirts the islands, and is ejected through the Florida Straits into the Atlantic in a mighty spurt 50 miles wide and a quarter of a mile deep. A tremendous amount of water is moved through this gap – something like 65 times as much as the combined flow of all the rivers in the world. Through the narrowest part of the Florida Straits – the Bimini Narrows – about 10 cubic miles of water is moved every hour at speeds of up to seven miles an hour. At this stage the Gulf Stream is moving so fast that it does not merge with the surrounding colder water, but pushes hundreds of miles north.

Later, like a great river, the Gulf Stream begins to slow down and meander, taking up a general circular motion. Tributaries turn south off southern Europe and northern Africa, finally rejoining the westward-flowing north equatorial current. Other tributaries continue in a general northeastern direction, some of them washing past the west coasts of Great Britain and Ireland until they die out north of Norway. The whole Gulf Stream has the shape of a wheel, with its hub the relatively motionless Sargasso Sea.

The northeastern-flowing tributaries have a profound effect on the climate of northwest Europe. They keep the ports of western Norway ice-free all winter, while the ones on the other side of the Atlantic and even those farther south in the Baltic freeze up. The Gulf Stream also keeps areas such as the northwest of Scotland mild in winter. But it is not just the stream that warms the area. The flow of warm southwestern winds blowing from the Sargasso Sea

and other temperate regions helps as well. Research continues into this complex aspect of land/sea inter-action, but it is thought that perhaps the Gulf Stream acts as a boundary current to prevent warm waters from mixing with the icy north seas, which would have a cooling effect on the air passing over them.

In the Pacific the northern gyre is made up of the northern equatorial current and the Kuro Shio current – the Pacific equivalent of the Gulf Stream. The Kuro Shio transports less water than the Gulf Stream – around 1766 million cubic feet per second off Japan – and reaches speeds of up to six miles per hour. Like the Gulf Stream it has an important effect on the climate of the northern Pacific, particularly of Japan, past which its flow is strongest.

The Kuro Shio current is the north-flowing extension of the northern equatorial current, starting its

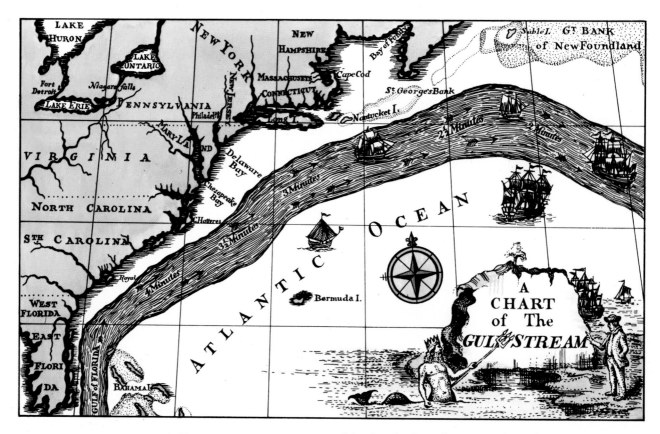

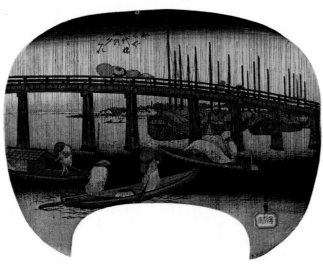

Above: the stormy weather so often part of the scene in Japanese paintings – such as this woodblock print by Hiroshige – is typical of regions affected by warm equatorial currents. In the case of Japan, it is the Kuro Shio Current off its southeast coast that helps determine the weather.

movement off the Philippines and moving along the coasts of Taiwan and southeastern Japan to finally turn east off the part of Japan called Honshu.

In the southern hemisphere the major currents are not nearly so well defined in terms of speed or volume of water transported. The direction of flow of the great gyres, of which there are four, is reversed. The three in the south Pacific, south Atlantic, and Indian Oceans have similar major characteristics. Water flows west in the southern equatorial current until it

hits land, then flows south to join the circumpolar current, the northern part of which carries the water across to the eastern land mass where it flows north to rejoin the equatorial current. In the Atlantic the Brazil current flows southward down the east coast of South America while the Benguela current flows north up the coast of Africa to complete the gyre. In the Pacific the East Australian current flows south while the Humboldt current flows north up the coast of South America, having a profound effect on fish population. In the southern Indian Ocean warm water courses down the east coast of Africa in the form of the Agulhas current, while on the other side of the ocean the cold West Australian current flows in a northerly direction.

Why is there no clockwise gyre in the northern part of the Indian Ocean to match those of the Atlantic and Pacific? According to the logic of surface current circulation such a system should exist. It does not because, for once, conditions on land dominate those in the ocean. The whole oceanic pattern in the Indian Ocean north of about 2° N is governed by the monsoons. Between April and October the Asiatic land mass heats up quickly, setting up a low pressure area to draw in winds from the southwest. This sucks along with it the eastward-flowing southwest monsoon drift current. Then from October until April there is a dramatic reversal. The land cools down swiftly, setting up a high pressure atmosphere so that the prevailing winds change direction to blow out to sea, dragging water away from the land. This water is deflected by the Coriolis effect to flow to the west in the same way as the northern equatorial currents in the Pacific and Atlantic Oceans.

50

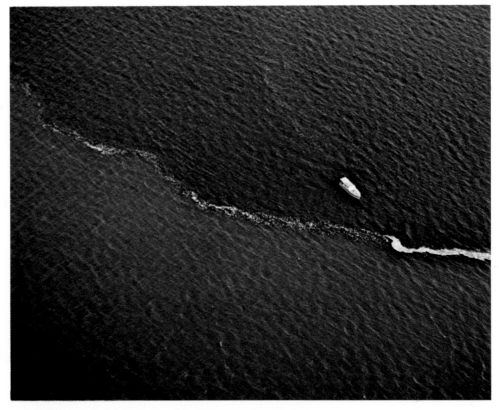

Opposite: an 18th-century chart of the Gulf Stream made by Benjamin Franklin for the guidance of mail packet captains sailing between Britain and the colonies. As the colonial deputy postmaster general, Franklin's advice was: "Don't fight the Gulf Stream."

Right: the meeting point of the Peru Current and the equatorial current. It is clearly marked by the contrast in the colors of the waters, the Peru Current being green – a sign of fertility – and the other being a dark blue.

The last of the great oceanic gyres is the circumpolar current in the Southern or Antarctic Ocean. This is the one place in the world ocean where the water can flow freely, unobstructed by land. Dragged around the world by fierce winds between 40° and 65° S, the current flows, with just one interruption, at a rate of 100 million cubic feet of water per second. The interruption is made by the Drake Passage, a 500-mile gap between the tip of South America and the Antarctic peninsula. The circumpolar current squeezes into this gap to move over 3000 million cubic

feet of water a second.

The circumpolar current does not exist in isolation. Winds, undersea ridges, and the occasional headland catch at it, sending it looping north. Because it is cold it has a high density, causing it to sink below warm water it meets drifting south, particularly at the Antarctic Convergence.

Whatever is known about oceanic circulation is complicated by important unknown factors. It has been learned, for example, that currents are set in motion and their momentum maintained by winds, although the exact mechanism by which a current is first made to flow by wind is still not fully understood; by the effect of the Earth's rotation, which is called the Coriolis effect; by changes in the level of seawater and differing pressure gradients when it is piled up

Below: the chart on the left shows the currents of the Indian Ocean during the winter and the one on the right indicates them during the summer. Unlike other sea currents, these are governed by the monsoons that strike on land, which causes them to reverse seasonally.

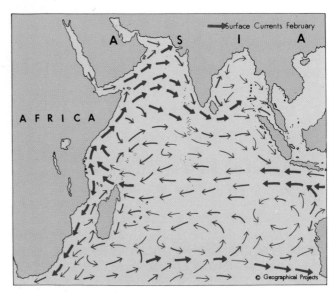

against the land and deflects; and by density differences. But each of these pieces of knowledge remains incomplete. Regarding the winds: they do not blow steadily from the same direction all the time. They swirl and move in different directions at different levels in the atmosphere at far greater speeds and in far more complex patterns than do the oceans which they drag in their wake. The Coriolis effect is not constant across the Earth's surface. Land masses against which the currents pile up do not have a regular shape. Therefore, the speed, volume, and direction of a major ocean current is changing constantly – from day to day, month to month, and year to year.

As complex as this is, there is even more to it because what has so far been described refers only to currents which extend on average to a depth of only 350 feet. But the depth to the abyssal plains is 10,000 feet and even there the water is moving. Sometimes the motion is so slow that it cannot be tracked by measuring the speed of flow, but only by plotting its temperature and salinity. Nevertheless, there is movement.

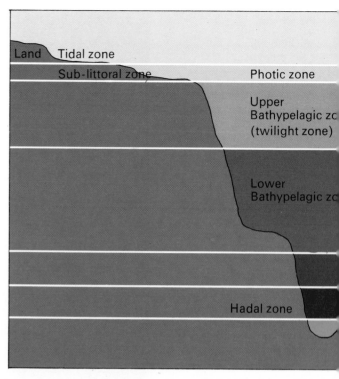

Above right: this chart of the mid-Atlantic floor, in which the vertical scale is much exaggerated, shows temperature variations at different depths of the ocean. Temperature and salinity have a great deal to do with creating currents. Cool salty water usually sinks, causing deep movements throughout the world's seas.

Right: *Ra II*, the papyrus boat in which the Norwegian anthropologist and explorer Thor Heyerdahl and seven others sailed from Morocco to Barbados, West Indies, in 1970. Heyerdahl was trying to prove that, once past the Canary Islands, the ancients of the Middle East could have easily drifted across the Atlantic on the North Equatorial (or Canary) Current to the New World. As he puts it: "To ride along in the Canary Current . . . to Middle America requires nothing but a support that will float."

Scientists have divided the layers of water in the ocean into five depth masses: surface, central, intermediate, deep, and bottom. The Atlantic Ocean provides a good example of the way in which these layers interact. The Weddell Sea, off Antarctica in the south, gets very cold in winter, sometimes below freezing point. It does not freeze solid because it is highly salty. This "heavy" cold water sinks and flows

north in great fingers along the ocean floor, traces of it having been found at 45 degrees north of the equator. This is called Antarctic bottom water. North Atlantic deep and bottom water forms off Greenland in the north, sinks, and makes its way south. It is not as dense as Antarctic bottom water and therefore flows on top of it all the way down to the South Atlantic where it reaches the surface at about 60° S

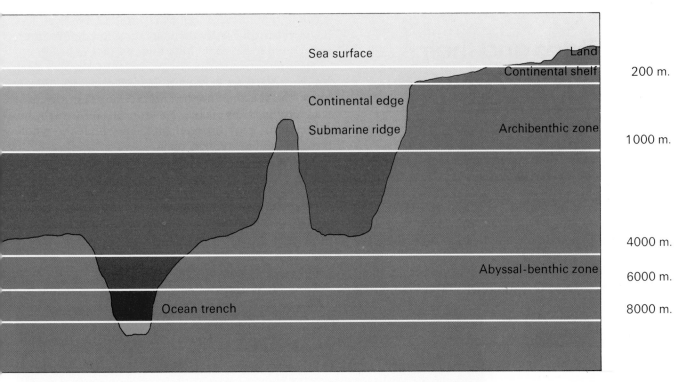

Sea surface   Land

Continental shelf   200 m.

Continental edge

Submarine ridge   Archibenthic zone   1000 m.

4000 m.

Abyssal-benthic zone   6000 m.

Ocean trench   8000 m.

Left: a satellite view of the Strait of Gibraltar. The surface current flowing into the Mediterranean from the Atlantic travels on top of the salt water flowing out. Below: a diagram showing how the Mediterranean water, made salty and dense by evaporation in the warmth, sinks and floats westward. The current is against a ship going west through the strait and with it going east.

at the Antarctic Convergence. Also at the Antarctic Convergence, Antarctic intermediate water cools during the winter and itself begins to sink and flow to 20° N where it mixes with North Atlantic intermediate water coming south. North and South Atlantic central water forms at around 45° N and S in the winter, sinks, and flows toward the equator. An added ingredient to this complex mix is provided by Mediterranean water. Cool salty water is made in the Mediterranean and this pours out over the sill at the Strait of Gibraltar into the Atlantic. It is replaced by less salty Atlantic water which flows through the straits on top of the outgoing water. A remarkable two-way flow is thus set up.

On the surface as currents and in the deeps, the waters are constantly sinking, rising, flowing, and mixing. There can be no absolute pattern to these movements, but greater knowledge can be – and is bound to be – gained. Oceanography is a young science and its marriage to meteorology younger still. But it has the advantages of modern technology. Armed with satellites, better ocean measuring devices, and

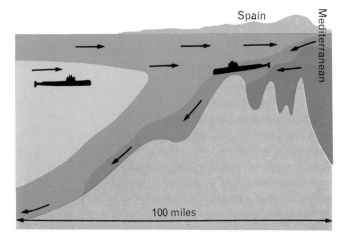

Spain

Mediterranean

100 miles

computers to analyze and predict, oceanography gives reason to suppose that there will be a breakthrough in knowledge and understanding soon. Unraveling the web of oceanic movement is one of the most fascinating and potentially one of the most rewarding activities left for the mind to challenge and conquer.

Tides and their Pattern of Motion

Tides and waves are the most easily seen evidence of the moving sea. More is known about tides than about many other ocean phenomena, and they have been observed for many centuries. When the Greek explorer Pytheas sailed to Britain in 325 BC, he formulated how spring, or high, and neap, or low, tides are related to the phases of the moon. The dictates of trade and war forced early mariners to study tidal patterns. For example, marauding Vikings mounting a cattle raid on a foreign coast would have been lost if, on returning with their spoils a few hours later, they found their boats high and dry on the beach.

The link between sun, moon, and tides has therefore been observed for centuries. However, people did not know the reasons for the interaction. The why of it was provided by Isaac Newton in 1687 with his principle of gravitational pull, on which the whole basis of tidal movement rests.

According to Newton's scientific finding, every single piece of matter on Earth is attracted by the gravitational pull of the moon. Even the land has "tides," but because it is a solid crust on the liquid mantle of the Earth, it moves very slightly. Fluids move so much more that there is a tidal range in any body of liquid, even in a cup of tea or coffee if you could possibly measure it.

To understand why tides vary in height and fre-

Above: this 8th-century stone relief from Gotland in the Baltic Sea shows two Vikings in a boat. All early seafarers had to learn about the tides.

Right: moon and water, the pull of the heavenly body causing the tidal pattern of the other. Although less powerful than the sun, the moon's effect is the greater because it is nearer.

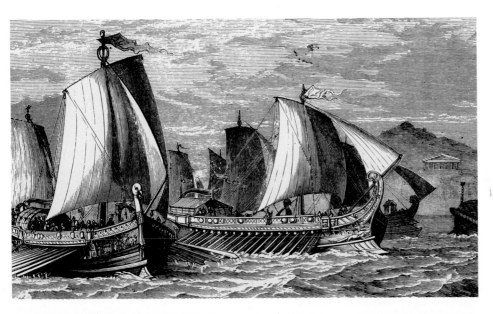

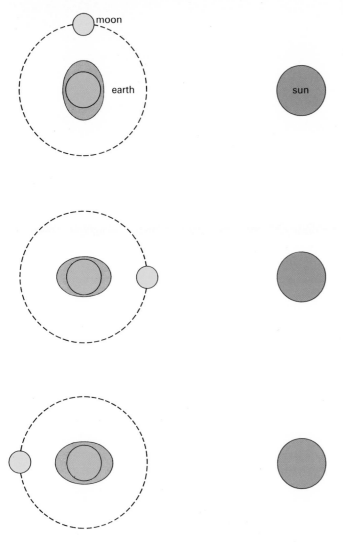

Above: the position of the moon and sun in relation to each other and the Earth affects the tides. When the sun and moon are at right angles (top), the larger and brighter sun offsets the gravitational attraction of the moon, and the range of tides in the Earth's waters is smaller. When the sun and the moon are on the same side of the Earth (center) or on opposite sides (bottom) their individual affect is more, and the tide range greater.

quency in different parts of the world, imagine a smooth-surfaced globe completely covered by water around which the moon revolves every 28 days. Each particle of water is attracted toward the center of the moon, but the farther away it is from the center, the weaker the attraction. So the waters nearest the moon are more strongly attracted to it and consequently are drawn out from the Earth's surface in a bulge. Because the Earth is rotating, centrifugal force is exerted. This opposes gravity, and this opposition keeps everything in balance. On the other side of the Earth farthest away from the moon, another bulge occurs. This happens because there is a weaker gravitational pull and the water tries to move away from the surface of the Earth.

The tidal pattern so established on the entirely smooth globe can be clearly seen to bring about two high tides each lunar day – one for the gravity-derived bulge toward the moon, the other for the bulge away from the Earth. Two low tides occur on those parts of the globe at right angles to the moon. The evening high tides on successive days tend to be more similar than the tides occurring on the morning and evening of the same day.

If the moon alone exerted a gravitational force, tidal variations would be constant over the lunar month. But the sun has its part to play too, although it raises tides to less than half the extent of the moon. This is because the sun, although having far more mass than the moon and therefore a stronger gravitational pull, is much farther away. When sun, moon, and Earth are in line at new or full moon, spring tides occur. When sun, moon, and Earth are at right angles at the first and third quarters of the moon, the forces exerted on the water want to cancel each other out. This produces neap tides.

The rotation of the Earth influences the times of tides. If the moon revolved round the Earth at the same speed as the Earth's own rotation, a high tide would occur at a fixed point on the surface each 12 hours. However, although the moon moves around

55

Right: the island of Saint Michael's Mount off the coast of southwest England. The natural causeway connecting island and mainland can only be used when the tide is low.

Opposite: a shoreline during high tide. The regular rise and fall of ocean waters is clearly noticeable. Tides in inland waters, on the other hand, are so slight as to be overlooked.

Below: wrecks near the Goodwin Sands, southeast England. Dangerous shoals like these are formed by tidal currents that pile up material in the sea.

the Earth in the same direction as the Earth is rotating, it moves just slightly slower. So the Earth has to rotate for an extra 50 minutes each day to bring the same point under the moon. This means that each of the two spring tides will be 25 minutes later each day.

The moon complicates things still further by not moving around the Earth in an exact orbit. It swings elliptically backward and forward in relation to the equator, setting the oceans oscillating in two types of tidal rhythm. In some areas the result is the pattern with which Europeans are familiar – two high and two low waters a day at roughly similar heights. These are semidiurnal oscillations. Elsewhere, such as in the Gulf of Mexico, there is only one high and one low tide each 24 hours and 50 minutes. This is a diurnal oscillation. In still other regions – the Pacific and Indian Oceans, for example – there are two high and two low waters a day, but the difference in heights of the morning and afternoon high waters is considerable. These are called mixed tides.

When sun and moon pull on the world ocean to create a bulge on either side of the Earth, it is not the bulge that is the tide. This can be proved by checking the time of a particular high tide against the position of the moon. If the bulge were causing the tide, the moon would be directly overhead. But the time difference between the passage of the moon and high water can be anything up to 12 hours and 25 minutes. This is not to say that high water will never occur when the moon is overhead but only that if it did, it would be purely coincidental.

What causes tidal flow, then, is not the bulge but the flow of water horizontal to the Earth's surface as it moves to make the bulge. When a body of water – in this case the individual oceans – is subjected to a regular disturbance – in this case the attraction of the sun and moon – it starts a rocking motion from side to side along a virtually motionless central line. The time taken for each rocking motion is determined by the shape of the coastlines surrounding the body of water, its depth, atmospheric conditions, and any

Above: a tide recorder, used to measure and record tidal movements.

Below: a special way of fishing in the Bay of Fundy in the horse-and-cart days. Because the tide range in this area of the northeast coast of the United States is so great, nets that collected fish during high tide could be easily emptied – on land – when the tide ebbed.

submarine features such as mountains and trenches. The time the rock takes is called the natural period of oscillation, and each ocean basin has its own natural period. When the natural period of oscillation is close to that of the tidal period a momentum builds up and high tides result. These are resonance tides.

There is very little tidal range in small bodies of water if these are cut off from the open ocean. That is why the tides in the Mediterranean – open to the Atlantic only through the narrow Strait of Gibraltar – are so small as to be almost unnoticeable. Coastal seas open to the ocean have a wider tidal range and many often get resonance tides. The rocking motion of the ocean is transmitted in the form of waves to the coastal sea which then sets up its own natural period of oscillation. If the natural period corresponds to the tidal period, resonance tides occur. This explains why there can be widely varying tidal ranges over quite short distances of coastline. At the center of the English Channel, for example, the spring tide range is about seven feet. Not far away in the Bristol Channel, the spring tide range at Avonmouth is 40 feet – one of the greatest in the world.

The single greatest tidal range in the world is in the Bay of Fundy between Nova Scotia and mainland Canada. The bay has a natural period of oscillation of about 12 hours. Its resonance tide, reinforced by the incoming Atlantic waters, normally rises 40 feet twice a day, reaching 70 feet at spring tides.

Tides can be predicted accurately for a given place once measurements of tidal range have been made at

that point over a period of time. On the basis of accurate predictions, the layout of ports can be planned, and the arrival time of all kinds of ships be scheduled. In a lighter vein, people building sand castles on the beach can calculate the time required to complete their architectural fantasies before the tide sweeps the structures away.

Any rise or fall in a body of water means that energy is being generated and theoretically is capable of being captured. Engineers are studying ways to capture tidal energy for generating power to keep modern society going.

57

What We Know about the Waves

Pick up the end of a piece of rope that is lying on the ground and give it a sharp flick. Waves run along its length but the rope does not move forward at all. It cannot, because it is being held at the end where the action started.

That is what waves in the sea do. They bob up and down without moving forward. But a wave of the height and speed that brings joy to a surfer on Eng-land's west coast may well have been generated by a storm in the South Atlantic over a week before. The storm has flicked the oceanic rope and sent a disturbance across the surface. The wave action has traveled, but no great mass of water has moved forward.

Waves are important. They are the main force in shaping our coastlines. They are what make some beaches sandy and others pebbly. They produce weirdly shaped coastal rock formations. They move vast chunks of coastline from one place – usually where the loss is most felt – and unceremoniously dump it somewhere else – usually where it is least wanted.

Scientists are as uncertain about how waves are generated as they are about the origin of currents. In both cases, this uncertainty exists because the study of waves and currents overlaps in oceanography and

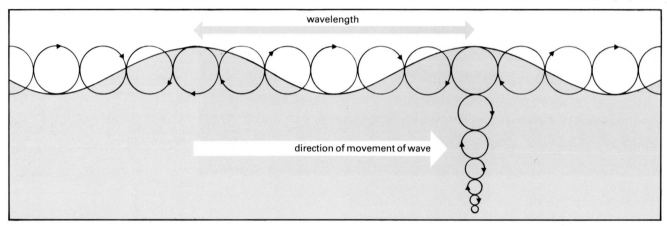

wavelength

direction of movement of wave

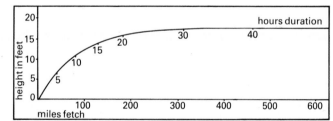

hours duration

height in feet

miles fetch

Above: a wave in deep water. Each water particle moves in an almost circular orbit whose diameter at the surface equals the wave height. Below the surface the orbital diameter diminishes sharply.

Left: the height of a wave depends on the duration of the wind forming it and on the length of the fetch – uninterrupted open water – over which the wind travels.

Left: Atlantic waves break-ing on an Irish beach. The shallow bottom at the shore slows and turns the waves so that they break parallel with the beach itself.

Opposite: water against rock. When waves batter a coastline, bits of rock they have previously picked up chip other pieces of rock off. The result is a constant wearing away of the land.

meteorology, and only in recent years have oceanographers and meteorologists come to the realization that they should relate their work more. It is fairly obvious that conventional surface waves are started by winds.

Wind blows across a stretch of water and produces ripples. In the oceans the wind does not blow steadily, either in direction or speed, but whirls around, picking up little bits of water and setting up a complicated chain of ripples which start traveling in the general direction of the wind. It is a case of cause and effect: as the wind keeps blowing, it finds it easier to make a bigger slope in the part of the water that it has already made a little slope in. The longer the wind blows at a particular speed, the greater the wave will grow in height and energy. Given a constant or increasing wind speed, the wave's ultimate height depends on the extent of the expanse of sea over which the wind has blown. The distance over the sea surface and the distance the waves travel under the influence of the wind is known as the fetch. As the waves build up, the distance between wave crests increases, and the longer this wavelength becomes the faster the wave will travel.

Remembering the rope analogy, it must be borne in mind that the water itself does not move forward. Rather, each particle of water in the wave goes through a circular orbit, the diameter of which is the same as the height of the wave. The particle moves upward and forward at the crest, then downward and backward through the trough. Throw a brick into a pond and it will start ripples. Immediately after, throw a rubber ball into the water and it will bob up and down without moving even though ripples are

Right: when waves break near shore, the greatest wave energy is concentrated on the short stretches of shore around a headland (A to B and C to D). Wave force (arrows) make the waves bend around the headland, causing erosion. Where the shore is recessed, wave force is diffused and material is deposited.

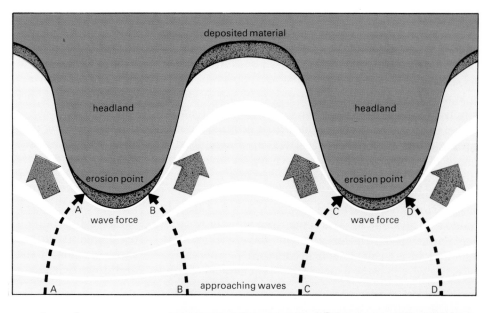

Below right: the surf. Waves break as they approach land because they are constricted by the shallowness of the bottom near shore. Breakers vary in form and force according to the shape of the bottom they move across.

breaking on the bank. It is the wave form that moves, not the water.

While the wind is blowing, the sea will look confused and angry. Under storm conditions the wave will get so steep that the crest topples forward and breaks in the open sea. These are the whitecaps or white horses that make such a dramatic sight.

Eventually the storm dies down, but the waves it has generated continue their rippling action. Away from the influence of the wind the wave changes shape – it becomes softer and more rounded. Its height decreases and the crest stretches out, often to a width of over a half a mile. The sea loses its confused look and an orderly procession of waves – now called swell – makes its steady progress toward the shore in long parallel rows.

Have you ever wondered how it is that waves nearly always break straight onto a beach, parallel with it? The swell comes in from the Atlantic into the English Channel, traveling roughly from west to east. Yet the waves break directly onto a beach that faces north in France and one that faces south in England. This is because swell is easily diverted by headlands, and can also be focussed on a spit of land in the sea.

As swell approaches land its characteristics change once the water depth is shallower than the distance from crest to crest. The swell begins to slow down, and this accounts for some of the beautiful sea surface patterns seen from the top of a cliff overlooking a bay. If the bay has a smooth sandy seabed deeper in the center than at the edges, the swell in the center maintains its speed while that on the edges slows down, producing a pattern that makes it look as if the swell has "fitted" itself to the shape of the bay.

As a wave runs into shore, the shelving seabed slows it down and it gets closer to the wave in front. But it is carrying just as much energy as ever and this energy has only one place to go – upward. Energy is transferred from the bottom of the wave to the top where the water particles speed up so that eventually the forward movement at the crest gives the wave a

height it can no longer support. It topples forward and breaks. When waves move across a gently sloping sandy seabed, their rise in height is gradual and the crests begin to curl and foam some distance out to sea. These are the ideal surfing waves to be seen off California on the west coast of the United States and – to a lesser and certainly more chilly extent – off Cornwall on the west coast of England. Where there is a steeply shelving beach, the swell pattern changes much more quickly and dramatically close to shore to produce foaming breakers.

Once waves have broken that is the end of their lives. But it is certainly not the end of their effect on our lives. The water that flows up a beach from a breaking wave has to go somewhere. Most of it returns to the sea, creeping under the incoming breakers. We can feel this when we stand on a shoreline and our feet tend to get pulled seaward from beneath us as shingle raps against our ankles. Where the incoming waves are strong the water turns and runs parallel with the coast in a longshore current until it reaches a place where the incoming waves are

weakest. It can then flow out to sea. If there is a gully or stream bed running up the beach, or a place where waves have broken across a promontory, the outgoing flow might be channeled into a strong jet called a rip current. This sand-discolored flow is dangerous to bathers and boaters.

As they progress along the shore, the currents generated by waves carry material from the land with them, especially on smoothly curving coasts. Sand and shingle are eroded and deposited out at sea or on another stretch of coastline. Beaches can be eroded quickly, which is why so many have a familiar pattern of groynes or breakwaters to prevent material being scoured away. We can judge the effectiveness of these simple wooden walls by looking at them at low tide after a storm. The sand will have piled up against one side but will have been scoured away from the other. Some beachgoers might even have had the unfortunate experience of having leaped over what was thought to be a two-foot high wall of sand with more beach behind only to plunge into a pool of water about six feet below.

Longshore currents, especially when they combine with currents set up by tidal flow, can be particularly destructive. At certain points along the east coast of England the land is being worn away at a rate of 17 feet a year. Much of the material is washed out to sea. The rest is redeposited, often in the form of long curving spits such as Spurn Head, Yorkshire on the northeast coast and Chesil Beach, Dorset on the south.

Both the longshore currents and the breaking wave itself can pack a heavy punch. A succession of big storm waves can temporarily strip a beach of its sand, or hurl it all inland. It will crack open cliffs. A second or two after the wave has thumped against a rocky shoreline there will be a whoosh, and a spout of water will fly skyward up the rock face. The wave will have compressed the air in the niche of a rock and the air will then explode to produce the spout. The niche will eventually succumb to the enormous forces continually being set up inside it, and the rock face will crumble.

Waves themselves have the power to unleash the

Right: Spurn Head on the North Sea coast of England. This pebbly sand spit extends for four miles across the mouth of the Humber. It was formed by longshore currents that deposited material there after eroding it from the nearby land.

forces of compressed air. As the crest curls over at the point of breaking, a pocket of air is often trapped inside. When the wave with its pocket of rapidly compressing air hits a cliff or sea wall, the air is released in a great spout of spray which carries sand, shingle, and even rocks high into the air. The lighthouse keepers at Dunnet Head, which stands at the entrance to the narrow Pentland Firth between the Orkney Islands and the mainland of Scotland, know the power of the waves full well. The lighthouse stands on top of a 300-foot cliff, yet often its top windows have been broken by stones that the sea below has flung up. This can be one of the angriest seas in the world when tidal currents meet storm-driven Atlantic waves.

The water flung up by breaking waves when they thump into a cliff can often confuse people into thinking that they have seen a truly enormous wave. But the biggest "true" waves occur in the open sea and only rarely. The height of a wave in the open ocean, averaged out clear around the world, is only about 12 feet. The biggest waves – in the Atlantic and Antarctic – get up to around 55 feet, but only for about nine weeks of the year.

There are the freaks, however, and they get very big indeed. To understand how they happen it is necessary to return briefly to the formation of waves. Although waves move in the general direction of the wind, the wind is swirling and eddying, and the same effect is created on the sea surface. A pattern of small waves of different heights, speed, and localized direction is set up, giving the sea surface a confused look. That confusion can now be sorted out in much the same way as electronics engineers sort a mass of confused radio or optical signals: a filter technique is used. An instrument records the distance between the crests of all the waves. The recorded signals are then played back through an electronic filter so that only those waves with a particular distance between crests are shown on the record. The process is then repeated to show another set of waves. Then by "tuning" the filter, a plot is produced which shows every type of wave – its speed, height, and direction – that makes up the confused sea. The technique is called spectral analysis and it can be invaluable in predicting the time at which storm waves are likely to reach shore.

Sometimes the mixed waves from a confused sea can combine. There is a great mix-up of energy and a huge wave builds up in the middle of a storm. It is not spread over a great distance and it disappears as quickly as it arrives. Storm waves of this type are probably responsible for many of the sea's tragic mysteries, accounting for ships that vanish without trace after being turned turtle and sucked under by a sudden huge wave.

The biggest storm wave recorded by a ship measured 112 feet from trough to crest. The trouble with measuring such giant waves from a ship is that

concern for survival can naturally overcome the concern for science. Today easier ways of measuring the height of freak waves has been provided by the installation of fixed oil production platforms in one of the worst stretches of water in the world – the northern North Sea. Because they offer a stable base from which to observe and make measurements, platforms such as these are providing us with a fund of knowledge about the movement of the oceans as well as playing a vital role in producing offshore oil.

Above left: a choppy sea like this is an excellent base for study of wave motion. By treating such a mixture of waves as simple patterns electronically, the complete spectrum of movements can be analyzed.

Above: the lighthouse at Dunnet Head, Scotland. Though it looks safe from any waves at its 300-foot height, it has had top windows broken by stones flung up by waves below. Such power comes from air compression by the waves.

Left: a fishing vessel battling storm waves in the North Sea. Storm waves are often huge, but their ferocity is great at any size. They can give even a good sized ship a severe buffeting.

63

Ocean Chemistry

Winds, waves, currents, and tides keep the world ocean in a state of perpetual motion. Water is an almost universal solvent. Anything that is capable of being dissolved will be dissolved by water. Even the glass that holds water will gradually be dissolved by its contents.

These two facts – the perpetual motion and the universal solvent – account for the saltiness of seawater. The oceans have continually scoured the continents and the seabeds, as well as collecting the dissolved minerals carried seaward by the rivers and the suspended solids that those rivers have pulled from the land. Once in the sea the dissolving and mixing power of the oceans insures that all the material is blended into the oceanic cocktail.

Seawater might be described as a dilute solution of

Right: this map shows the salinity of different parts of the world ocean. Salinity is measured in parts per thousand by weight (‰), which is indicated on the key.

Above: in every wave and in every part of the world's seawater lie the same elements that exist on land. Some are only traces, and all are much diluted.

all the elements that exist on earth. However, it is such a dilute solution that scientists leave their academic options open. "We shall state that seawater contains all naturally occurring elements on the assumption that one day we shall have the sophisticated measuring and analytical devices available to prove our hypothesis," they say.

Analysis is difficult because more than 99 percent of the dissolved matter in seawater is made up of only 11 major constituents. Of these, two dominate. Chloride takes up 19.35 grams per kilogram and sodium 10.76 grams per kilogram. At the end of the 11-element list comes fluoride at 0.001 grams per kilogram, yet it is considered one of the major constituents. By this it can be seen that the minor constituents and the concentrations are extremely small, but in terms of their total weight some of them seem to represent enormous riches. Gold, for instance. There are around nine million tons of gold in solution in the world oceans, and this gold is more or less universally distributed. So there would seem to be a marvelous opportunity for somebody simply to suck it out of the water on the local beach. That is what the German chemist Fritz Haber thought he could do when he calculated that the German debt from the 1914–18 war could easily be paid off by processing seawater to extract the precious metal. He had gone wrong in doing his sums about the concentrations of gold, however. The metal occurs in concentrations only of 0.000004 milligrams per kilogram of seawater. To extract the tiniest heap, millions and millions of tons of seawater would have to be processed and millions of tons of worthless elements be disposed of.

The total concentration of all the dissolved salts in the oceans is what determines seawater's salinity, and salinity contributes to density. It is a tribute to the dissolving and mixing powers of water that the saltiness of the world ocean is fairly consistent.

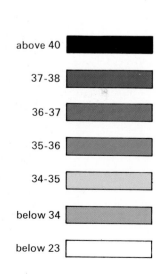

above 40	■
37-38	
36-37	
35-36	
34-35	
below 34	
below 23	□

Above: a person can float with ease – in fact, can hardly sink – in the Dead Sea because of its high salinity. This degree of salinity occurs because all the fresh waters flowing into the landlocked sea are quickly evaporated in the extreme heat. Dense salty water keeps the body up.

Salinity is expressed in terms of parts per thousand (‰) and the average value of seawater is 35‰. About 97 percent of the world ocean has a salinity value of between 33° and 37‰. For the other three percent, extremely localized conditions affect salinity. Where it is very warm, water evaporates from the sea and leaves salts behind to increase salinity. When seawater freezes, the ice formed is fresh water and the unfrozen water left has a high salinity. If it rains a great deal, the seawater becomes diluted and therefore less salty. In coastal areas where the sea is almost enclosed and the weather cool, fresh water running from the land decreases overall salinity. The Baltic Sea, for example, can have a salinity concentration of as little as 7.2‰. Conversely, in an enclosed sea in tropical regions, saltiness can reach very high levels. In the Red Sea salinity builds to nearly 40‰, and even higher in the Dead Sea. It is possible to float so easily in the Dead Sea's warm waters because the water is so dense that it will support a load far more readily than other seas or any body of fresh water. Scrubbing off the salt crystals after a swim in the highly saline Dead Sea can be painful, however.

What is interesting and useful to the scientist is that although the salinity of seawater may differ in various parts of the world ocean, the proportions in which the elements that make it up occur vary only

very slightly from place to place. In fact, for oceanographic work it is safe to assume that there is an absolute relative constancy of elements. This makes for fairly simple chemical analysis to determine salinity. In a sample of seawater, the elements of chlorine, bromine, and iodine can be thrown out of solution by adding silver nitrate. The precipitate formed can be measured to determine the amount of chlorine present in the sample and, since the proportions of the other salts are known, the overall salinity of the sample can be determined quickly.

It is not vitally important to know the salinity of a sample of seawater just for its own sake. What is important are the clues that accurate salinity measurements provide to the movement of water masses. Although the speed and direction of surface currents can usually be measured directly by a simple meter, many currents – particularly subsurface flows in deep water – move so slowly that no instrument could possibly record their speed. It has been calculated that Antarctic bottom water, for example, has not been at the surface of the Atlantic for 750 years.

The way in which the movement of a great ocean bottom current is tracked is to measure its density, since this characteristic distinguishes it from surrounding water flows. Density cannot be measured at sea and therefore has to be calculated from a number of factors, the chief of which are salinity, temperature, and depth. Staying with the example of Antarctic bottom water, we have already seen that this has a high salinity because so much of the Southern Ocean's fresh water content is removed to form ice. It is also very cold. The dense, cold water sinks beneath the less dense water in the Antarctic and flows slowly northward along the ocean floor. In the mid-Atlantic, then, the oceanographer can obtain a sample of water from near the ocean floor and by measuring its salinity and temperature can identify it as Antarctic bottom water.

The oceanographer also has the tools to measure other variables in the chemical make-up of seawater and these also provide valuable clues to its distribution and fertility. In addition to its dissolved solids, seawater has a dissolved gas content. Gases become dissolved in a number of ways. When rain hits a sheet of water it creates little bubbles which trap air, for instance. Rivers foam and spray into the sea. Chemical reactions in the water itself create gases. Four main groups exist. Inert gases – nitrogen, argon, helium, neon, xenon, and krypton – get into the seawater directly from the air and through river flows. Oxygen is introduced in the same way as well as being a product of the process of photosynthesis by marine plants. Carbon dioxide is produced by air/sea reaction, land run-off, and decaying plants and animals. Other gases are produced by all these methods as well as being a result of air pollution.

Oxygen, nitrogen, phosphorous, and silicon are four of the essential elements for sustaining marine life and the study of these nutrient chemicals by

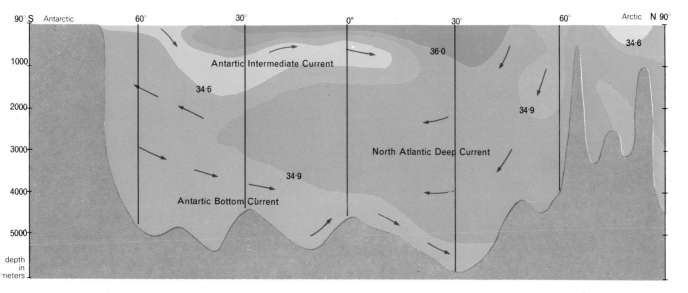

90° S Antarctic   60°   30°   0°   30°   60°   Arctic N 90°

1000

36·0   34·6

Antarctic Intermediate Current

34·6

2000

34·9

3000

North Atlantic Deep Current

4000

34·9

Antarctic Bottom Current

5000

depth
in
meters

Above: deep water currents in the Atlantic between the Antarctic (90°S) and the Arctic (90°N), with salinity figures. Salinity helps to determine currents.

Left: pack ice in the Antarctic. The extreme cold freezes fresh water to form ice, thereby leaving the remaining water highly saline and dense.

Below: marine scientists at work. Salinity is one field of investigation.

marine biologists is one way toward the more efficient harvesting of living marine resources. The alkalinity and acidity (pH) of seawater also varies with depth and local conditions and its measurement, together with those of temperature, salinity, and dissolved gas content further enhances the accuracy of analysis of water movement. Naturally occurring radioisotopes have recently aided the oceanographer's analytical armory. Carbon 14 isotopes decay at a fixed rate over a very long period, and by removing them

from a seawater sample and counting how many remain, the age of the sample can be expressed with a high degree of accuracy.

The study of currents, tides, waves, and marine chemistry forms the basis of the science of oceanography, which is one of the most exciting activities open to inquiring minds. Through oceanography, scientists grapple with the complexity of the oceans to further our understanding of its mysterious processes.

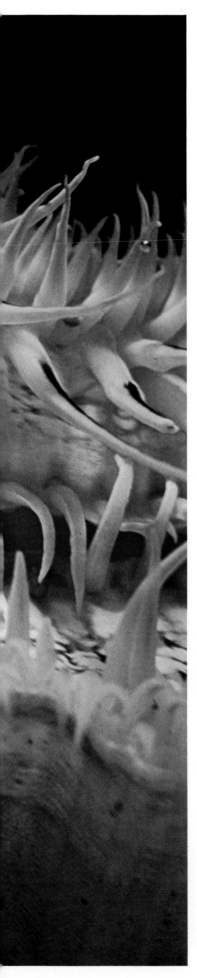

Chapter 4

Extracting Ocean Secrets

Until nearly the end of the 19th century, human knowledge had not particularly concerned itself with what happened beneath the sea surface. It was the increasing interest in fish as a major food source and the laying of telegraph cables on the sea floor that prompted the need to know more about the depths of the oceans and the creatures that inhabited those depths.

The first major oceanographic expedition was by HMS *Challenger*, and it pioneered an era of systematic study of the oceans. All the research expeditions have revealed how difficult it is to make things work at sea; to determine exact position; to "see" underwater; and to measure such things as temperature, current speeds, and the exact depth of the seabed.

In retracing the routes and work of those early expeditions, we can take a look at the way in which equipment has been developed to keep pace with the ever-increasing demand for knowledge of the oceans. Some of that equipment has played its part in one of the greatest scientific discoveries of our time.

Opposite: for many long years scientists argued whether anything could exist beyond a certain point in the ocean. When proof came, it revealed life in the deepest parts.

Explorers and Ocean Knowledge

The world ocean has kept its ancient secrets for so long partly because the study of it on a systematic basis is such a recent science. For example, only over about the past 50 years have comprehensive attempts been made to plot the depths of the oceans. By as

tides and relate springs and neaps to the phases of the moon.

Magellan embarked on his epic circumnavigation of the globe in 1519 partly because Charles V of Spain had promised him a share of the profits resulting from the discovery of exploitable South Sea Islands. Captain Cook ostensibly was sent into the Pacific so that scientists could observe the 1769 transit of Venus from Tahiti, but he had sealed orders from the government to find the "Great Southern Continent" and to explore the coast of New Zealand for Britain's possession. His discoveries led to the addition of Australia to the British Empire. During his epic voyages he made two invaluable but widely differing

Left: depiction of Ferdinand Magellan entering the strait now named after him. Explorers like him expanded human knowledge of ocean size and currents, but not on a scientific basis.

Below: an early ship's chronometer. This instrument helps determine longitude at sea from time, using the international time standard at Greenwich, England. Captain James Cook advanced the science of seafaring by his use of the chronometer.

recently as 1914 there were records of only 6000 deep ocean soundings.

The awareness that a better understanding of the complete oceanic system can lead to increased material wealth has been the main spur to the burgeoning of oceanography as a science. As is the case for so many other fields of science and technology, the greatest advances in our knowledge of the oceans have resulted from the need to capture or colonize new areas of the Earth or to steal a technological march on an enemy. It seems always to have been so. Brave Pytheas sailed from France and out of the Mediterranean into the unknown Atlantic in 325 BC not only to see what lay on the other side of the Pillars of Hercules, but also to investigate the source of tin that had been making its way onto the Greek market from the north. He found it in Cornwall, England. On this journey of exploration he made some valuable contributions to marine knowledge. He devised a fairly accurate way of fixing latitude; he sailed around the British Isles; and he was one of the first to describe

contributions to the art and science of seafaring. He showed that scurvy, the scourge of the sailor, could be completely eliminated by the provision of a balanced diet. He also proved that it was possible to obtain accurate longitudinal positions by use of the recently developed chronometer – in terms of accuracy, the 18th-century horological equivalent of today's quartz crystal digital watches.

Although for thousands of years commerce, territorial gain, and war dictated the growth of the ocean sciences, much of the information gained was kept secret. Carthaginian navigators were threatened with beheading if they divulged details of optimum sea routes, and Spanish rulers of the 16th-century, whose mariners discovered the advantages of the Gulf Stream and the north equatorial current, kept the knowledge to themselves. In any case, very little had ever been written down about navigational techniques, optimum routing, or any other finds about the oceans. What was transmitted at all was by word of mouth. Even today, skippers of trawlers fishing the rich grounds of the North Atlantic have their favorite catching areas which they try to keep to themselves. The way in which they rig their fishing gear, the exact speed at which they tow their nets, the subtle changes of course they make while the trawl is on the seabed – all these pieces of knowledge are still jealously guarded secrets.

In the purely scientific world of today, there has developed a remarkable clubbing together of oceanographers, with free exchange of all but the most sensitive information. Cold wars may grow colder – or heat up alarmingly – but the camaraderie of the oceanographers flourishes. The story may be apocryphal, but it is said that at the height of the Cuban missile crisis in 1963 when United States warships were preventing Soviet ships from gaining access to Cuban ports, a routine international oceanographic mission was being undertaken in the Atlantic. An American research ship met a Soviet research ship in mid-ocean. Scientists on both sides knew each other and, crisis or no crisis, they acted in friendship. The two ships hove close to each other, greetings were exchanged, and the Soviets sent over some choice fish. The Americans reciprocated by sending back cans of beer. The Soviets looked puzzled by these until the American seamen indicated, by mime, the way to open a ring-pull can. The beer had been well shaken on its journey between the two ships, and when a Soviet seaman pulled the ring, there was a loud pop. Suspicions aroused, the Soviets threw themselves to the deck, waiting for what they thought was an American bomb to go off. But the incident ended with understanding restored.

International cooperation in the ocean sciences was fostered by Charles Wyville Thomson, who led the world-famous *Challenger* expedition. This was the expedition that first awakened wide interest in what lay beneath the waves. Until then, scientific observations had been largely confined to the surface with the

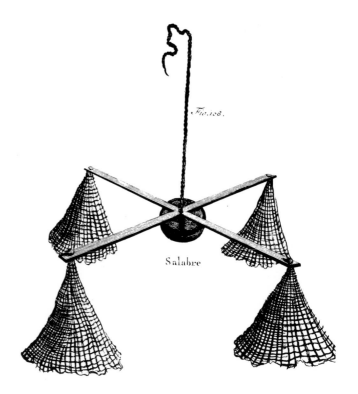

Above: an illustration of the dredging device used by Count Luigi Ferdinando Marsigli, as it appeared in his book *The Physical History of the Sea* (published 1711). This naturalist and geographer, who founded the Italian Institute of Science and Arts, was a pioneer in the field of oceanography. He used this dragnet to collect marine animals, one of the first scientists to do so.

almost sole aim of improving navigation. In the 1670s, for example, Sir Edward Boyle had proved that pressure in the oceans increased with depth. The 18th-century at least paid a little more attention to the undersea. Count Luigi Ferdinando Marsigli published *The Physical History of the Sea* in 1711. He was one of the first to use a dredge to collect marine animals, and he also constructed a meter to measure currents in the Bosphorus. Swedish botanist Carolus Linnaeus devised a classification system for living creatures which gave much impetus to the search for new marine life. Alexander von Humboldt, who laid the foundations for modern meteorology, also appreciated that the atmospheric processes were inextricably

Right: two species of fish discovered by Charles Darwin on his voyage in the *Beagle*. Both are of the Scorpaenidae family and both are found in the Galápagos Archipelago. Darwin's fame rests on his Theory of Evolution, but his work in marine biology would have gained him a reputation on its own.

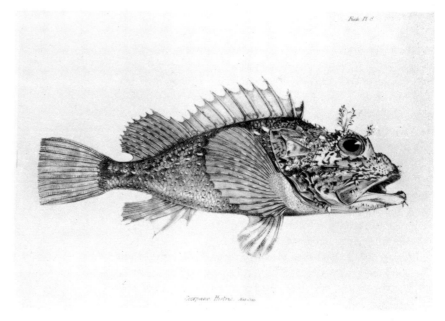

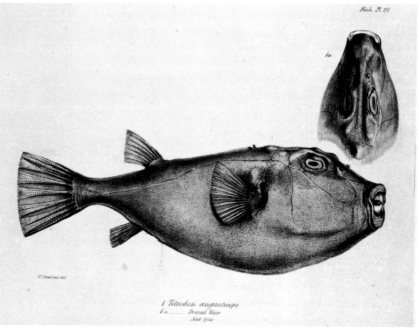

entwined with those of the oceans. He made measurements of the water temperature off the coast of Peru and documented what mariners had noticed for hundreds of years – that there was a cold current sweeping north along the west coast of South America. Humboldt, after whom this current was named, formulated the theory that the water originated in the Antarctic. He also studied the marine life that teemed in the cold water. It was left to later students of the area to show that the low temperatures were caused by upwelling of the subsurface waters as the surface waters were dragged away by the wind. This upwelling water is rich in nutrients and supports massive stocks of fish.

Humboldt's work was carried out at the turn of the 19th century, and from then on oceanography began to be taken seriously. The Royal Society of Great Britain, which until then had taken scant notice of oceans, began to devote more time and attention to oceanic subjects. Charles Darwin, who had read Humboldt's work, made his 1831–1836 voyage as an unpaid naturalist on the *Beagle*. The theories he began to formulate on that voyage changed the course of human thought, and their towering importance tends to overshadow some of the contributions he made to marine biology.

Benjamin Franklin, the American statesman whose wide interests included science, made his contribution by mapping the Gulf Stream. Another American, Lieutenant Matthew Fontaine Maury, saw the great potential for gaining oceanographic data by having the navigators of commercial ships make routine observations of winds, temperatures, and current speeds and directions. His organizational genius in obtaining and disseminating the information was calculated to have saved shipping lines millions. As a direct result of his work, for example, sailing times from Britain to California were reduced by as much as

30 days by making optimum use of currents and winds.

The last quarter of the 19th century saw more and more interest focussing on the underwater. This was stimulated by the increasing importance of the fishing industry, which demanded fundamental biological studies, and by the laying of subsea telegraph cables, which required data about the nature of the deep ocean floor, ocean currents, and the marine life that might affect the cable.

The results of expeditions mounted to provide some of this information promoted a growing interest in the oceans and helped establish the right climate to launch the greatest-ever single voyage of oceanographic study: the *Challenger* expedition.

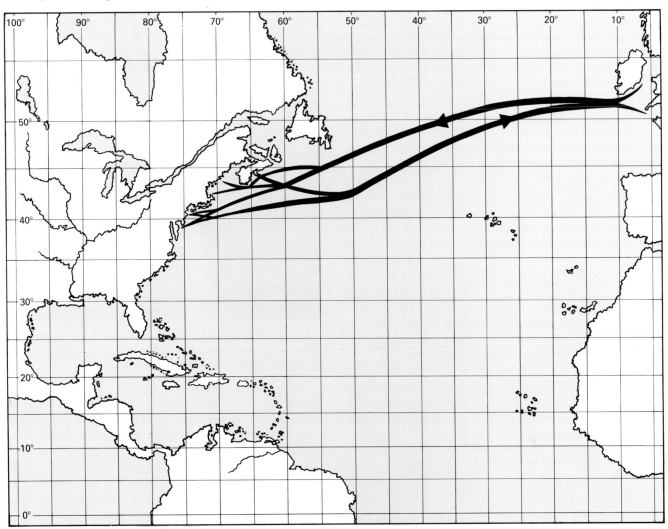

Above: a chart of the best commercial routes across the Atlantic, based on the one published by the American oceanographer Matthew Maury in 1855. He worked them out after a scientific study of currents and winds.

Left: a drawing of the search for a fault in the Atlantic telegraph cable in 1865. The cable had to be recovered from the seabed for inspection. The need for accurate oceanic data in laying cable promoted study of the seas.

Pioneer Expedition of the *Challenger*

By all accounts, HMS *Challenger* was a foul ship to live in, let alone in which to carry out delicate and precise laboratory work. A 226-foot naval corvette with sails and auxiliary steam power, it was slow and prone to roll heavily – through 46 degrees in one direction and 52 in the other. These adverse conditions faced *Challenger's* scientists during a three-and-a-half year circumnavigation of 68,890 miles, covering over 140 million square miles of ocean.

The concept of the *Challenger* expedition was not seized upon with great enthusiasm by the British government. Funds were allocated because, it was argued, the cost of the voyage would not be much more than that of keeping the ship afloat and the crew paid. The enthusiasm and drive that was responsible for getting the epic voyage approved was generated largely by two men – William B. Carpenter and Charles Wyville Thomson.

Carpenter was a biologist. He was also the type of administrative scientist who, when it becomes necessary to extract funds for a favorite project, will wheel, deal, cajole, and threaten with all the skill and determination of the most hard-headed businessman. In 1868 he suggested to Thomson that they collaborate on investigations of life in the deep oceans. Thomson was at that time professor of Natural History at Queen's College, Belfast, and a leading marine biologist.

The two men wanted to undertake deep sea dredging for marine life to prove that life did not stop at a depth

Above: the dredging and sounding equipment on H.M.S. *Challenger*. After use, the equipment was reeled in by two steam-powered drums near the mainmast.

of 1800 feet in the ocean, the point of view widely accepted at the time. This theory had been made in 1840 by Edward Forbes of the University of Edinburgh, who seemed totally to ignore the earlier finding of the Arctic explorer John Ross. On his search for a northwest passage in the Arctic, Ross had developed a grab that brought up worms, starfish, and other live animals from depths of over 18,000 feet.

Left: the two scientists who were largely responsible for the *Challenger* expedition – William Benjamin Carpenter and Sir Charles Wyville Thomson (far left). The idea originated with Carpenter, a naturalist and physiologist. He sought and gained the co-operation of Thomson, an influential marine biologist, who directed the ship's scientists.

Above: a watercolor drawing of H.M.S. *Challenger*. The ship, which was lent to the Royal Society of London by the British Navy for the scientific ocean expedition, had an auxiliary steam engine but mostly used sail.

Forbes' influence as a founding father of marine biological studies was so great, and Ross' findings were so little publicized, that the idea of a lifeless zone beneath 1800 feet had persisted.

Thomson, who had studied with Forbes, had accepted the theory, but began to change his mind in the 1860s when more and more evidence to the contrary came to light. For example, the Norwegian marine biologist Michael Sars had shown Thomson animals dredged from 2700 feet, and other creatures had been found clinging to a cable brought up for repair from 7000 feet beneath the Mediterranean.

Carpenter persuaded the Royal Society, of which he was then vice-president, to approach the British Admiralty to let him and Thomson sail on a hydrographic survey ship doing routine work in the North

Atlantic. He also extracted some funds from the Royal Society to pay for equipment.

The two sailed in August 1868 on HMS *Lightning* – a most inappropriately named ship, according to Thomson, who described it as the oldest and most cranky paddle steamer in the navy. The weather was bad and the opportunities for dredging limited, but from the abundance of life obtained from depths as great as 3000 feet, it was clear that the Forbes theory was poised for demolition. What also excited and yet puzzled the two men was a discovery made almost by accident. Various makes of thermometer had been attached to sounding lines on the *Lightning*, not primarily to measure ocean floor temperatures but to test how pressure affected their performance. Carpenter and Thomson found that the temperature at various levels of the ocean varied and that this variation was dramatic over quite small areas. This challenged the universal acceptance that all temperature below a certain depth in the ocean was

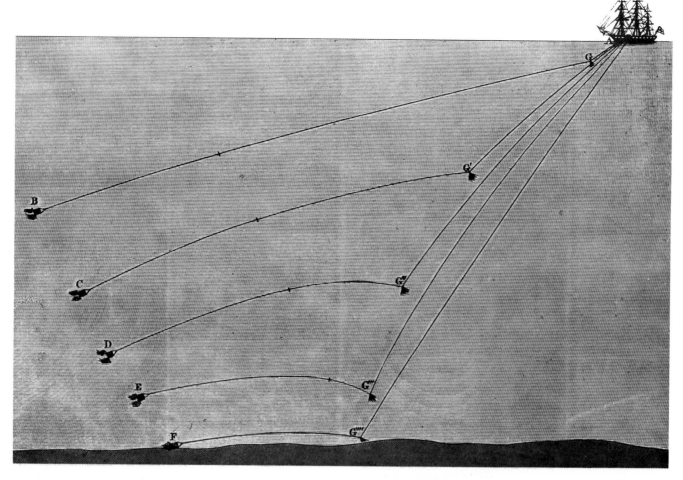

constant, varying only slightly with latitude.

The interest that these biological and physical revelations aroused made it easier to obtain sanction for additional voyages so that more data could be gathered to consolidate the *Lightning* findings. Two voyages were made in the HMS *Porcupine* in 1869 and 1870. Both were highly successful. They established once and for all that life exists in the deep ocean, and they put Britain in the forefront of oceanographic discovery. They also sparked a controversy about oceanic circulation that was to rage for years.

Carpenter promoted the idea that a major voyage of exploration was necessary to maintain Britain's academic lead, and gained backing for the *Challenger* expedition in April 1872. Some idea of the enthusiasm the scientists brought to the project can be given from the fact that the ship was converted and equipped in eight months, sailing from Portsmouth in December 1872. The ship's guns had been taken out and laboratories installed in their place, winches were bolted to the deck, and scientific equipment loaded aboard.

Under the leadership of Thomson, *Challenger* sailed to the Canary Islands, zigzagged the Atlantic, and then rounded the Cape of Good Hope to head

Opposite: two contemporary drawings showing work aboard the *Challenger*. In the one on the far right a scientist is emptying a water sample bottle. In the one on the left two crew members are unloading the dredge net.

Left: one of the two well-equipped laboratories on the *Challenger*. This one was designed for the study of sea animals and plants. There were four naturalists and one chemist on the ship's team of scientific investigators.

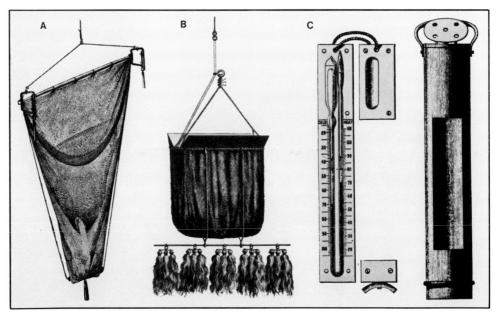

Opposite: the stages of travel of a weight pulling the *Challenger's* dredge to the sea bottom. As the weight moves along the line toward a toggle, it carries the dredge down, reaching bottom when it reaches the toggle.

Left to right: the trawl net, the dredge, and the thermometer used on board the *Challenger*. The equipment on the ship far outshone the ship itself.

south almost to Antarctica before reaching Australia in March 1874. There scientists and crew had a two-month break. For the rest of 1874 and for the early part of 1875 the ship explored the network of seas off Southeast Asia, and then sailed into the north Pacific. It swung south, calling at Hawaii and later at Venezuela before passing into the Atlantic via the Strait of Magellan early in 1876. It arrived back in England on May 24, 1876.

The *Challenger* voyage was the culmination of the 19th-century movement toward acquiring greater knowledge of the oceans. Using equipment that was often untried and remarkably crude by present-day standards, the ship's crew worked hard to recover samples for the scientists to study. Dredges were pulled along the ocean floor to capture sediments, and this enabled the first map of the geology of the sea floor to be drawn. The samples were analyzed and classified by John Murray who was Thomson's chief scientific assistant. His work in the expedition can

be seen as the start of modern marine geological studies.

If there had been any lingering doubts among the members of the marine academic community about the existence of life in the deep ocean, the *Challenger* finally settled them. By dragging trawls at the end of up to eight miles of hemp line, thousands of living samples were brought to the surface. Plankton nets were also towed at various depths. In all, the *Challenger* expedition discovered an impressive 4417 new species of sea animals and plants and 715 new genera. Some 77 samples of seawater from various depths were obtained. Work on these by expedition chemist J Y Buchanan, followed up by William Dittmar, formed the basis of modern chemical oceanography.

Only in one aspect did the expedition disappoint the waiting scientific world. The measurements of temperature, specific gravity, and current speed and direction taken on the voyage made no useful contribution to settling the controversy raging within the

hallowed portals of the Royal Society about the true nature and cause of oceanic circulation. Carpenter, who had not wished to join the *Challenger* expedition although he had helped get it sponsored, was the most vociferous defender of one of the schools of thought. Moreover, the *Challenger's* scientific crew did not include a physicist trained to take full advantage of the wealth of data obtained from observations of temperatures and of current movement.

However, when the expedition reached home there were no immediate clouds to shadow the shining glory of the achievement. Thomson was knighted and people flocked to see the ship. Trouble began only when it came to classifying and analyzing the thousands of samples and masses of data that had been obtained, and to producing the report of the voyage. Thomson, as professor of Natural History at Edinburgh University, was a dedicated scientist who simply wanted to see the best scientific interpretation made of the results of the voyage of discovery. He was hardly prepared for the scientific establishment's wrangling over, and the government's parsimonious attitude toward the work that had to be done. There was argument about whether the work on specimens should be carried out at the British Museum or at

Top: a drawing of a dredging operation being carried out from the *Challenger* on the Parramatta River in Australia. Some of the scientific gear used on the world's first oceanographic expedition was somewhat primitive.

Below left, below, and opposite: original drawings of four new specimens of marine life discovered by the *Challenger* and published in the *Challenger Reports*. They are a microscopic organism, and a crab and two fish.

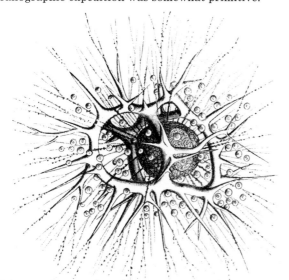

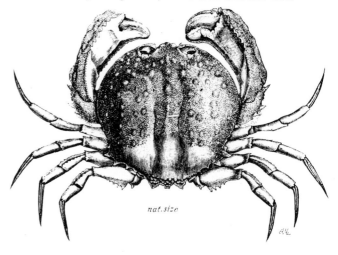

nat. size

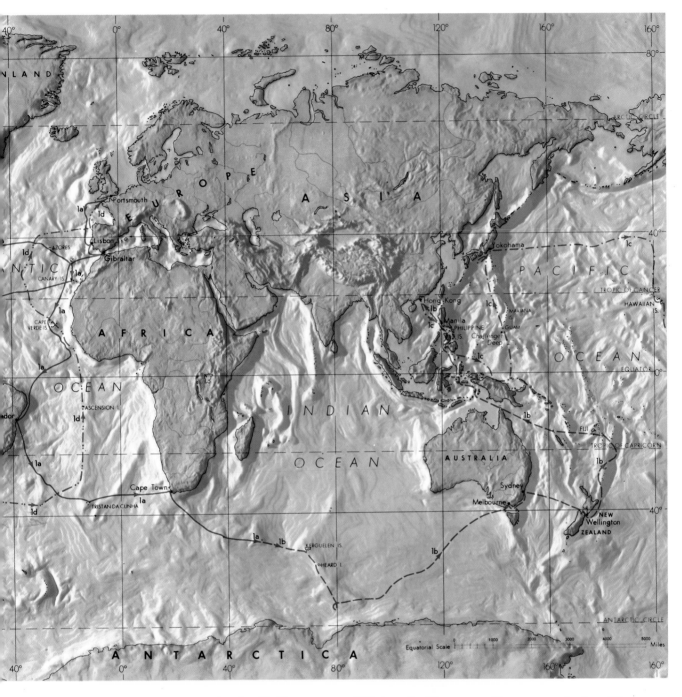

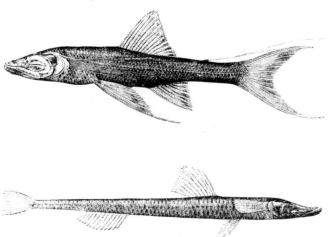

Above: the route of the voyage made by the *Challenger*. The expeditionary ship circled the Earth, visiting every ocean except the Arctic. Compared to many, the enterprise was not so costly in lives – two men died during the trip.

Edinburgh. There was disagreement when Thomson, on the basis of using the best talent available, asked foreign scientists to carry out the work. There was continual bickering with the government, not only about more funds to support the all-important task of writing the report and to pay the authors, but even about the layout of the pages. Thomson had hoped that the report could be kept succinct, amounting to perhaps 16 volumes, and wanted them all published within five years. Faced with the squabbling and the prospect that financial support for the completion of the work would not be forthcoming, Thomson collapsed under the strain in 1881 and died in March 1882 only six years after *Challenger's* triumphant return. The last reports on the voyage were not published until 1895 – 19 years after the voyage ended. The total report filled 50 volumes.

Learning about the Sea's Secrets

Other nations were quick to respond to the challenge in the ocean sciencies that had been laid down by Britain, and marine biology seemed to be the subject that fired the scientific imagination. In the United States Alexander Agassiz developed sampling techniques to the extent that, only a few years after the *Challenger* expedition, he was responsible for bringing up in one haul from over 10,000 feet more specimens of deep sea fish than the *Challenger* had collected in three-and-a-half years. Such fruitful hauls were obtained by using steel cables for dredging instead of the bulky hemp rope used by the *Challenger*, and by the development of a dredge that would operate even if it fell upside down on the seabed. Agassiz initially financed much marine biological research work himself, conducting voyages on the steamer *Blake* in the Caribbean Sea and the Gulf of Mexico. Later he was in charge of expeditions on what is generally accepted to be the first vessel purpose built for oceanographic exploration, the *Albatross*. The ship was built in the 1890s by the then US Commission of Fish and Fisheries to undertake basic research into fish stocks – another example of commercial considerations providing the impetus for the pursuit of knowledge.

Toward the end of the 19th century scientists and governments began to realize that fish catches could be improved by a closer study of the environment in which the main commercial species lived, as well as of the plant and animal food on which they preyed. To

Above: Alexander Agassiz, one of the founders of oceanography, watching the arrival of a deepsea trawl on the *Albatross*. A marine zoologist and mining engineer as well, he specialized in study of coral reefs.

be able to plot the migratory patterns of fish would be to know where to deploy fishing fleets. However, before this could be done, it was necessary to understand the reasons for their migration; to understand the currents that they used on their long journeys; and to be able to record accurately temperatures, salinity, and the other oceanic variables to which particular species of commercial fish and their food were sensitive.

Left: a group photograph taken on board H.M.S. *Discovery* in the Antarctic on Christmas in 1926. The two-year expedition of the *Discovery* added greatly to the knowledge of whales. The research program was directed by Stanley Kemp, an Irish naturalist.

Aside from Agassiz in the United States, scientists elsewhere and particularly in Scandinavia studied fish stocks. They all added to the total sum of knowledge about the shape of the ocean basins, the circulation of their waters, and the distribution of their plant and animal life. As an example, studies of the biology of the Antarctic whale by Britain's *Discovery*, and the result of subsequent voyages by it and other ships, made the Southern one of the best-known of all five oceans. Knowledge about the life of the whale was also increased by the work of Prince Albert of Monaco, another self-financing oceanographic pioneer. He used his royal yachts – *Hirondelle I* and *II* and *Princess Alice I* and *II* – to good purpose by equipping them to carry out extensive oceanographic work in the Mediterranean and North Atlantic. Intrigued by the pieces of a giant squid found in the stomach contents of a whale killed by the crew of *Princess Alice I* off the Azores, he initiated a program of research into whales. The results showed which types of food were favored by different species of the sea mammal.

World War I gave a serious setback to the progress of academic studies of the oceans which, at the outset of hostilities, were being actively pursued by France, the United States of America, Germany, Norway, Russia, and Britain. After the war other nations, notably Japan, joined in. There followed an era of

Above: a fine mesh triangular net used for collecting specimens from great depths. It is part of the equipment on board *Princess Alice*, the yacht from which Prince Albert I of Monaco did a great deal of research.
Below left: *Princess Alice* on its first expedition in 1892.

systematic study of the physics, chemistry, biology, and geology of the oceans and their floors. Each nation, it seemed, was intent on gathering masses of data over the widest possible areas.

War came again and academic studies were abruptly halted. But with so much of World War II fought at sea – and especially beneath the surface – scientists worked feverishly to develop equipment that would help surface hunters seek out submarines. Sonar equipment was one result. Anyone who has watched war movies featuring submarines has seen at least one scene in which the crew of the submarine sits edgily and silently listening to the ominous ping of the enemy's searching sonar.

Above: Prince Albert. Known as the "scientist prince" because of his marine research, he wrote many books and encouraged wider interest in oceanography. He also established a marine museum in Monaco in 1910.

After the war sonar and other equipment was put to peacetime use. The electronics revolution had as profound an effect on oceanography as it did on so many other aspects of everyday life. The 1950s saw the realization that the study of the oceans was far too big a subject to be tackled by single ships from individual nations, whose efforts were often duplicated. International expeditions of several ships were mounted, establishing a trend that continues today.

Techniques for Studying the Seas

Merely to sail a ship in a straight line across the oceans can be a difficult, sometimes hazardous, task. When the ship has to stop in mid-ocean to launch and recover equipment, the difficulties are compounded.

It may seem odd, but one of the most basic problems of working at sea is simply to know where the ship is. The problem of fixing location is common to every aspect of oceanic activity, from research work to drilling for oil.

Within sight of land a ship can fix its position accurately relative to known landmarks, using simple instruments like theodolites and sextants. Electronic versions of these instruments are used today to speed the task. In the open ocean, beyond the sight of land, however, the problems begin. Celestial navigation, in which the known position of sun, stars, and moon makes it possible to compute latitude and longitude, was used for hundreds of years, but its accuracy was limited to measurement in miles. Today, when an oil rig needs to move back to drill an abandoned well again, it is necessary for the crew to know the position of that well to within a few feet. Modern technology makes this possible.

The most widely used method relies on radio waves to give an accurate position. Waves of predetermined frequency are transmitted from shore stations, and the difference in arrival time – phase difference – of the waves is measured by a receiver on the ship. Lines are drawn on a chart through points of equal phase difference. The distances between these lines are known as lanes. By identifying which lane the ship is on, the navigator can translate this information into latitude and longitude. The farther away the ship is from the shore stations, the greater the lane widths become, which reduces the accuracy of the fix. Distance also decreases accuracy because the velocity of radio waves is affected by temperature, humidity, and atmospheric pressure. So the farther they have to travel, the greater the likelihood of interference. With vessels of all types today needing accurate positioning hundreds of miles from shore, still better methods have had to be found. The most successful has been by means of satellites, which can give pinpoint accuracy of position. It works like this: as a satellite orbits the Earth, its trajectory is measured by a receiver on the ship. A computer aboard ship uses this information in conjunction with known facts such as the ship's speed, the height of the receiving antenna, and other information, and produces a simple print-out of the ship's position in latitude and longitude.

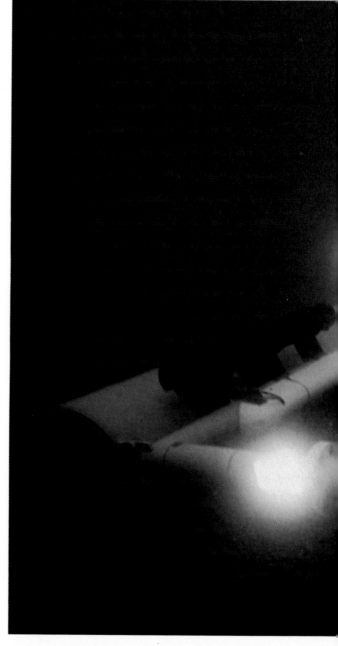

A whole new set of problems is presented in working at sea when equipment must be lowered beneath the surface, whether it is a thermometer or a small submarine. All marine equipment must be made tough enough not only to take the hard knocks it will inevitably get through constantly being launched from the deck of a pitching and rolling ship, but also to withstand what the oceans themselves will try to do to it.

Pressure is one of the problems. We can get an impression of the pressure exerted by water when we dive into a swimming pool and feel the pounding in the eardrums. On land at sea level, pressure is 14.72 pounds per square inch, known as one atmosphere. In seawater that pressure increases by one atmosphere for every 33 feet of depth. At its deepest points, therefore, the ocean exerts a pressure of seven tons per square inch. Equipment such as a deep ocean camera must be encased in an extremely strong shell

Above: a hydraulic A-frame in use to lower a submersible over a ship's side.

Left: a submersible with powerful lights in action in deep ocean waters.

Below: a diagram of the hyperbolic system of fixing position by radio waves. It requires three or more land-based transmitting stations and a shipborne receiver. The lattice is made up of lines joining points of equal phase difference. The receiver sets position by measuring the phase difference of the radio waves on the lattice and giving readings that can be plotted on a chart.

Bottom: the receiver of a medium-range radio positioning system.

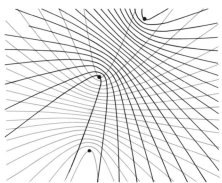

● *Fixed ground station*

to keep it watertight under such enormous strains. It must also be corrosion-resistant and capable of withstanding extremes of temperature when, as is often true, it is launched from the deck of a ship in the tropical sun to plunge to the ocean's near-freezing depths.

Seeing underwater is another problem. Even in the clearest waters very little light penetrates much below 2000 feet. Where there is any turbidity because of suspended sand, mud, or dense animal and plant life, the oceans can become very dark just a few feet beneath the surface. Lights help for short-range observation, but over any distance it is somewhat like trying to drive a car in thick fog, with the headlight beams being reflected straight back in the face.

Navigation, pressure, darkness, corrosion. The simplest task of obtaining measurements at sea is fraught with logistical and technical problems. Scientists are constantly at work to overcome them.

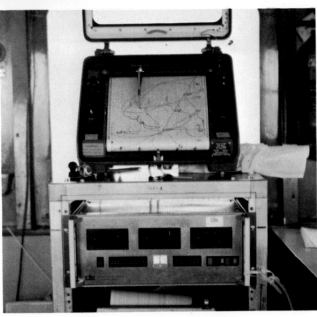

Problems of Ocean Exploration

When HMS *Challenger* made its way round the world on its scientific expedition, one of the biggest complaints of the crew was against the benumbing tedium of lowering equipment into the ocean depths and winching them back again in a routine that could take hours. Boredom would give way to frustration when, as so often happened, thermometers would be found smashed, nets ripped and empty, and other equipment damaged in other ways.

Much of the seemingly slow progress that has been made in the building up of knowledge of the movement of the oceans and of their living and mineral resources can be attributed to the simple logistical problem of having a ship sit for hours at a time dangling equipment on the end of a long piece of rope. Take, for example, one of the most basic measurements, that of seawater density which enables very slow-moving deep ocean currents to be identified. Formerly it was necessary to obtain data on the salinity and temperature of a sample and analyze them in relation to the depth at which the sample was obtained. This involved having a string of containers on a wire at fixed points to collect water. A brass weight or "messenger" slid down the wire to trip a mechanism that closed the two ends of the containers to trap a water sample and also to activate reversing thermometers to record temperature. Samples from known depths were then analyzed to determine the degree of salinity. This time-consuming technique was superseded by the electric salinometer which compared the electrical conductivity of the captured sample with a control sample of seawater whose salinity had been accurately calculated.

That still left the problem of lowering water containers into the water from a ship and laboriously recovering them. An added disadvantage was that data could only be retrieved at fixed points determined by the positioning of the containers on the line. The development of electronic techniques solved the problem in the shape of the continuous salinity/temperature/depth (STD) probe. Lowered from a ship the STD probe measures the three factors continuously on its downward journey and converts them to electrical impulses which are relayed up the cable and, after filtering, recorded on charts. Other instruments such as current meters may be fitted to the same cable, dramatically increasing the amount of data obtained from one deployment of an instrument package.

The ship has to stop to obtain the STD probe data, but other instruments have been developed that can

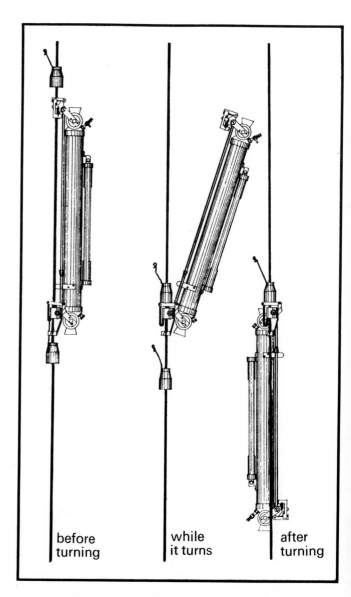

before turning | while it turns | after turning

Above: the Nansen bottle for taking water samples. It is basically a metal tube that has valves at each end and that is fitted with reversing thermometers. As many as 20 bottles in a series are lowered on a wire. A "messenger" weight released from the ship travels down the wire and trips the first bottle, making it reverse. As it reverses, valves at each end close and trap a water sample.

record important oceanic variables while being towed by a ship at full speed. Temperatures at known depths are recorded by a device known as an expendable bathythermograph (XBT). This is a streamlined probe attached to a coil of fine wire. Temperature is measured and transmitted up the wire as the XBT falls. Since the rate of fall is known, depth is determined simultaneously. Operation of the device is so simple that it can be fitted to merchant ships on which, unlike the *Challenger* sailors, the crews welcome the change in routine that launching the device represents. The XBT probe is simply abandoned after its transmissions have been recorded. Through it, the gathering of the important oceanographic variable of temperature is no longer confined to a handful of specialized ships. As a result, the number of temperature measurements has been increased by a factor of

Left: measuring density of water by means of an STD (salinity/temperature/depth) probe from a ship. The STD works electronically to take these important measurements that tell oceanographers a great deal about currents.

Below: recovering data-gathering equipment from the sea. These buoys were released by an electronic signal (see next page) that made them float to the top, carrying the equipment with them.

thousands – at least along shipping routes. The use of merchant ships to obtain environmental data is by no means a new concept. Since the 15th century, scientists have begged ships' captains to record the phenomena they observe every day. Even today much of the information that goes to produce the latest European weather forecasts is obtained from the daily broadcasts made by fishing vessels in the North Atlantic.

Another way to obtain more oceanographic data without having to deploy expensive ships is to operate the recording instruments remotely. At its simplest this means stringing instruments along a cable that is anchored to the seabed and kept afloat by a buoy which floats on the surface. A series of buoy instrument packages is launched by a ship over an area, the instruments record data over a period from

turns to face the prevailing current, and this directional data is also transmitted to the ship.

The self-recording current meter operates on the same principle except that data is recorded within the instrument onto magnetic tape, usually a cassette of the same type used on domestic music equipment, in digital form. On recovery the cassette may be interfaced with computer equipment so that data is quickly processed to give average current speeds and directions. The latest current meters do not even depend on magnetic tape but have a built-in computer memory logic system. The instrument is taken from the water and plugged unopened into a processing unit that produces a print-out of current speed and direction. These instruments have to be opened only once every four years, and then only to change the batteries.

The next step in oceanographic data gathering is to dispense with the services of the ship almost completely. In the southwestern approaches to the British Isles is now found the UK Data Buoy One (DB1), with a discus-shaped hull over 20 feet in diameter and a 24-foot high mast bristling with antennas and lights. The data is retrieved not by ship but by telemetric link with the shore. The buoy stores up a mass of data

Above: an acoustic release on a mooring line. On receiving a coded signal from a ship, the release detaches itself, allowing the mooring line to be carried up by buoys.

weeks to months, and the ship then picks them up. Surface buoy arrays are vulnerable, however. They can become detached from the instrument package if the cable is chafed by the action of heavy waves, or – and this is the most common problem – if they are removed by passing mariners.

To counteract these problems oceanographers now bury their data gathering equipment beneath the sea surface. The configuration of the array is much the same as that used for surface buoy equipment. The main difference is that between the array of instruments and the anchor – a concrete weight or some heavy chain – on the seabed is an acoustic release. The equipment array sits beneath the surface for its allotted time, recording data on tape. When the time comes for it to be recovered the ship approaches and sends a complex coded signal through the water. This is received by the acoustic release and a shackle is activated so that the link between the equipment and the anchor is parted. The whole array then becomes buoyant and floats to the surface where it is recovered by the ship. Equipment of this type may be left in the oceans for months at a time, thanks to the development of self-recording instruments, and in particular to the self-recording current meter.

The kind of current meter used on board ships is read directly. The instrument is suspended from the ship by a cable which carries an electronic signal. The meter has a rotor that turns in the current, and the rate of rotation is translated into a voltage that is transmitted up the cable to the ship to give the speed of current. A vane on the instrument means that it

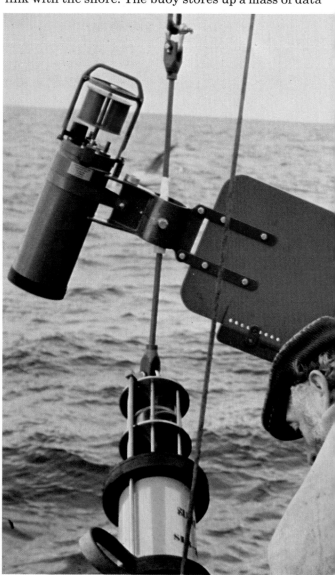

including winds, atmospheric pressure, tides, wave heights, and current movements until a shore-based radio station calls it up. The buoy then transmits all its stored data by high frequency radio for subsequent processing ashore. Buoys of this type will steadily replace lightships around Great Britain and other countries, and in addition to marking shipping lanes, will record meteorological and oceanographic data. The use of large data buoys of this type, which need to be serviced for battery changes and repairs only a few times a year, will vastly improve the effectiveness of weather forecasting and marine data gathering.

But by far the most sophisticated data gathering technique of all is that in which the measuring instruments do not even come into contact with seawater. This is what is happening as another spin-off from the space research program. Satellites orbiting the Earth are now able to use remote sensors to identify sea states, current direction, distribution of biological resources, oil slicks, icebergs, sheet ice patterns, and a host of other phenomena. Spin-off from space research and further development of small seagoing computers will give oceanographic data gathering programs an enormous boost in the coming years.

Above: an artist's impression of the Nimbus 7 satellite operated by the United States space agency NASA. This craft uses advanced sensors to gather oceanographic and atmospheric information and to research into pollution.

Above: the UK Data Buoy One (DB1) on site, moored some 150 nautical miles off the southwest coast of England. It is collecting data about possible oil fields.
Left: deploying a self-recording current meter. The system houses a mini-computer, batteries, crystal clock, magnetic compass, digital encoders, and memory bank.

Cameras for Use under Water

Just by looking upward on a cloudless night you can see features in the universe that are millions of miles away: planets, stars, and whole galaxies. Arm yourself with a cheap telescope and you can examine in some detail the craters and plains of the moon.

On Earth, we have to go to expensive lengths to look at features of our own planet that are only a couple of miles under water. The oceans are dark, so that right on our doorstep we have a huge mysterious area that we have only begun to look at in any detail.

There are three ways to look at the secrets of the ocean: by cameras sent down from the surface, in a diving suit, or in a submarine. For simple, quick-look observations of underwater features, television, movie or film cameras are used.

Deep-sea cameras have to be encapsulated against the pressures of the ocean depths. They are usually mounted on a frame that has electronic flash units and/or powerful lights fitted. The array is lowered from the ship and, as it approaches the sea floor, is activated by one or more devices. The simplest of these is a signal from the surface that turns on lights and starts the film. One of the few advantages of working in complete darkness is that the shutter speed and lens aperture can be fixed. Another method is to use a simple probe slung from the bottom of the array so that when it touches the seabed it triggers off the lights, flash gear, and camera, and makes the film advance at preset intervals. A more sophisticated

Left: a diver photographing the coral and fish in the Red Sea. Some of the pictures taken underwater are strikingly beautiful as well as highly informative.

Below: this deepsea camera contains a pinger that monitors the depth of the unit continuously. When the pinger indicates that the camera has reached the required depth, it is activated.

Above: interpreting photographs obtained by stereoscopic cameras. The two prints are viewed side by side through a special viewer that produces a three-dimensional effect.

Left: two stereoscopic cameras mounted on an SMT-2 submersible used for inspections.

unit has an echo sounder mounted alongside the camera. This sends an acoustic signal to the seabed which echoes back to the ship. An accurate plot of the depth of the camera array can be maintained and a signal sent to activate lights, flash gear, and film. Simple triggers are also used to obtain photographs of deep sea animals. The trigger is baited and when a fish takes the bait it activates the lights and the preset focus camera. Echo sounders and infrared lights are also used. When an animal swims across the beam of sound or light, the camera is triggered off to start shooting.

Distinguishing the scale of seabed features from a flashlit, fixed focus photographic print is not easy for even the most experienced interpreter. This work has been simplified by the development of stereoscopic cameras. These are simply two cameras mounted side by side with the angle of view of their lenses slightly out of synchronization. When triggered they take pictures simultaneously, and the resulting prints are placed side by side under a special viewer to give a remarkable three-dimensional effect.

A development from somewhat sinister military intelligence operations has been the low-light level TV camera. Because light is attenuated so highly by seawater, the range obtained by a TV/light combination is often so short as to be practically useless for direct observation. However, low-light level units make full use of the tiniest source of ambient light – sometimes even that given off by a passing fish – to send live pictures back to the surface for direct observation and for recording onto videotape.

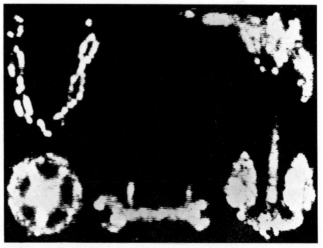

Above: images seen on an underwater acoustic television system, in this case observed in zero visibility conditions. The revolver (top right) was completely buried in silt. Also seen are a chain, pulley wheel, wrench, and anchor.

Left: a diver equipped with an acoustic television camera. The acoustic TV system is designed to locate and inspect submerged objects in turbid conditions and zero visibility.

Sea Probes by Sonar Instruments

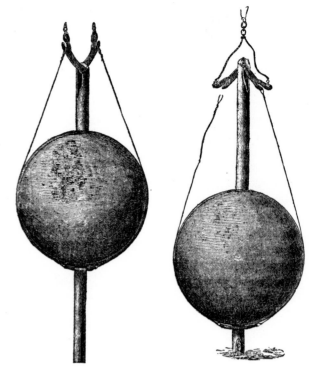

When the first attempts were made to discover the true depths of the oceans, some extraordinary figures appeared on charts. They showed the deep ocean floor to be many miles deep – or even bottomless. The method of depth measurements was crude at the time. Measurements were made using a lead line. A rope with a lead weight on the end was paid out from the ship and the amount of line that paid out before the weight touched bottom and the line went slack gave an indication of depth. This method worked well enough in shallow water. In deep water a cannonball on the end of a marked line was hurled over the side, and it was thought that when it touched bottom the line would go suddenly slack like the line did in shallow depths. The line never did go suddenly slack, and so for some time it was thought that the oceans were practically bottomless. What was happening was that the weight of the line itself was making it continue to pay out long after the weight had touched bottom.

In the 1850s navy officer Matthew Maury, the American oceonographer, initiated a program that led to depths being recorded more accurately. He instructed ship's crews to observe carefully the rate at which the line paid out. When the speed decreased, said Maury, it was an indication that the weight had touched bottom. The method was successful, and from the results he obtained Maury was able to produce his bathymetric chart of the North Atlantic in 1854. It included results from over 150 soundings made from depths greater than 6000 feet.

It was not until the 1920s, however, that fast and accurate measurements of the depths of the oceans could be obtained, and this was because of the development of the echo sounder.

The echo sounder is a simple example of one of the best methods of "seeing" underwater by using sound. Seawater is a good medium for the propagation of sound, and an enormous amount of marine exploration work is based on the principle that the right kind of noise underwater will make the sound bounce back when it hits something. The echo can be recorded and displayed.

Sound travels through water at about 4940 feet per second – five times faster than through air. At its most basic, then, an echo sounder consists of a hammer and a watch: the hammer is struck against the hull of a ship and a watch records the time taken for the sound to return as an echo. By use of the known speed of sound through water, the depth can easily be calculated.

Modern echo sounders work on this simple principle. An electric current is passed across ceramic or crystal-based materials which contract slightly and make a sound of a known frequency. The sound

Top: a 19th-century deepsea sounding device, consisting of a hollow tube inside a cannon ball. When the tube struck bottom, the drop in tension on the line tripped the hinged brackets and released the ball. The tube, with a sample of the seabed, was reeled in.

Left: a chart from an echo sounder, showing how it acts as a fish-finding device. The bottom echo is electronically separated from any fish echoes above it and both show up on the chart with a white line between them.

transmitted by this transducer hits the seabed and returns as an echo. The echo is received by the transducer, reconverted into electrical energy, and displayed. One type of display has a belt-driven stylus that traces across a slowly moving roll of electrosensitive paper. The stylus marks the paper as the amplified echoes are transmitted through it, and a scale on the chart shows the depth from which the echo has been received. Another form of display shows the depth in digital form – a simple read-out of feet or meters below the keel of the ship. Digital sounding means that hydrographic surveying (charting the seabed) may now be conducted speedily and automatically. In its most sophisticated form, the data is

water in infinitely variable positions from straight ahead to astern. The area of sea and seabed swathed by the beam of sound is far greater than that covered by an echo sounder – about 1000 million cubic yards of seawater can be scanned in about two minutes. The use of sonar has revolutionized fish hunting techniques.

Another form of sonar is used to obtain scans of the seabed which, as the illustration shows, are uncannily like photographs. They are produced by side scan sonar. The transducers are mounted in a casing called a fish, which is towed beneath the surface by a survey ship. Narrow beams of sound are transmitted by the fish in a fan shape at right angles to the

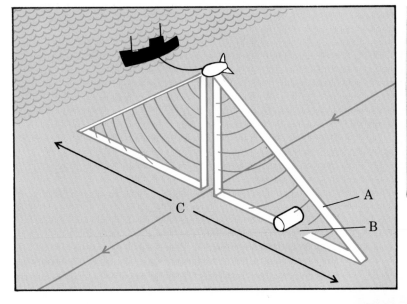

Above: this picture, produced by a side scan sonar unit, shows an old sunken vessel.
Left: a side scan sonar. A "fish" holding transducers is towed by a ship. The sound beams (A) spread out at right angles to the direction of travel (C). Objects make acoustic "shadow" (B), whose echoes are traced by the unit.

fed into a plotter that draws a contoured chart, with allowances made for the height of the tide.

The echo sounder is used to record more than depth alone. In 1933 a British herring fisherman noticed that his depth recorder chart was picking up some interference. Echoes were appearing above the firm black line that represented the seabed, and these echoes were from fish swimming in midwater. Thus was born one of the most effective aids in the search for fish.

The echo sounder searches only the narrow area of sea and seabed directly beneath a ship. During World War II it was essential to have some means of searching greater areas of water so as to be able to find submarines. The Allied Submarine Detection Investigation Committee, a wartime organization of scientists and engineers, succeeded in developing such equipment, which became known as Asdic after the committee's initials. The equipment is now known as sonar, from Sound Navigation and Ranging. It gave another dimension to underwater exploration as well as serving a wartime purpose. Sonar acts like a searchlight in that a beam of sound can be trained under-

Right: putting a side scan sonar unit, called a towfish, into action. It will reveal objects on the floor of the sea.

91

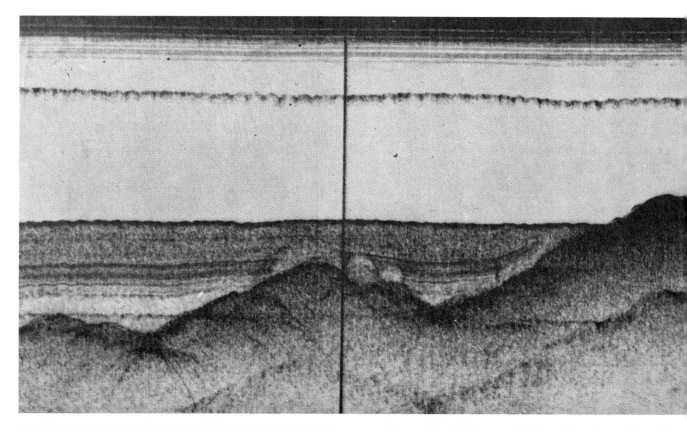

Above: air guns being deployed from the side of a seismic survey vessel. The guns – four in this case – have been mounted on a cradle to be lowered by a hydraulic crane. Seismic surveys of the ocean floor and subsurface geology are important to exploration for oil, gas, and ore resources. Probes of sea floor sediments today often utilize air guns because of their powerful sound beams.

direction of travel. When the beam hits an object standing clear of the seabed, an echo is returned and an acoustic "shadow" is left directly behind the object, just as if a light had been shone on it. A graphic

recorder on the ship records the returning echoes. It is fascinating to watch a "picture" of the seabed and its features – perhaps a long-forgotten shipwreck or an oil pipeline – slowly build up. Side scan sonar has many uses, among them the survey of the seabed for obstructions that may slow down the laying of an oil pipeline; the search for wrecks; the inspection of underwater installations; and the making of acoustic maps of wide areas of seabed. The definition that can be obtained with side scan sonar is so high that a seabed telephone cable can be picked out hundreds of feet beneath seawater to show distinctly on the record.

Sound may also be used to look at features beneath the sea floor. The lower the frequency and the greater the power at which a sound beam is transmitted, the greater distance it will travel – though at the cost of the definition of the features it picks out. Sound beams can be transmitted straight through the water column to penetrate the seabed. They reveal a vast amount of detail about geological strata and are invaluable in delineating sand and gravel deposits; in finding buried pipelines and cables; and in showing the best sites in which to position oil rigs and platforms. The most commonly used acoustic energy source is the airgun, utilizing compressed air. It can produce useful acoustic energy reflections from depths in excess of 20,000 feet beneath the sea bottom. The energy passes at different speeds through rocks of different densities, and a complete geological picture can be built up with the aid of computers that filter out spurious or double echoes. This is the technique used to find the most likely types of sedimentary rock that may contain trapped hydrocarbons.

Sound waves do not always have to be transmitted

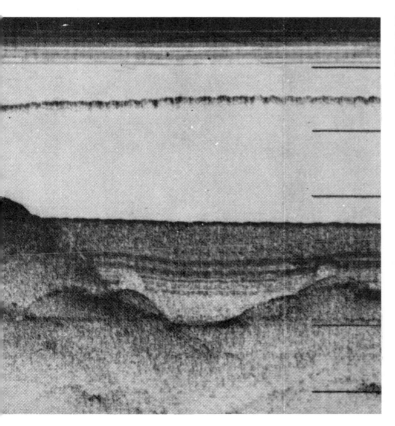

Left: a vertical profile of shallow sediments on the sea floor. Such a profile is obtained by a side scan sonar which displays a continuous sub-bottom profile by processing signals from a sub-bottom profiler attached to the towfish.

from, or just beneath, the surface to aid marine surveyors and engineers. They can also be used as underwater beacons and navigation aids. Floating oil rigs, for example, are fitted with pingers – named after the sound they make – so that the rig may quickly be spotted on the sonar of a submarine. The basic item of equipment used in underwater navigation is a transponder. This is an electronic device that receives an interrogatory acoustic signal which triggers it to make its own acoustic response. Transponders are fitted to temporarily abandoned equipment such as oil wellheads so that when the time comes for the well to be redrilled or the wellhead to be recovered, the ship or rig can call up the transponder when in the near vicinity. Transponders are also used to tell the crew of an underwater vehicle their exact position. Two or more of the devices are placed on the seabed, perhaps up to five miles apart. A sonar on the vehicle activates a transmitter which interrogates each transponder in turn and notes the time taken for the answering signal to arrive. By computing the difference in the time that the signal arrives from each transponder, the vehicle is able to determine its position relative to the known mooring point of each transponder.

In the underwater world, where light and radio waves have little penetration, sound is used to give us eyes.

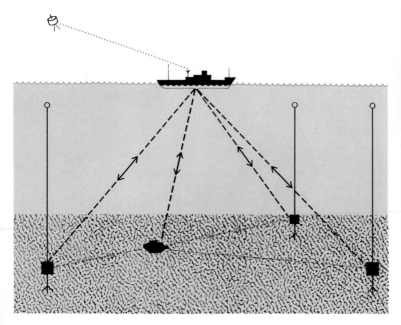

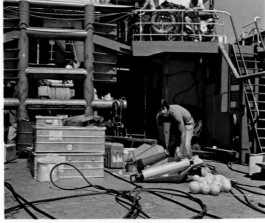

Transponders (**above** and as black squares in diagram **left**) are interrogated by a submersible and "read" on an echo sounder to fix relative position. Precise position needs interrogation of both submersible and transponders.

The Theory of Continental Drift

"Laws of nature are human inventions, like ghosts. Laws of logic, of mathematics are also human inventions, like ghosts. The whole blessed thing is a human invention. The world has no existence whatsoever outside the human imagination. It's all a ghost, and in antiquity was so recognized as a ghost, the whole blessed world we live in. It's run by ghosts. We see what we see because these ghosts *show* it to us, ghosts of Moses and Christ and the Buddha, and Plato, and Descartes, and Rousseau and Jefferson and Lincoln, and on and on and on.... Your common sense is nothing more than the voices of thousands and thousands of these ghosts from the past." So said Robert M. Pirsig, author of the book about science for nonscientists called *Zen and the Art of Motorcycle Maintenance* (1974).

Ghosts of geological thought stifled the development and acceptance of a theory that explains the very formation of the oceans: the theory of continental drift. Such ghosts arise because science is not just the collection and classification of facts. It is also a creative activity, and as subject as painting or musical composition to quirks of fashionable acceptance, of personality conflicts, and of competing schools of thought. So it was that the ghosts of the hard-and-fast opinions of eminent scientists made it hard to change opinion about the world ocean, no matter how many facts were amassed to back up new hypotheses. But the breakthrough has been made, and it is encouraging that today we have come to terms with a major upheaval in thought.

In the earth sciences, fact had been piling on scientific fact in formulating a theory that would support them, and it all came to a head in the early 1960s. What was underway was a complete revolution in ideas about the structure and evolution of the Earth. The more facts that were discovered, the closer they came to providing the evidence to support a hypothesis postulated in 1912 by Alfred Wegener, a German meteorologist and geophysicist. His explanation is called the theory of continental drift.

Wegener maintained that 200 million years ago the continents were all joined in a single vast land mass

Below: the Alps, photographed from space and showing Lake Geneva and the Italian lakes. According to the theory of continental drift, the European Alps were forced up when the Eurasian and African plates collided.

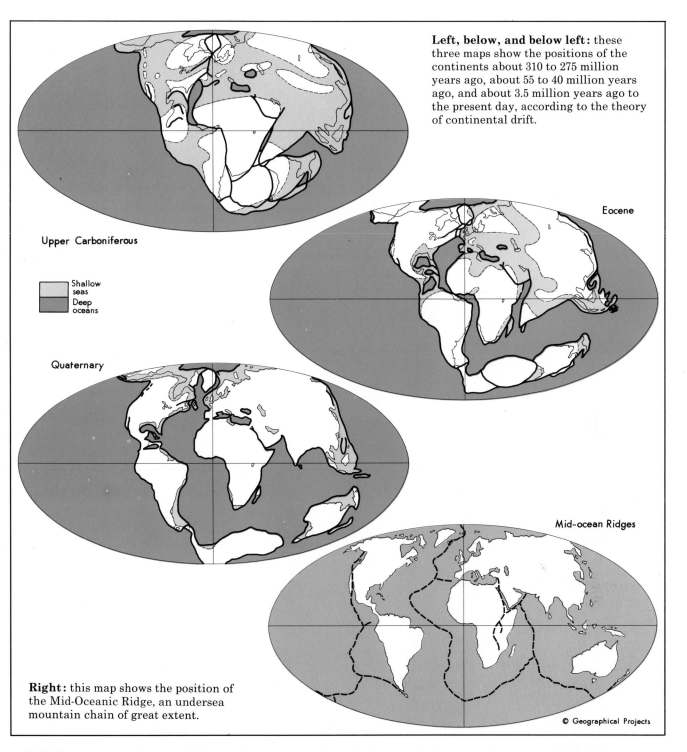

Upper Carboniferous

Shallow seas
Deep oceans

Quaternary

Eocene

Mid-ocean Ridges

Left, below, and below left: these three maps show the positions of the continents about 310 to 275 million years ago, about 55 to 40 million years ago, and about 3.5 million years ago to the present day, according to the theory of continental drift.

Right: this map shows the position of the Mid-Oceanic Ridge, an undersea mountain chain of great extent.

© Geographical Projects

called Panagaea, and two thirds of the Earth was covered by one great ocean. This can be seen if you cut up a world map and piece the continents together to make one supercontinent. The pieces fit even more exactly if you cut around not just the coastlines of the continents but also around the outlines of the continental shelves, the extensions of the land lying underwater.

According to the theory, great rifts appeared in the supercontinent starting in the Gulf of Mexico 190 million years ago. This pushed North America and Africa apart. Next to move was South America, followed by Antarctica. Later the northern end of North America split from Europe and the Atlantic

Ocean was fully opened up. At the same time Australia separated from Antarctica. About 50 million years ago India detached itself from Africa and shifted toward Asia, which it hit at such speed and with such force that the Himalaya mountain range was formed by being forced upward by the collision. Elsewhere smaller movements were in progress: Italy, for example, was driven into Europe with such a thump that it raised the Alps.

Wegener put forward a mass of evidence to support his views but they were not well received. The fact that he was not an accepted member of the international scientific establishment made it the easier to dismiss him. Scientists preferred to stick to the

Right: a computer drawing of the location of earthquakes over several years. The pattern agrees with the theory that the outer crust of the Earth is divided into a number of plates that are constantly moving. Accordingly, the earthquakes occur where disruption is the greatest – at the edges of the plates.

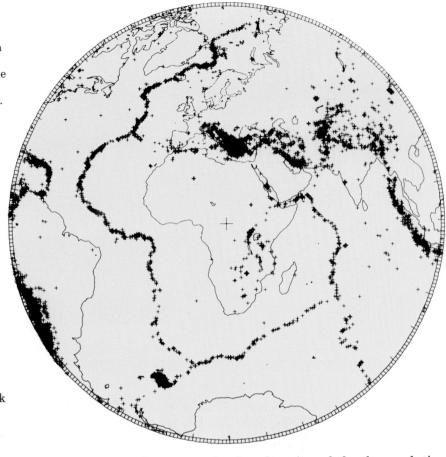

Below: this world map shows the main shield areas (brown), the areas of mountain upheavals (black narrow lines), and ridges of the mid-ocean (dark gray thick lines). The shields are remains of ancient ranges now eroded down to their roots.

then generally held theories about the Earth's structure and evolution – or rather they were prepared to look at what had happened on the Earth but seemed content to remain in doubt and mystery about how it had happened. Certainly it was thought that there was no room for Wegener's outlandish theories.

It was amazing how long it took for the revolution in thought about the Earth's structure to become fully accepted. Even in the early 1960s talk in favor of the continental drift theory in American universities could cause the protagonist to be laughed at, and in some extreme cases, to be discharged. In the

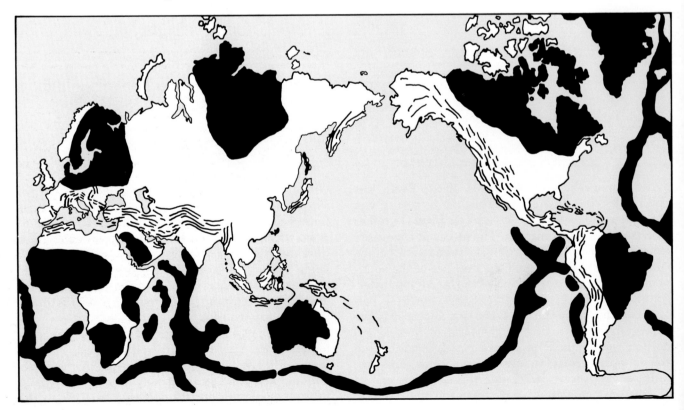

mid-1960s scientific papers were still being published explaining, with all the paraphernalia of mathematical equations, that mountain ranges were formed not by the collision of land masses but by the shrinking of the Earth's crust, which made mountains by a "crinkling" movement.

The facts, many of them just observations that posed questions, were nonetheless piling up as though waiting for the theory into which they could be fitted. The growth in knowledge of the formation and structure of the ocean basins, spurred by new techniques and equipment, threw up many of the questions and later provided many of the answers. It is worth looking at the flowering of the theory of continental drift and plate tectonics in the late 1960s in terms of some of those questions.

Question: Why does the Earth's longest mountain range, the 40,000-mile Mid-Oceanic Ridge, run more or less down the centers of the great oceans, roughly following the outline of the continents, and with a huge median rift valley? The exception to the mid-ocean course is in the Pacific, where the range runs into western North America to form the Rocky Mountains. The point at which the ridge runs ashore is at the notorious earthquake belt of Southern California. In the north Atlantic as well, the ridge breaks the surface to form Iceland, another area of intense volcanic activity as the plentiful supplies of hot water

Above right: a close-up picture of red-hot lava sliding down the unstable coastline of the island of Surtsey into the Atlantic Ocean. Volcanic activity in the Mid-Atlantic Ridge, which was formed where a crack in the Earth's crust separates the great American and Eurasian continental plates, threw up the new island in 1963. It is in one of the world's most active volcanic regions.

Below: this diagram illustrates the obvious geographic fit of South America and Africa. The continuity of glacial deposits and mountain ranges between the two land masses supports the continental drift theory.

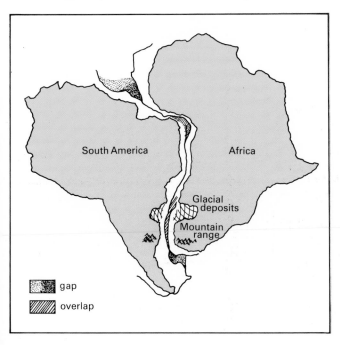

that gush through cracks in the surface testify. In 1963 more dramatic evidence of the geological liveliness of the Mid-Atlantic Ridge was supplied off the coast of Iceland. A crew member of a fishing boat noticed the sea beginning to steam, and then to boil. Within 24 hours an entirely new island – called Surtsey – had reared up from the ocean depths. Why is the whole of the Mid-Ocean Ridge rift valley an area of intense volcanic activity?

Question: Why should the major ocean trenches occur so close to land? The oceans deepen gradually the farther they are from land, so the trenches ought to be in the middle. But the ocean basins seem to be the wrong way round, for in their centers are towering mountains while at the edges are plunging trenches. Some of them close to the Philippines and Japan are as deep as 35,000 feet. Like the rift valley of the Mid-Oceanic Ridge, the deep ocean trenches are centers of earthquake activity. Scientists observed and recorded these facts, but the why eluded them.

Question: Why are sediments on the deep ocean floor so much thinner than was expected before they were effectively sampled? The predictions of their thickness were based on the belief that the ocean basins had been there since the beginning of the Earth and had been steadily accumulating sediment. Could it be that the oceans were relatively young? If so, this fact would upset a number of accepted assumptions.

Question: Why was the Earth's magnetic field found to be about one percent stronger than expected over the Mid-Atlantic Ridge when it was surveyed in the early 1960s? Why too did a "weak-strong" pattern of magnetic stripes appear across the ocean floor?

Mid-ocean ridges, near-shore trenches, thin sediments, fluctuating magnetism, earthquakes. All observed as facts, all raising questions, none with ready answers. For the why and a logical link to the facts, scientists – particularly in the United States – began to look again at Wegener's theory and to apply their new-found knowledge to it. On closer inspection, it worked. Its advocates may have been branded as geological heretics at the time in the mid-1960s, but they showed that the continents had in fact moved apart. Even more, they demonstrated that the continents are still moving. The surface of the Earth is in a continuous state of destruction and rebirth. This is the theory of plate tectonics, and it answers most of the questions posed above.

Plate tectonics in the most simple terms means the construction of the Earth's surface by the movement of great plates of material. The inside of the Earth is

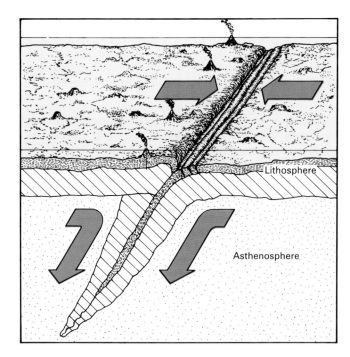

Below: pillow lava is usually found near volcanic vents and fissures formed when the ocean floor spreads out from the mid-ocean ridges as huge crustal plates. It gets its rounded shape because it cools rapidly in the cold sea.

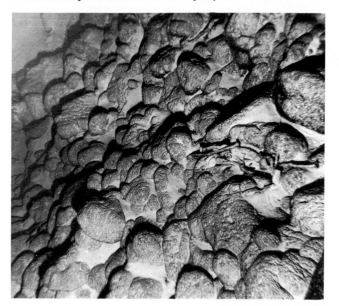

molten but a 40-mile thickness of surface rock is "frozen" like a coating of solidified mud on a football. This is the Earth's crust, and cracks in it divide it into islands or plates. These plates are disturbed by the heat within the Earth. They nudge each other and molten material keeps bubbling through the cracks to solidify and form new surface rock. This molten material is pouring out of the great rift valley that runs down the Mid-Oceanic Ridge and it spreads across the ocean floors. The rift valley represents one of the edges of the plate system. As one plate moves away from another, new material from the center of the Earth wells up to fill the gap. This explains the appearance of Surtsey on that night in 1963 when the sea boiled: the Mid-Atlantic Ridge rift valley had produced new material in a spectacular burst of energy to create an entirely new island. It also explains the instability of that area of southern California where the ridge makes its landward lurch and on which Los Angeles and San Francisco uneasily perch.

This flood of new material does not mean that more

dark stripes: magnetized (north)   light stripes: magnetized (south)

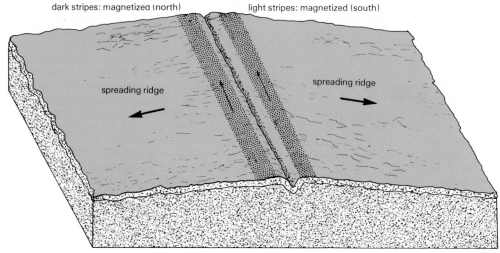

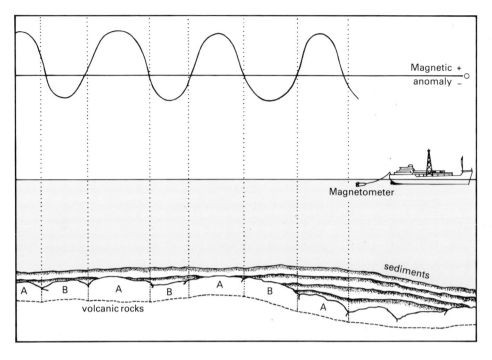

Left: how a trench is created. The edge of one plate collides with another and cannot absorb the shock by buckling. One therefore plunges into the asthenosphere below the other and is destroyed in that hot layer. The impact produces a trench as well as volcanoes and islands.

Right: study of the ocean's magnetic pattern. A magnetometer towed by a ship measures the small magnetic variations of the ocean crust. If the rocks were magnetized in the Earth's present-day polarity (A), they reinforce the magnetic effect. If the polarity were reversed when the rock solidified (B), they reduce the magnetic effect. Both these effects result in so-called magnetic anomalies, represented by the curved line.

material is being added to the Earth's surface because there is a counterbalance. As one plate moves toward another, their leading edges do not overlap. The edge of one plate disappears beneath the edge of its neighbor. As the plate bends it creates an ocean trench, with deep earthquakes and a string of volcanoes on the far side of the trench.

That explains ridges and trenches. What had also been observed was that the Mid-Oceanic Ridge did not always proceed in an orderly line down the ocean floors. Instead it tended to zigzag, sliced by transform faults. These occur where plates do not move away from each other as at the Mid-Oceanic Ridge, nor slide under one another, as at the trenches. Instead they just slip alongside each other in parallel motion.

The facts were fitting together under the mantle of the original Wegener theory. This also allowed an explanation for the unexpected thinness of the ocean sediments: the ocean floors are in fact younger than the diehards had proclaimed, since they are made up largely of new material pumped out from the Mid-Oceanic Ridges.

What about the magnetic anomalies? That turned out to be the final evidence in the argument for the theories of continental drift and plate tectonics. Periodically the Earth's magnetic field changes and the poles reverse. This last happened 30,000 years ago and the reversal remained for about 2000 years. In terms of geological time, these reversals of magnetism occurred fairly often and the "stripes" shown by the magnetic survey of the deep ocean floor near the Mid-Oceanic Ridges were a perfect record of the reversals. It had already been shown that molten rock retained the prevailing magnetic field of the Earth as it cooled. Fred Vine, a young scientist at Cambridge University, showed that the striped magnetic pattern of oceanic rocks was in fact the geological record of successive reversals of the Earth's magnetic field. As molten material spewed away from each side of the Mid-Oceanic Ridges, it adopted the prevailing magnetic field of the Earth as it cooled down. At the next change in the magnetic field, the latest batch of cooling material would freeze that polarity into the rock. So stripes of "normal" and "reverse" magnet-

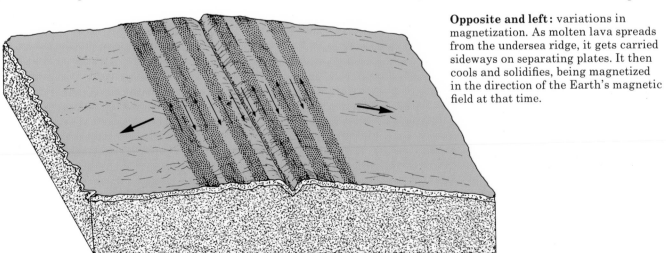

Opposite and left: variations in magnetization. As molten lava spreads from the undersea ridge, it gets carried sideways on separating plates. It then cools and solidifies, being magnetized in the direction of the Earth's magnetic field at that time.

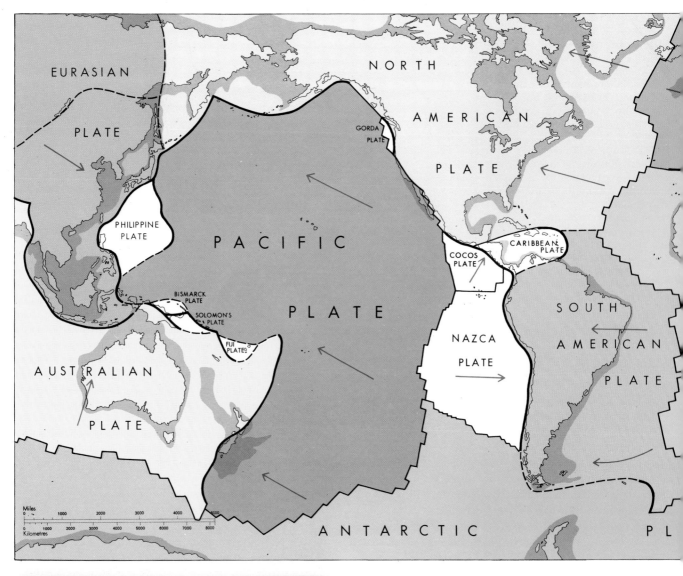

Left: the Red Sea and the Gulf of Aden taken from a spacecraft. Both bodies of water were formed by a shift in the ocean floor. If Ethiopia (left) and the Arabian peninsula (right) were joined, the fit would be excellent.

ized rock drift away from the Mid-Oceanic Ridge.

The reversals of magnetism shown spreading from the underwater ridges could not have been observed if the material had not been molten as it was produced, and this was the evidence needed to prove that new material is continually being created from the center of the ocean basins. It was also shown that the material spreads evenly on each side of the ridge so that the pattern on either side is symmetrical but in a mirror image. It was also possible to put a geological date on the growth of the ocean floors, since the times of magnetic reversals had already been calculated from rocks on land.

From all this evidence scientists were able to move quickly to draw a map of the various plates that make up the surface of the Earth. Their edges can be defined by looking at the Mid-Oceanic Ridge systems, the trench system, and other areas of persistent volcanic activity. Basically, there are six major and a number of smaller plates into which the Earth's surface is

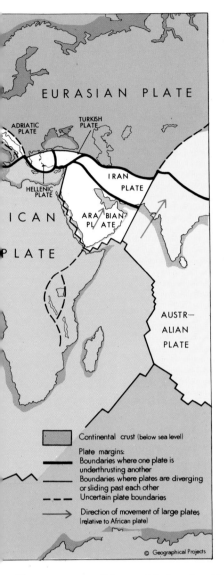

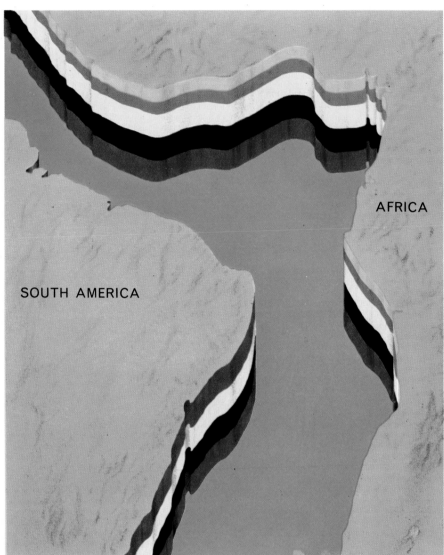

Above: a map of the world showing the plates of crust that carry the Earth's continents. They are in constant motion because of the spread of molton material from deep within the mantle, especially at the mid-ocean ridges.

Above: this diagram shows clearly how well the northeast coast of South America fits into the Gulf of Guinea on Africa's northwest coast, supporting the continental drift theory. Rock strata of the two are also similar.

divided. These are shown in the map on this page.

In spite of all the progress, toward the end of the 1960s there remained a sizeable number of unconvinced scientists. *Glomar Challenger* tipped the balance. *Glomar Challenger* is a remarkable drill ship 400 feet long, fitted with a 142-foot high drilling derrick. It can drill in water depths of over 20,000 feet to recover cores under 3000 feet of sediment.

Glomar Challenger went into service in 1968 on the deep sea drilling project which was formulated by the Joint Oceanographic Institutions for Deep Earth Sampling. This is a remarkable example of international cooperation. The major ocean research institutes throughout the world are invited to participate in each of the ship's cruises so that the skills of the world's top marine scientists can be combined in the detective work undertaken by the vessel.

The research ship first set out to prove the theory of sea floor spreading which had been developed from observing the magnetic stripes running parallel to

the Mid-Oceanic Ridge. This was done by drilling deep into ocean floor sediments. Before *Glomar Challenger* went into action, American researchers had put forward a daring theory which laid down a time-scale for magnetic "events" throughout the world, based on the fact that the same sequence of magnetic striping occurred at the ridges of different oceans and ought to be able to be correlated.

This theory had to be tested. *Glomar Challenger* did so by drilling deep into the sediments overlaying basement rock to obtain cores. These were analyzed to check the ages of fossils that had built up on the ocean floor directly after its formation. Besides sampling close to the ocean ridges, the ship drilled thousands of miles away from them in what, if the theories were to be proved, should be the oldest parts of the ocean floors. In those areas drilling was the only way in which to date the rocks because there was no distinct pattern of magnetic striping. Either there had been no reversals of the Earth's magnetic field

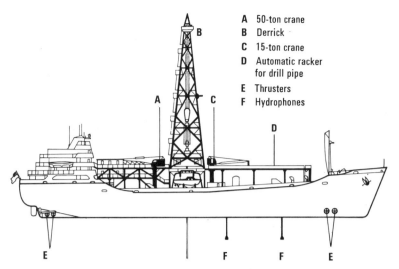

A 50-ton crane
B Derrick
C 15-ton crane
D Automatic racker for drill pipe
E Thrusters
F Hydrophones

at that time, or the magnetic record had just faded away.

Making a complete traverse of the South Atlantic at about 30° S on the third leg of the series of cruises, *Glomar Challenger* proved what had been postulated. The age of the sediments away from the axes of the Mid-Atlantic Ridge was found to increase, showing that the sea floor had a spreading rate of nearly an inch a year.

Hundreds of boreholes later, *Glomar Challenger* is still adding evidence to the Wegener hypothesis. If

Below: a view down the center of the drilling derrick of the *Glomar Challenger*. In 1970 the ship was responsible for discovering the oil fields in the Gulf of Mexico.

Above left: the *Glomar Challenger's* principle features. **Above:** the stern thrusters of the *Glomar Challenger*. The ship has thrusters at both bow and stern. This arrangement enables the research vessel to move sideways so that it can remain precisely over the drill site in the open sea.

any proof is needed that the study of the oceans is a young and dynamic subject, the development of the theories of continental drift and plate tectonics provides it.

Far more than mere academic knowledge has come from the detective work undertaken by scientists pursuing the continental drift theories. New equipment and techniques have had to be developed to gain scientific evidence, and some of these will help to

develop marine resources. *Glomar Challenger*, for example, pioneered many techniques which are in commercial use today. The ship has a satellite navigation system which gives automatic accurate positioning via an on-board digital computer. Weather satellites transmit photographs direct to the ship for the meteorologist to use in forecasting weather and sea states – which is important when there is over 20,000 feet of drill pipe to haul to safety on board if a bad storm blows up. To drill in great depths precludes the use of conventional mooring systems such as chains, wire rope, and anchors. *Glomar Challenger* remains on station while drilling with no anchors at all. Sonar equipment is used for this. An acoustic beacon is dropped to the sea floor and sends sound back to the ship where it is picked up by hydrophones (underwater microphones) fitted in the hull. Because of the distance between the various hydrophones in the hull, each sound wave front is received by one hydrophone before another. These differences in arrival time are fed to the computer soon after the ship arrives on station, and this gives the reference for the ship's exact position. If wind or waves start to drive the ship off position, the distance between the sea floor beacon and each hydrophone naturally changes. The time difference is recorded and is instantly noted by the computer which sends a corrective signal to activate the ship's main propellers and thrusters mounted at bow and stern. It is brought quickly back onto station before dangerous kinks appear in the drill string.

Above: a geologist examines core samples in the laboratory on the *Glomar Challenger*. Some of the cores taken from the seabed were 140 million years old. They helped scientists explain the nature of the ocean floor.

Left: the 400-foot-long *Glomar Challenger* on station over a drilling site. Built in 1968, it was named after the original *Challenger*. Unlike that ship, however, it was purposely designed for oceanographic research.

Another pioneering technique developed on *Glomar Challenger* enables a deepwater borehole to be reentered after the drill string has been removed to replace a blunted drill bit or to move off station when a storm blows up. Underwater sound is again utilized. A cone is positioned over the borehole, having been fitted with acoustic beacons. An interrogatory acoustic device is fitted to the drill string that is to reenter the hole, and by monitoring the sound waves received on the surface the drill is steered into the cone and funneled back into the borehole.

These and other techniques developed during *Glomar Challenger's* triumphant series of voyages will probably makes its expeditions as important in the development of the study of the oceans as those of its famous namesake 100 years before.

Atlantis: Mystery of the Deep Sea

Scientific evidence is mounting for the existence of a sunken continent in the mid-Atlantic Ocean. Archaeological finds and, more recently, underwater investigations have confirmed the strong possibility of a once great civilization on the island of Crete. Are either of these the remains of the fabulous lost Atlantis, that flourishing region described by the Greek philosopher Plato as a paradise on Earth?

The legend of the lost continent of Atlantis has persisted since Plato first wrote about it in about 335BC. He told of a remarkably advanced culture of peaceful, pleasure-loving people on a huge island continent "beyond the Pillars of Hercules" (the Strait of Gibraltar). The account described in detail the splendid capital city of Atlantis, but the book fragment ends with the total destruction of the island and its inhabitants in a cataclysm that made it sink beneath the waves in only 36 hours. Plato claimed the story was true, but there are many inconsistencies and implausibilities in it.

This account generated debate and spawned thousands of books and articles over the centuries, seeking answers to whether the story was fact and where exactly the huge continent had been. At the turn of the century, the whole legend came into new focus.

not disappeared. How could such a theory hold up?

Proponents of the idea that Crete and Atlantis were one argue that it has been established that a catastrophic volcanic eruption partially destroyed the island of Thera, near Crete, between 1500 and 1450 BC. Study of the Krakatoa disaster of 1883 showed that the damage from a tremendous eruption extends over unexpectedly great distances. Therefore, they say,

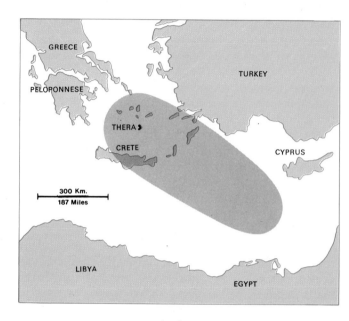

Above: a map of the eastern Mediterranean showing the extent of the layer of ash produced by the Thera eruption.

Below: Ignatius Donnelly, a successful politician who wrote an influential book on the lost Atlantis.
Below left: Donnelly's map of Atlantis and its empire from his popular book, *Atlantis: the Antediluvian World*.

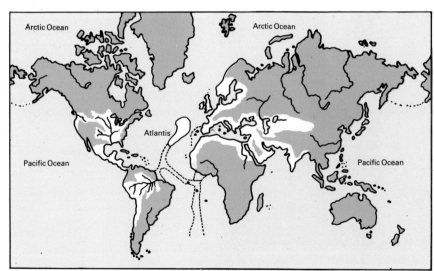

Excavations on Crete, the mountainous island at the southern end of the Aegean Sea, unearthed the ruins of a glorious civilization that flourished about 4500 years ago. Scholars called the people Minoans, named after one of their greatest rulers, Minos. The link between Crete and Atlantis was proposed in 1909 by a scholar at a leading Irish university. However, Crete is not in the Atlantic, it is not huge, and it has

the Minoan civilization on Crete had been totally destroyed by the eruption on Thera. They go on to say that the legend could have been describing the disappearance of Atlantis in an historical rather than a geographical sense, which could then apply to the destruction of a whole people, their civilization, and their culture.

As for the argument that Atlantis was in the

Atlantic Ocean, it has been pointed out that there is another geographical feature called the "Pillars of Hercules" by the ancient Greeks: two promontories on the south coast of Greece *facing Crete*.

Many Atlantis experts, however, denied the Crete-Minoan theory and insisted that the great civilization was located in the Atlantic. They found reinforcement for their argument in the theory of continental drift, now widely accepted as the explanation of how the Earth's land masses came to be in their present positions, and by sightings of underwater formations off the island of Bimini in the Bahamas in 1968.

Atlantists claim that there is evidence of roads, walls, a great arch, and pyramids deep under the surface of the water around Bimini. An expedition on board a bathyscaphe later gave support to the existence of a sunken continent when they reported seeing flights of steps carved into the continental shelf off Puerto Rico. Enlisting the theory of shifting tectonic plates, Atlantists argue that Bimini and the region of the eastern Caribbean could once have been dry land. Geologists and oceanographers refute this. Many agree that a continent might once have existed in the mid-Atlantic, but if so, they say it was long before there was human life on Earth. Scientists also claim that there is no trace of ruined cities beneath the waters, and that everything the Atlantists identify as the work of a former civilization can be attributed to natural rock formations.

Whatever the claims and counterclaims, the legend of the lost continent of Atlantis still intrigues scientists as well as nonscientific investigators. Will technology finally provide the answers to this ancient mystery?

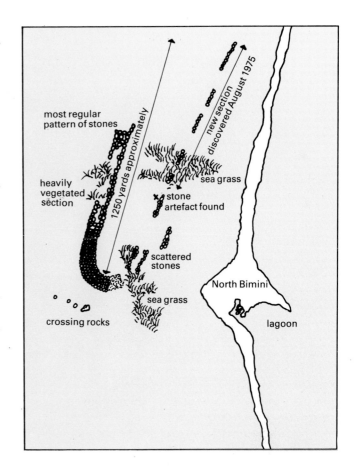

Above: a map of the Bimini Road drawn after aerial and underwater surveys carried out by an American group.

Below: a volcanic eruption on the new island of Surtsey. Atlantists believe that the kind of rumbles that created Surtsey in the Atlantic could also have destroyed Atlantis by a cataclysmic upheaval.

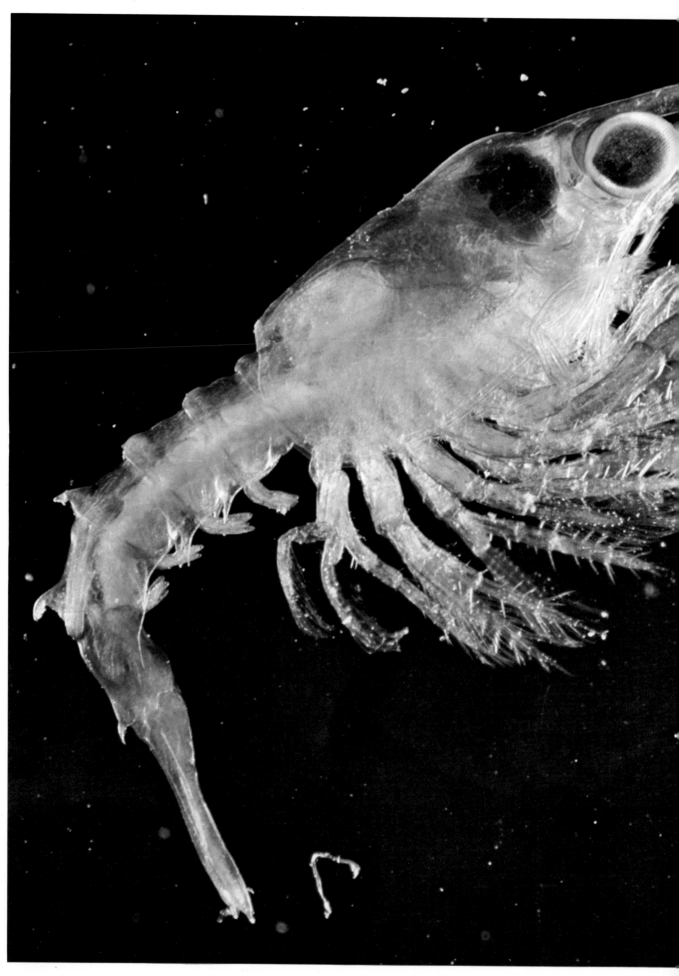

Chapter 5

Life
in the Sea

Life in the sea begins with phytoplankton –
microscopic plants that rely for growth on the process
of photosynthesis. They are preyed upon by the
zooplankton, tiny organisms which in turn are the
basic food of commercially important fish. Jellyfish are
also plankton, although it is hard to think of these
unformed blobs of matter as animals. Nonetheless,
jellyfish together with corals and sponges are the first
rung on the marine evolutionary ladder. Next come
starfish and then worms, followed by two great classes
of marine invertebrates. One comprises the mollusks,
which start with lowly mussels and clams and end at
their most advanced with the mysterious giant squid.
The other great class is the crustaceans – crabs,
shrimps, and lobsters – whose land-based relatives
are the insects.

Opposite: the teeming life of
the world's ocean contains many
colorful and strangely shaped
organisms. The crustaceans are
among the most varied sea
creatures.

How Food Chains Affect Ocean Life

Properly exploited, the oceans could provide all the food human beings need for many years. In seeming contradiction, the oceans are deserts in terms of productivity compared with fertile land areas.

These facts do not make a paradox. The oceans can indeed supply all the food we need – but only because of their immensity rather than their fertility. The oceans contain enough protein to feed us all, but they will have to be exploited in a far less haphazard way than they have been up to now. Until quite recently we have not thought carefully enough about the exploitation of the protein resources of the sea because it has been all too easy and we have eaten only the marine foods that we like – such as haddock, cod, herring, scallops, and shrimp, to name a few. To feed us all in future, however, we cannot afford to be so choosy. We must look for protein in bulk, something that is easily exploited and easy to process. Where to start?

In the sea as on land, there are food chains. Plants are eaten by little animals, bigger animals eat little animals, humans eat mostly the big animals. That is putting it at its simplest. Along this line some protein is "lost" to human consumers because of energy transfer. This means that when a big fish eats two small fish, the protein value of that big fish does not represent two times the protein value of the two fish it ate. Much of the protein value that the big fish has accumulated from its prey is burned up merely to keep it alive, rather than in making it grow.

Little is known about the efficiency of this food chain, but a rule-of-thumb calculation puts it at about 10 percent overall. In other words, it takes 10 pounds of plant life to support one pound of small fish, 10 pounds of small fish to support one pound of big fish, and so. This is a crude extrapolation. Efficiency also depends on the environment in which the organisms live, especially as to whether there is a plentiful supply of food at all times; the age of the animal, young fish needing more food and growing faster than older animals; and the exact point in the food chain at which measuring the efficiency of food energy transfer starts. The last factor creates a problem. For example, the largest mammal in the world is the blue whale. It does not eat fish at all, living almost exclusively on a tiny crustacean which in turn feeds on planktonic plants. That puts the blue whale low down on the food chain and makes it difficult to assess its energy transfer efficiency. How do we assess ourselves in an energy transfer table? Humans eat shellfish, which is low down the chain, small fish, big fish, and whales.

We complicate the formula still further by taking fish such as herring, fairly high in the chain, turning them into fish meal and feeding them to chickens, which have a high energy transfer efficiency, and then eating the chickens. All these variables combine to show that there is not so much a simple food chain as a rather complicated web.

It can be taken as said that there is great loss of energy efficiency when a big organism eats a smaller one. Isn't it then the logical solution to world feeding problems to get as low down on the food chain as possible to maximize sources of protein? In other

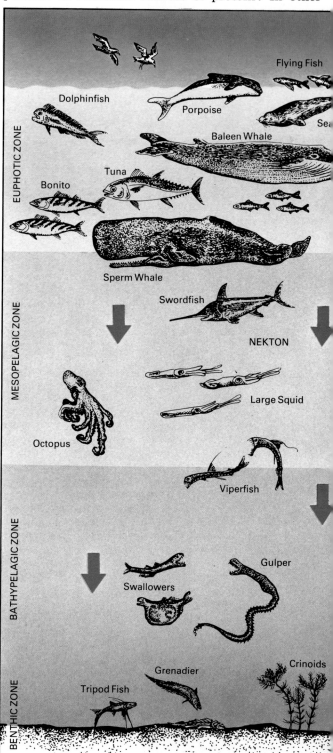

words, to cut out the inefficient intermediate food and go for the primary source. Unfortunately, it is just not that simple. The most basic form of life in the ocean is phytoplankton. To say that we should eat these microscopic plants because they are the first link in the food chain and in order to eliminate the inefficiencies of later energy transfer is like saying that we should cut out beef and eat grass. The problems of harvesting and processing the "grass" of the oceans into an even remotely palatable form would be too great. We need the intermediary – fish.

This is not to say that we can ignore the patterns of growth and distribution of phytoplankton just because we cannot eat it, any more than a farmer can afford to ignore the rate at which the grass grows in the pastures. Phytoplankton is the most widely distributed and abundant form of plant life on earth, although this is because of the sheer size of the oceans. All marine life depends on it. So in order to unravel the marine food web, the patterns of distribution of food fish, and the most efficient ways by which to exploit them for the overall human good, it is necessary to have a full understanding of the grass of the sea.

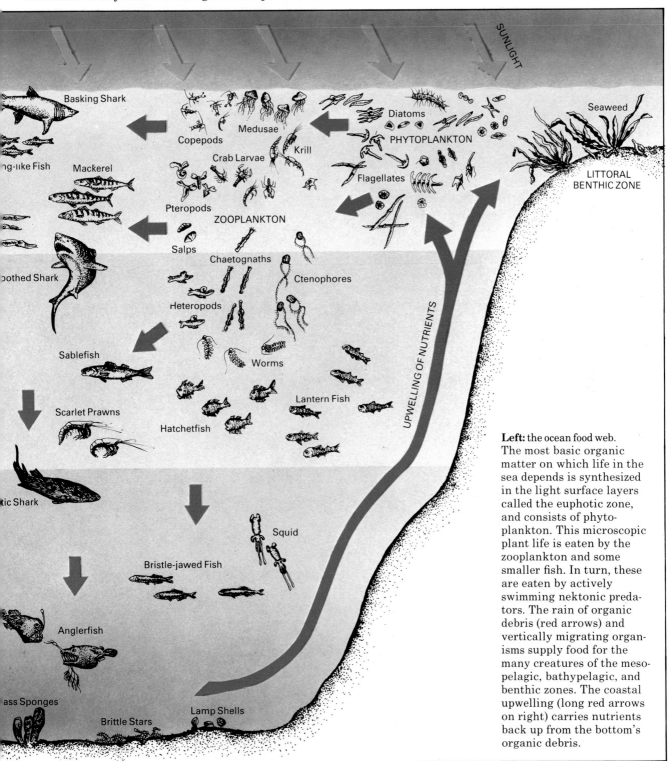

SUNLIGHT

Basking Shark

Copepods
Medusae
Crab Larvae
Krill

Diatoms
PHYTOPLANKTON

Flagellates

Seaweed

LITTORAL
BENTHIC ZONE

ng-like Fish
Mackerel

Pteropods
ZOOPLANKTON

Salps

Chaetognaths

oothed Shark

Ctenophores

Heteropods

UPWELLING OF NUTRIENTS

Sablefish

Worms

Scarlet Prawns

Lantern Fish

Hatchetfish

ic Shark

Squid

Bristle-jawed Fish

Anglerfish

ass Sponges

Brittle Stars

Lamp Shells

Left: the ocean food web. The most basic organic matter on which life in the sea depends is synthesized in the light surface layers called the euphotic zone, and consists of phytoplankton. This microscopic plant life is eaten by the zooplankton and some smaller fish. In turn, these are eaten by actively swimming nektonic predators. The rain of organic debris (red arrows) and vertically migrating organisms supply food for the many creatures of the mesopelagic, bathypelagic, and benthic zones. The coastal upwelling (long red arrows on right) carries nutrients back up from the bottom's organic debris.

Phytoplankton and Zooplankton

Life in the sea begins with plankton. It is the basis of all oceanic life. Until a century ago, however, plankton was largely ignored by biologists: only in 1887 did Victor Hensen originate the name, taking it from the Greek word meaning "wandering." This is descriptive of these plants and animals which have limited independent powers of movement and which drift almost powerlessly around the oceans at the whim of waves, currents, and tides. They have existed on earth for 180 million years.

Among plankton, the most primitive form is phytoplankton, the grasses of the oceans. Individually a microscopic plant, phytoplankton occurs in such quantities that the mass lends color to great stretches of ocean. Northern European waters, for example, often have a greenish-brown hue that makes them look as if they are muddy. In fact, that mudlike substance is floating life – millions upon millions of diatoms, the biggest single group of plants that makes up phytoplankton. Tow a fine silk net through the brownish waters and it will come up full of greenish-brown slime. Carefully drop some of that slime onto the slide of a microscope – and a new and beautiful world opens up. Exquisite shapes in perfectly symmetrical circles and chains, with patterns of rods and spikes in an infinite variety of designs – and all made

of glass. Diatoms are single cell organisms with an external skeleton of silica, and the processes by which this substance is obtained from seawater and then built up by the plant are a mystery. Neither is it understood why the seawater does not dissolve the silica.

Inside the diatom's silica shell is a single unit of living protoplasm which interacts with the sea outside through slits and pores in the skeleton. The

Above: a bloom – the sudden mass reproduction of plant plankton – seen from a ship. Blooms sometimes extend for miles and are so dense that they lend the color of the plankton to the sea surface.

Left: a photomicrograph of diatoms. One of the most widespread of the phytoplanktons, diatoms possess silica skeletons. Some of them are remarkably beautiful.

and it can withstand cold or – in intertidal areas – even drought. When things get better, it comes to life and starts reproducing again.

It is essential that these microscopic plants have such remarkable attributes of growth, reproduction, and survival because they are the very basis of all life. Annual phytoplankton production, expressed in terms of carbon, is probably greater by as much as three times that of land plants. They account for at least a half of total plant production on Earth, making them the most abundant form of plant life. Nothing, it seems, has been left to chance in the evolution of this most basic form of existence. Even the delicate and beautiful glass shapes that show up under the

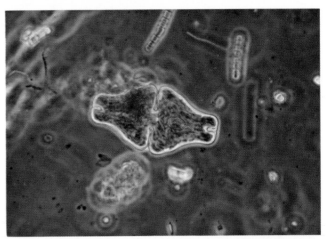

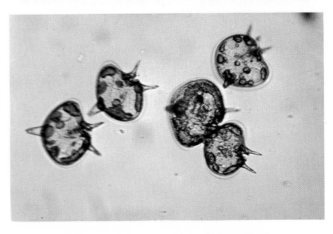

Top: one of the freshwater species of phytoplankton. There are several thousand species of marine plant life.
Above: the dinoflagellate *Peridinium depressum*. Some of this group of phytoplankton are plants, some are animals.

transparency of the silica means that the cell can absorb the light essential for the diffusion of gases and nutrients into the cell.

Reproductive division is a simple matter for most plants because their cell walls are soft. The rigid glass walls of diatoms would make the processes of growth and division extremely tricky were they not so cunningly constructed. The cell wall is built in halves, with the bottom half fitting neatly into the top like a pillbox. As the diatom reproduces, the halves of the box separate, and each half forms a new bottom through secretions.

Given the right conditions, diatoms reproduce at a tremendous rate, sometimes up to four times a day. But it follows that with "new boxes" continually being formed to fit "old lids," the cells get progressively smaller. Once they shrink to a certain size, either reproduction stops or the diatoms die quickly. What stops the species from becoming extinct is a process that bears the rather splendid name of auxosporulation. In this little understood type of sexual reproduction, certain cells bulge out from between the halves of their glass pillboxes in a swollen lump. This lump separates and grows a new shell, forming a plant that is back to original size and strength. The whole reproductive process then begins again. If the conditions for reproduction are not right – if the sea is too cold or if there are insufficient nutrient salts, for example – certain cells insure the perpetuation of the species. They do this by wrapping themselves in a thick overcoat of silica, which serves not only as protection but also gives the plant the requisite weight to sink to the sea floor. There it rests, for months if necessary, in a state akin to hibernation,

microscope are formed that way for a purpose: diatoms have to float in the oceanic layer in which light penetrates – the eutrophic zone. This is essential for the process of photosynthesis. The high density of silica, which could theoretically cause diatoms to sink to the ocean floor, is overcome because the shapes have a large ratio of surface to volume area. This provides buoyancy as well as making the nutrient exchange with the surrounding medium more efficient.

The next link in the oceanic evolutionary chain is still planktonic, but marks the dividing line between

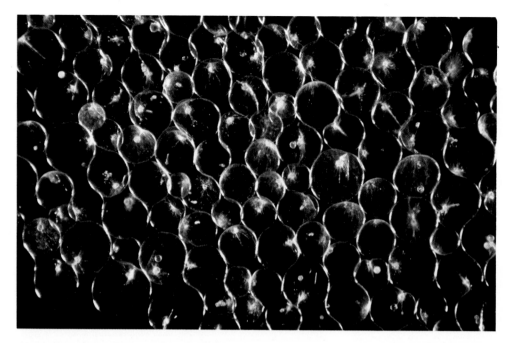

plant and animal life: dinoflagellates. They have been in existence for some 70 million years and, like the smaller diatoms, are single-cell organisms. They have two flagella or whips which give them their name. These whips are threads of trailing protoplasm that constantly lash, giving rise to the Greek prefix of their name that means "terrible." The lashing gives them a small degree of propulsion, perhaps a few inches. Some dinoflagellates eat diatoms and even fish eggs; others rely entirely on the process of photosynthesis for growth.

One remarkable attribute of certain species of dinoflagellates is bioluminescence, in which the organism itself lights up. In the open sea or sometimes even in harbors the sea will glow. Sometimes even the imprints left by feet in wet sand at night will suddenly shine, or breaking waves burst with light, or a leaping dolphin shimmer eerily. How this species of dino-flagellate – *Noctiluca scintillans* – emits this pale cold light is the subject of much study. Why it does it is simply not known. It is another one of the mysteries of the seas.

Dinoflagellates are microscopic but their presence is sometimes easily observed – in the case of the genus *Coccolithophorea*, as a good sign. This genus thrives in warm tropical waters where other planktonic forms do not flourish because of the comparative lack of nutrient salts. How they manage so well is still a matter for conjecture – another mystery. However, some species of the genus extend to northern waters and particularly to the North Sea and Norwegian Sea, where they reproduce at such an enormous rate

that they can be found in concentrations of 100 million to a couple of pints of water. This has the effect of giving the sea a milky-green color which is greeted with enthusiasm by herring boats because it means that large shoals of herring will appear to feed on the plankton.

In contrast, the telltale signs of the presence of other forms of dinoflagellate are greeted with dismay by fishers. When the right conditions of heat, light, and nutrient salts coincide to cause an explosion in the reproduction rate – a "bloom" – of plankton, the water turns a tomato color in the phenomenon called red tide. Under these conditions the plankton make such a demand on the oxygen content of the water that they drive the fish away or cause them to die through respiratory failure because their gills become clogged with the plants. Sometimes the effects can be even more serious: one dinoflagellate species under red tide conditions produces a strong neurotoxin which kills marine life. When it accumulates in shellfish

that feed by means of filtering, it quickly kills people who eat the shellfish.

The families and orders of plant plankton are small compared to those of animal plankton, which represent the greatest diversity of marine life. This diversity makes zooplankton an exciting field of study for marine biologists because they know that on every cruise they are likely to discover a new class – or even a new species – of animal.

The lowest form of animal life in the ocean is the single-cell Protozoa group, most of which are microscopic but some of which can be seen with a powerful magnifying glass. For all their primitive state, the animals display great beauty. One of the most abundant orders of Protozoa, the Radiolaria, has a glass skeleton like the diatom, but even more intricate: delicate spheres within spheres, fluted urn shapes, and hundreds of radiating spines. All types have radiating spines that trap food, as do the Foraminifera, the other important Protozoa, which has a lime skeleton divided into communicating chambers varying in number according to the species.

Dominating all the zooplankton are the copepods which are the single most important group of animals in the sea. They often make up 70 to 90 percent of a zooplankton sample and in one group alone – the calanoid copepods – there are over 1200 species. The size range is great – from microscopic to a quarter of an inch in length – but whatever the size, for the first time in our examination of planktonic life, we are able to discern shapes that correspond fairly closely to those of animals with which we are more familiar. Copepods are crustaceans, the insects of the sea. Superficially they have a flealike appearance. Typically they have a pair of long antennae and up to six pairs of legs on the oval part of the body behind the head, each pair of legs joined at the base. The animal swims by flicking these legs and its antennae.

The copepod is a vast consumer of diatoms, eating

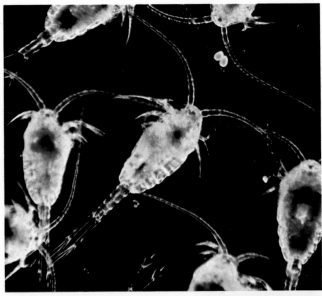

Above: copepods of the *Temora* species. Although of the crustacean subclass, copepods do not have a carapace. They are vastly numerous and vitally important in the food chain of the ocean.

Right: the dinoflagellate species *Ceratium tripos*. Its tripodlike shape and tough spiky covering are its main protections against enemies that prey on it.

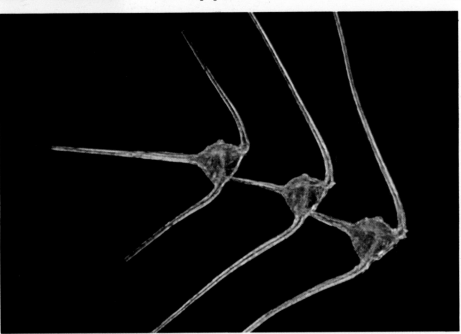

half its own weight of the little plant plankton in a day. It does this by filtering seawater. Because of its profusion it is an important food source for many larger animals: herring, for example, feeds on one of

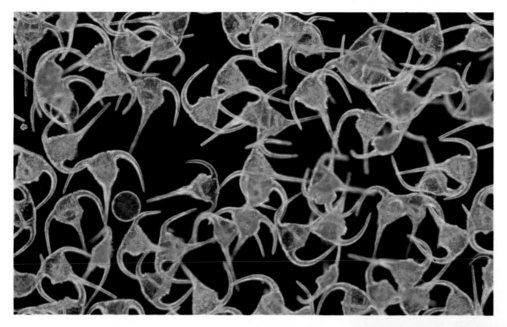

Left: a close-up of dino-flagellates. This marine plankton has a light-sensitive spot by which it can tell where the light is strongest, so being able to swim toward it and keep near the surface.

Below: the arrow worm, a minute zooplankton. Arrow worms have transparent bodies that make them difficult to see in water. They move by flexing trunk and tail.

the most economically important copepods, the *Calanus finmarchicus*, which forms dense shoals off northwestern Europe. Sardines are wholly dependent on the smaller copepods, and seabirds like the albatross and stormy petrel rely on them for a large part of their food intake.

Because the diatom-copepod-fish/bird link is more complicated than this description indicates, it is a good example of the way in which the oceanic food chain must be regarded more as a web than a chain. Copepods themselves are eaten by other plankton and particularly by arrow worms (Chaetognatha). Ranging from hairlike creatures of less than one 16th of an inch long to pencil size, arrow worms are vicious little creatures with spiny heads and hooklike teeth. Stabilizing fins and a tail by which they can steer means that they can dart quickly over limited distances to secure their prey. When that prey is the smaller copepod it is devoured whole. There are some 70 species of arrow worm with life spans ranging from six weeks to two years.

The idea of a simple "big-fish-eats-smaller-fish-eats plankton" is again confounded when we look at another crustacean plankton, the euphausiid. Of the 90 known species just one – the *Euphausia superba* – plays a vitally important part in sustaining the largest mammal on earth, the baleen whale. With this link the simple view of the food chain is quickly blurred, because the big fish – the baleen whale – lives exclusively on a diet of *Euphausia superba* or krill – plankton rather than a smaller fish. The mighty whales browse on krill, which can grow to lengths of 2 inches, filtering them from the water through the plates in their jaws. In the Antarctic krill drifts in huge shoals through which the whale can lazily swim,

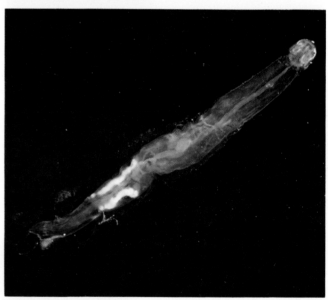

sucking them in. Because the number of whales in the Antarctic are well known, and because analysis of one whale's stomach contents is an easy task, it is possible to make the simple but extraordinary extra-polation that over a four-month summer feeding period whales will consume 150 million tons of euphasiid. Fish and birds also take their toll of the krill stock, but still the great swarms of the animal appear, giving the sea a reddish color. Euphasiids contain large amounts of Vitamin A and this is stored in the livers of many fish. Other less welcome substances are also accumulated by krill and passed on to fish where they are further concentrated. The euphasiid, for instance, is thought to play a large part in the oceanic distribution of such products of fallout as radioactive zinc.

All the plankton we have so far looked at is distinguished by exquisite shapes at best and unusual shapes at worst. As we approach the top end of the planktonic scale in terms of size, formlessness prevails, and the largest plankton of all is the shapeless jellyfish.

Jellyfish are grouped with plankton because they drift with the winds and currents, although some have a limited ability to propel themselves. Although the *Physalia* or Portuguese Man-o'-War is known as the most notorious jellyfish of all, it is strictly speaking not a jellyfish but a siphonophore. It has a crestlike sail which is inflated by a gas-filled float so that the animal can tack along with the wind, trailing its deadly tentacles which can be over 30 feet long. Plankton or fish coming into contact with them are paralyzed by an injection of a poison-filled, threadlike whip. The tentacles then contract and the prey is drawn into one of the animal's flexible central mouths. The Portuguese Man-o'-War delivers a particularly painful sting to swimmers, as do other species of jellyfish, but the sting from the common jellyfish (*Aurelia aurita*) cannot be felt. This is the species usually seen in large colonies around European shores where it makes swimming unpleasant rather than dangerous. It is difficult, however, to distinguish between the common jellyfish and some of its nastier cousins, so the best rule is to keep clear.

Not all the tentacles on jellyfish are used for capturing prey. Some are used for digestion, others for reproduction. The male *Scyphozoa* jellyfish leaves his sperm near the female, which are drawn into her by a form of chemical stimulus and fertilize her eggs. The hatched eggs develop into tiny polyps which attach themselves to the seabed. The polyps form buds from which young jellyfish grow and break away after reaching a certain size.

So far we have considered plankton in an environmental vacuum, but the areas in which it lives and

Above: the *Euphausia superba*, largest of the species of a shrimplike marine animal that is commonly known as krill. The 10 species in the Antarctic exist in such huge numbers that they are the main food for fish, seals, birds, and whales.

Left: the Portuguese Man-o'-War, a jellyfish with a powerful sting. It is most frequently found in the warm parts of the North Atlantic. Occasionally strong winds cast thousands of them onto coastal shores.

115

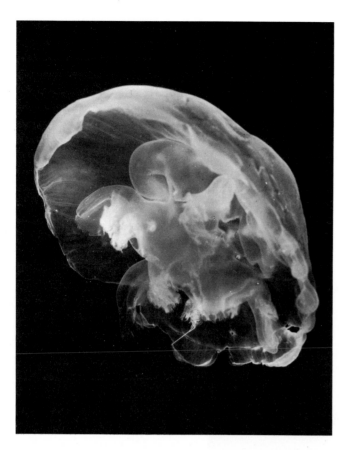

Above: the *Aurelia aurita* or moon jellyfish. It has the typical umbrella shaped body of its species, which it pulsates slowly in order to move. The moon jellyfish is the one most common to Europe and North America.

Right: this diver is collecting specimens of phytoplankton in the Caribbean. His job is speeded and eased by the use of a motor-driven tug that pulls him along.

breeds, and its seasonal health, are vital factors in the make-up of the whole oceanic life web. Plankton is now studied closely by fisheries scientists because the life cycles and distribution patterns of plankton tell us so much about commercially exploitable fish stocks. For example, when a herring fishery in the North Sea remains empty of fish for a season, the void can very often be attributed to a lack of food which, in the case of herring, is the copepod *Calanus finmarchicus*. If there are not enough *Calanus* to support the herring, it is reasonable to assume that there are not enough plants – diatoms – to support the *Calanus*. So we have to look at the reasons for the lack of diatoms to explain the failed fishing grounds. That takes us back still further into the food web to an examination of the conditions needed to support phytoplanktonic existence.

Basically, the conditions needed by diatoms as the grass of the oceans are the same as those needed by grass on land: sunlight from above and nutrients from below. In northern European waters at the end of March there is a bloom of diatoms. Their numbers will have quadrupled during the month. This is due first to the lengthening days and the increasing strength of the sun, which encourages the process of photosynthesis. Second, there is a rich supply of nutrient salts in the surface layers of the sea that have built up during the winter when there were few plants to absorb them. One possibility for the failure of a fishery immediately presents itself: if the nutrient salts are not present the diatoms will not reproduce at a rate fast enough to supply adequate food for zooplankton of which the herring's favorite food is one. Variations in normal weather conditions and slight changes in the flows of currents and tides could cause nutrients to be dispersed.

During April the diatoms start exhausting the nutrients in the surface waters and the plant crop experiences a decline until August. In September there is a secondary but lesser bloom of diatoms. This can occur because of the interaction of the warm and cold layers of water at a sharply defined point called the thermocline. During the summer there is a layer, perhaps 60 feet thick, of warm water. The top part of this warm water layer becomes thin in nutrient salts; plants and animals die and sink through the thermocline to the cold water below. Toward the end of summer the surface layer begins to lose heat until its temperature is the same as that of the lower layer. The thermocline disappears and the two bodies of water mix so that there is a renewal of the nutrient

content of the sunlit upper layer. There is still enough strength in the sun for photosynthesis to take place, so the diatoms can bloom again. They diminish in October and reach a minimum in December.

Zooplankton gorges on diatoms in the spring and summer, multiplying at a tremendous rate. Numbers fall off in the autumn, however, which contributes to the secondary bloom of diatoms because there are just not as many animals eating them.

The richest fishing grounds in the world are in the temperate waters of the north Atlantic and near Antarctica. This is because diatoms flourish best in cooler waters. The richest crops of diatoms are to be found in shallow waters of the continental shelf where light often penetrates to the seabed to allow photosynthesis right through the water column.

Above: the big plankton net on the *Discovery*.
Below: patterns of vertical migration of zooplankton. White stars represent the copepod *Calanus finmarchicus*; yellow asterisks, the jellyfish *Cosmetira pilosella*; and the red asterisks, the shrimplike *Leptomysis gracilis*.

There is also a greater abundance of nutrient salts there since these substances are carried to the seas by rivers. This is why some 90 percent of world commercial fisheries are based on or near the continental shelves.

Plankton migrate vertically as well as horizontally on a daily and seasonal basis. It chooses the degree of light intensity and temperature which suits it best in much the same way as a sunflower will turn its face to the sun. This in turn influences the movement of fish. Some five hours after sunset, for example, herring are abundant just below the surface where they have pursued the vertically migrating *Calanus*. It was this simple pattern of movement that long sustained the great drift net fishery. Boats would string walls of netting in the surface layers of the sea at night to catch the herring at their feeding.

To gain more knowledge of the distribution of fish and so to be able to catch them more efficiently, we must know more about plankton, their basic food supplies. A full understanding of plankton can only be gained by studying factors such as ocean temperature, salinity, and currents. This means that the fisher, the marine biologist, the chemist, and the oceanographer have common aims in their study of the oceans.

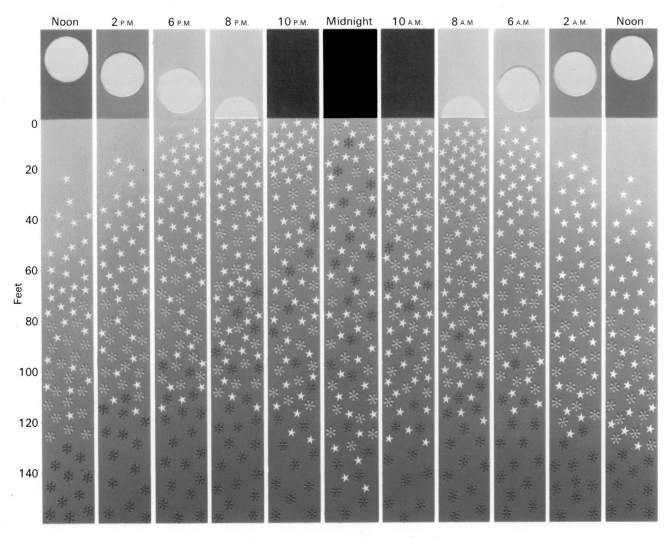

Second Step of Evolutionary Scale

It is hard to think of corals and sponges as animals, but they are. Australia's vast Great Barrier Reef is made up of the skeletons of millions of coral polyps. A simple bath sponge is a skeleton.

Strangely enough, corals belong to the Coelenterata, the same group as contains siphonophores such as the Portuguese Man-o'-War. In cool northern waters, corals do not build reefs but form dense undergrowth on the sea floor, sometimes over many miles. The two most common of this type are the sea fan and a variety called, rather gruesomely, dead men's fingers because of its shape. Both types branch across the seabed, and each has a flesh-covered skeleton supporting thousands of tentacled, anemonelike polyps. The tentacles sting prey. In some species this sting is powerful enough to cause a human diver excrutiating pain.

The true corals – Madreporaria – are those that build reefs. They can only live in temperatures of around 70°F and above and are therefore restricted to tropical latitudes, usually on the western sides of oceans. Besides catching passing plants and animals with their stinging tentacles, most corals have single-cell green plants around their mouths and tentacles on which they live in times of shortage of other food. In some corals the polyps depend completely on these plant cells and can live only in shallow waters where

Above: the soft coral known as dead men's fingers. Soft corals, which usually grow treelike, are not true corals.

the plants can carry on the process of photosynthesis.

The great reefs and atolls that Hollywood films have romanticized as the very symbol of the Pacific are formed by the continuing division and subdivision of a polyp in its limestone skeleton. As new polyps form, the old ones die and the reef grows ever outward, increasing in length to form huge natural features. Australia's Great Barrier Reef is more than 1000 miles long.

One of the most beautiful sights in the Pacific is that of a coral atoll, a near-perfect circle of coral topped by palm trees and surrounding a limpid blue lagoon. Atolls are often hundreds of miles from land and rise through thousands of feet of water. Yet coral only grows in the top 200 feet or so of water where the temperature is right and where, in the case of those animals completely dependent on plant life, photo-

Left: a true coral with polyps expanded for feeding. The polyps are supported by a hard chalky skeleton which is covered by an often beautifully colored layer of flesh. The reef-building species live mainly in tropical seas.

synthesis can take place. So how are the atolls formed? Charles Darwin, on his famous *Beagle* voyage between 1831 and 1836, took time off from amassing evidence for the theory that would change the course of human thought on human evolution to put forward some simple ideas on the formation of atolls. His theory was strikingly simple – perhaps too simple for many scientists since it started a controversy that rumbled on for over a century. Today, however, armed with the results of deep ocean drilling and advanced seismic studies, Darwin's theory has been proved. In essence it is as follows:

First, there is an extinct volcano soaring thousands of feet up to penetrate the sea surface. Around the top of the volcano, colonies of corals anchor themselves, growing from perhaps 200 feet deep to just beneath the surface. The volcano sinks slowly back into the Earth's crust, but the coral grows faster than the rate at which the volcano subsides so that there is always a fringe of living coral at the surface. It is supported by thousands of feet of dead coral beneath. Eventually the volcano sinks far below sea level, leaving only the circular ring of coral to show where it has been.

Sponges cannot build reefs. Sponges, in fact, do not do much of anything. They are among the simplest of marine animals with no sense organs, no nervous system, and no means of digestion. The sponge used in the bath, if it has not been replaced by the synthetic variety, is the skeleton of the creature and is made up of a substance called spongin. Bath sponges come mainly from the Mediterranean, the Gulf of Mexico, and the Caribbean.

The sponge spends its energy in generating enough water current through its body to filter out organic matter, bacteria, and plankton. It is essentially a pump, powered by pulsating cells that draw in water through tiny pores to their protoplasmic collars which collect the food. The water then passes into a central cavity and out through a few larger holes. The cells that make up a sponge can often be completely independent of each other so that if a piece of sponge is ripped off it will reconstitute itself into a "new" organism. This characteristic means that it is relatively easy to cultivate the commercial species simply by cutting up a large animal and "planting" the pieces. If the cutting is done scientifically, sponges of the most suitable commercial shape can be grown.

Above: a sponge seller in Athens. Bath sponges mostly come from the horny sponges of the species *Spongia*. They live only in warm seas and in relatively shallow waters. The Greek islands are the center of the sponge industry.

Below: the tube sponge of the Caribbean waters. There are also sponges with bushy, fingerlike, or treelike shapes.

Echinoderms and Plentiful Mollusks

Mention life in the sea to most people and they think of fish – the vertebrates. But in terms of numbers of species, fish and mammals are far outweighed by the invertebrates. Corals, sponges, and jellyfish are the most primitive of the invertebrates. Next come the echinoderms or spiny skins. In this group we find for the first time animals with a degree, although limited, of independent locomotion. Starfish, for example, have hundreds of suction pads along each of their five arms. Water is drawn in through a sieve in the animal's central hub and is pumped to the arms, which extend. The suction pads fix to a surface and when the starfish contracts one or more arms, it is able to pull itself along. The animal also uses its suction pads to deadly effect in consuming bivalve mollusks such as mussels. It wraps itself around the two halves of the shell and exerts continuous suction until the shellfish opens. The starfish then turns its own stomach inside out, pushing it into the shellfish and devouring the flesh completely.

The muscular ability of the starfish, as well as its fairly complex nerve cell system, is shared by other echinoderms such as the sea urchins and sea cucumbers. But the range of movement of all these creatures is limited, mainly because of their shape. The next

Above: a starfish feeding on a bivalve. Starfish occur in many colors, the most common being yellow, orange, pink, and yellow. Others can be blue, green, or purple.

rung on the evolutionary ladder holds groups of animals far better equipped to move around. They have heads, segmented bodies, and crude swimming apparatus. These are the annelid worms, which on land include the earthworms and leeches. The marine type is the bristle worm, of which there are over 60 known species. They are capable of swimming by means of oarlike projections from each of their segments: these are the bristles that give the creatures their name. The bristle worm has eyes, big jaws to devour plankton, and a nervous system which includes a cord running the length of the body from a

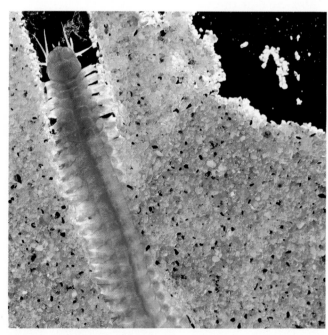

Above: a bristle worm known as mussel, clam and pile worm in North America and ragworm in Europe. Like their land relatives the earthworm, they are used as bait.
Left: the fanlike crown of this bristle worm gives it the name of fanworm. The animal builds itself a tube to live in.

Above: this colorful fanworm looks like a red flower in the sea. It is of the species that has adapted one of its crown's tentacles to act as a stopper to the entrance of the tube it builds as a home. It never leaves this tube.

rudimentary brain. Some bristle worms do not swim, but build themselves tubes of sand or mucus on the seabed in which they wait for passing prey with their heads poking out from their burrows.

From the bristle worms developed two enormous groups of marine invertebrates – the mollusks and crustaceans.

Descriptive marine biology is a relatively new science. Only in recent years have there been the facilities in terms of equipment and techniques to study the flora and fauna of the oceans in any detail. So when the name Mollusca was coined for a marine phylum it was probably fairly accurate for the limited number of species known at that time. Those species probably had at least a common superficial physical resemblance. As more and more species were discovered, they were lumped under the general name "mollusk" because they shared biological characteristics, even though they had no physical resemblance. Mollusk means soft-bodied from the Latin *mollis* (soft). This does not mean that a mollusk's body

is necessarily softer than that of other marine animals, but that it has a soft body inside a lime shell. Not all mollusks have hard outer shells, however. The class known as Cephalopoda, which includes the squid and octopus, either have no shells at all or have a vestigial shell inside their bodies.

However, the class of mollusk that most closely conforms to the popular conception of the organism is the Pelecypoda. These are the bivalves and they comprise the most numerous species of mollusks (over 10,000 are known). They include oysters, mussels, clams, scallops, cockles, and other commercially exploited species. Bivalves were probably the first marine animals to be eaten by humans, who picked them up from the seashore.

The bivalve has a hinged shell and buries itself in sand or attaches itself to a rock by threads to prevent itself from being washed away by waves. Typical of the type that leads a buried existence is the cockle. It feeds by sucking seawater through an inhalant syphon that is extended through the shell opening to get it clear of the sand in which the creature is buried. Seawater is sucked into the body through the gill and passes to the mouth where the food particles – plankton and organic detritus – are strained out and diges-

Left: the *Pecten maximus*, known as the great scallop or Saint James' shell. There are about 300 species of scallop living throughout the world's waters at depths from shallow to about 300 feet. Scallops are the only bivalves with enough independent motion to be said to swim. They do this by closing their valves quickly, forcing the water out to drive themselves forward or back.

Below: the file shell, seen open underwater. Bivalves often keep their shells open when they are at rest and there is no disturbance. They snap the shells shut at any sign of danger.

ted in the viscera. The strained water, unwanted material, and excreta pass through the anus to an exhalant syphon which, like the inhalant, projects above the sand. The cockle opens and closes its shell with two muscles and can propel itself by means of a foot that protrudes through the shell opening.

When mollusks breed they discharge thousands of larvae that float with currents and tides. Some species, including cockles and clams, settle to the bottom, grow hard shells, and burrow, but are able to disinter themselves to move to more favorable sites if they want to. Others, such as oysters, attach themselves for life to a hard surface. Mussels, on the other hand, neither bury themselves nor require a hard surface to fasten to. Once past the larval stage the mussel settles to the seabed and attaches itself to any convenient surface – even other mussels. It does this with threads known as byssus which it discharges through a groove in its muscular propulsion foot. The threads glue the animal to its chosen surface – and it takes a strong tug to dislodge it. The animal is able to move around by spinning out more byssus and moving its foot.

Another bivalve, the scallop, has a far more effective way of moving around. It is literally jet-propelled, and uses its unusual means of motion to get out of the way of starfish which it sees approaching with a series of eyes around the edge of its shell. The scallop simply opens its shell, draws water in, then closes the shell swiftly, expelling water through gaps on either side

of the hinge. It must be an uncanny sight when a shoal of the animals whizzes backward through the water.

One particularly persistent predator on bivalves is another mollusk: the common whelk. It has the rather unpleasant ability to bore through the shells of such animals as oysters and mussels in order to devour their flesh. The whelk is a member of another of the great classes of mollusk – the gastropods. On land their equivalents are the snails and slugs.

The gastropods are characterized by their conical or spiral shells, as seen in whelks and limpets, but like their landbased cousin the slug, some may have no shells at all. One common feature is the snail-like flat muscular foot by which the animal moves around the seabed and over rocks.

Right: a view of a limpet from the underside, showing its muscular foot curling over as the mollusk tries to turn over. The shell of a limpet looks like a flattened tent.

The best-known gastropods, such as whelks, periwinkles, and limpets, live mostly in intertidal areas, and although they have the ability to move, they do not have the speed to keep up with the ebb and flow of the tides. So they have to be tough enough to withstand being dried out, perhaps twice a day, and they have to retain enough moisture for respiration, since they breath dissolved oxygen in water and cannot breathe air. The periwinkle is a good example of a gastropod with remarkable staying powers: some species are able to survive exposure to dry air for well over a month. Gastropods, along with other animals living in the intertidal range, also have the greatest tolerance for temperature and salinity of all ocean creatures. In temperate waters they have to withstand temperatures ranging from freezing to perhaps the high seventies Fahrenheit. The salinity content of the water, usually so constant in the open ocean, varies widely at the shore because fresh water runs in from the land to dilute the seawater content of the pool. Conversely, the sun may evaporate the water to an extent that salinity increases to as much as 300 parts per thousand (in the open sea it is between 35 and 36‰). Gastropods have adapted to withstand these extremes and to live and breed in one of the most biologically hostile of all marine environments.

Right: a periwinkle in the process of feeding. Called winkles for short, these gastropod mollusks are vegetarians, rasping their food to pieces with a horny tongue. The common winkle is the largest and has long been eaten as food by humans.

Below: a street oyster stall in England, probably at the turn of the century. Once a poor person's food because of its abundance, the oyster is now a high-priced delicacy.

Mollusks are not confined to the seashore. Elephant tusk shells, named after their shape, are one of the species of the class Scaphopoda found at depths of well over 20,000 feet. Some bivalves, which have adapted for life at over 30,000 feet, have large guts so that they can make the most of the poor nutrient content of the deep waters.

As with so many other forms of marine life there are many species, even classes, of mollusks yet to be found. In 1952, for example, the German research ship *Galathea* dredged up 10 specimens of a species of mollusk belonging to a group which, from fossil records, had been thought to have become extinct 350 million years before. It was a tiny limpet, living on the ocean floor at a depth of over 10,000 feet. Because it had a segmented body, it was seen as another link between the annelid worms and the mollusks and crustaceans.

Even humble mollusks, then, can contribute to our better understanding of the mysteries of the oceans and life on Earth.

Mollusk Family's Advanced Species

It is hard to think of a squid as a mollusk. Most mollusks have hard outside shells and are slow-moving creatures that cling to intertidal rocks for survival, scooping or sucking in food that drifts past. Not so the squid. It is very fast moving, lives in the open sea, and has no shell. It is also a vicious predator on the largest animals in the sea, and a mysterious animal capable of invoking fear. The crew of the *Kon-Tiki* raft during its epic expedition in the late 1940s slept with

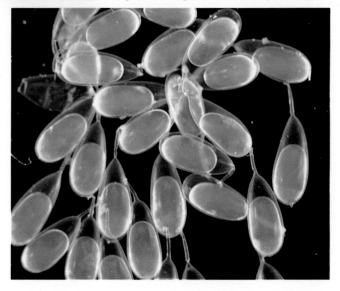

arms swiftly take the fish to the parrotlike beak.

Just how big is the biggest squid? Dead specimens recovered from the stomachs of sperm whales have been known to be up to 55 feet long, with 20-foot bodies and 35-foot tentacles. There are stories of squid so big as to make the 55-foot long specimen seem puny. By extrapolation from the thickness of pieces of tentacles found in whales' stomachs, it would seem that there might be squid some 150 feet long weighing 50 tons. J D Starkey, a naturalist, said that he once walked the length of a ship studying a squid that had "moored" itself alongside. The ship was 175 feet long.

Squids move fast. They have been seen to overtake a ship making 12 knots. Small squid can fly. One of the crew members of *Kon-Tiki* was struck in the chest by a squid traveling fast about five feet above the surface of the water. It was one of a shoal being pursued by dolphins, which are among the squid's major preda-

Left: the squid is a mollusk without a shell, though the ancestral remains of a shell can be seen at the end of its tapered body. Squids feed on fish, crustaceans, and smaller squids, paralyzing them with venom carried in two pairs of salivary glands. There are about 350 species that range in size from tiny to huge.

Below: a string of squid eggs, each one encased protectively in its own transparent capsule. The strings are attached to rocks by the female after they are laid. It is believed that one species of squid dies after spawning.

knives alongside in case a squid should probe an eager tentacle into their cabin. Once they looked out at night to see a pair of plate-size eyes staring at them out of a vast luminescent head. The squid slowly moved away and the *Kon-Tiki* crew carefully checked their knives before they slept.

Squids are the most advanced of the mollusks and a species of the Cephalopod class which includes octopus and cuttelfish. As mollusks, squids are low on the evolutionary scale, but the large varieties are formidable creatures perfectly well equipped to do battle with the most advanced marine mammals. Squids have 10 tentacles. Two of these are longer than the rest and are used to make the kill: they shoot out from the body and their club-shaped ends grasp a fish with the aid of closely set rows of suckers which draw the prey back to the shorter, grasping arms. These

tors. A squid swims by jet propulsion. It takes water
into its body cavity, contracts powerful muscles, and
shoots out the water through a funnel-shaped nozzle
(the hyponome) beneath its head. This gives the animal its speed. For direction it uses two lobes of skin
near its tail controlled by directional balancing
organs in its head that are similar to the inner ear,
the balancing mechanism, of human beings. When
moving fast the squid zooms backward. It has very
well-developed eyes and a sense of smell and taste.

All in all it has a formidable armory: good eyesight;
smell and taste; speed; and powerful equipment with
which to catch and kill. But there is more. Some
species have light organs on the ends of their tentacles

Above: a fanciful illustration in a Paris journal of
1906 showing a fishing boat
being attacked by octopuses.

Left: a common octopus, the
species that is plentiful in
the Mediterranean. It also
lives in tropical and subtropical waters of Africa
and the Atlantic side of
America.

Below: an underside view of
a common octopus. Its
suckers do not have the
horny rings seen in the
suckers of squid.

which they use to lure their prey. Others will flash
sparks to blind their predators, or produce a confusing
trail of a dark substance called "ink."

Squids are an important source of food for many fish,
particularly salmon, swordfish, and tuna. Marine
mammals also take their toll of the population:
dolphins, seals, and sealions devour the animals. The
toothed whales, especially the sperm, are the greatest
single species of predator, and over 28,000 small squid
have been found in the stomach of one sperm whale.

The octopus, the other well-known member of the
Cephalopod family, has been maligned for ferocity in
fiction and film. Most people must have seen films in
which a giant octopus has attacked a diver or even a
ship. Yet in contrast to the squid, the octopus is a shy
animal and at its biggest has a span of 32 feet, of
which only 18 inches is taken up by its body. The
eight-armed octopus lacks the fast movement of the
squid and does not have such well-developed eyes. It
makes up for these apparent deficiencies by having a
better developed brain, which enables it to sense
passing prey through taste and smell organs. Many
species can secrete a powerful toxin with which to
stun its victims.

The Crustaceans

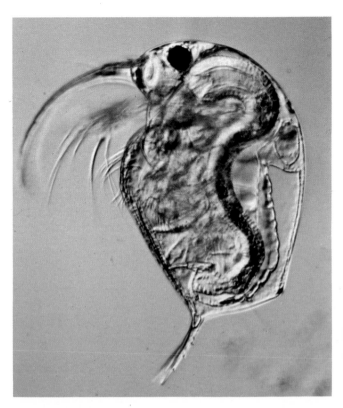

Giant squids have given rise to many tales of mysterious and horrible sea monsters. Similar tales have been recorded of monster crabs – and perhaps with some justification. One Japanese spider crab that was captured had an overall span of over 12 feet from claw tip to claw tip.

Crabs are arthropods, so named from the Greek words meaning "jointed legs." Arthropods are represented on land by the insects and spiders, and these are the highest form of invertebrate. In the sea, however, the advanced insects are poorly represented: marine evolution seems just not to have advanced as far. The most advanced forms of arthropods are the crustaceans and these are among the most widely spread and successful of all marine creatures, ranging

Above: the water flea, one of the lower crustacea and an important component of freshwater plankton. It is a tiny organism whose transparent carapace encases it fully.

Above: the *Gigantocypris*, an ostracod known as mussel or seed shrimp because of its resemblance to a mollusk.

in size from barely visible plankton to those monster Japanese spider crabs.

Besides its jointed legs, a crustacean is characterized by having its skeleton on the outside of its body – an exoskeleton. This is a horny cuticle, jointed to give the body a certain degree of flexibility. Other features are two pairs of antennae and pairs of legs on the body segments, usually for walking, but in some species adapted for swimming.

There are five main groups of crustaceans. The branchipods – brine shrimps, fairy shrimps, and water fleas – live mainly in fresh and brackish water. The second group, the ostracods, look very little like the conventional idea of a crustacean. Their exoskeletons or carapaces enclose the tiny animals almost entirely so that superficially they resemble bivalve mollusks. The antennae and limbs emerge from the skeleton and

the largest species, the quarter-inch *Gigantocypris*, has huge forward-looking eyes.

The third group, the copepods, were dealt with earlier in this chapter in the section on plankton. But it is worth emphasizing again that these tiny animals – the largest of which is about a quarter of an inch – occur in such vast swarms that entire fisheries depend on them.

Surprisingly enough, the barnacles that coat the hulls of ships and intertidal rocks are the fourth group of crustaceans, even though the adults resemble limpets and lead an immobile existence. The free-swimming or planktonic larvae of the many species of barnacle attach themselves by a gland in the head to a wide variety of bases: ships, rocks, fish, turtles – even the teeth of dolphins – where they swiftly grow, enclosing themselves in their conical shells. One species goes as far as to attach itself to a lump of floating oil and, as it grows, pushes out a float to keep itself on the surface. Barnacles feed by pushing their legs out through an opening in the shell and scooping in passing food.

The fifth main group of crustaceans is the best known. This consists of the decapods which include the familiar shrimps, prawns, crabs, crayfish, and lobsters that form such an important part of the world shellfish catch.

Popular names of the more common decapods can be highly confusing. Some examples: brown shrimps are shrimps but pink shrimps are really prawns; Dublin Bay or Norway prawns or scampi are a species

Right: an acorn barnacle feeds by catching passing food in its feathery appendages, which it then withdraws into its shell. It eats tiny animal and plant organisms.

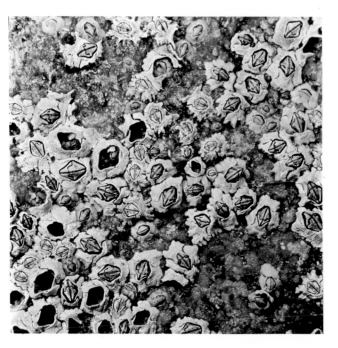

Above: acorn barnacles exposed at low tide. They survive out of water for some time by holding water inside them with a muscle that pulls the shell plates together.

Below: a deepsea shrimp with a minnow in its clutch. This species, *Systellaspis*, has light-producing organs.

of lobster; in the United States shrimps and prawns are both usually called shrimps; many authorities feel that what we call crayfish should be called crawfish since crayfish are small freshwater animals somewhat like shrimp.

The differences between prawns and shrimps are few. Shrimps are usually smaller than prawns, and close inspection will show that some species of shrimps have round bodies while prawns are more oval in section and often have a sharp kink across the back. Generally speaking, however, it is probably best to let confusion reign.

There is no confusion in distinguishing between the most advanced forms of the decapods – lobsters and crabs – but only in their adult form. As larvae all crustaceans resemble small shrimps, and they drift in the top layers of the oceans in planktonic form. In adulthood, however, most species of crabs and lobsters spend their lives on the seabed, and they are remarkably well-equipped for this lifestyle.

The common lobster, for example, is bluish-black in color and has powerful jaws, two even more powerful claws, and four pairs of walking legs as well as small paddles or swimmerets on its abdomen. This impressive set of equipment is completed by a tail fan with which the animal can flick itself backward to beat a hasty retreat from predators.

Like nearly all crustaceans, lobsters shed their shells at regular intervals. It is all very well to be armor plated, but when the inner body grows, something has to give. In the case of crustaceans it is the outer shell that peels off, or molts, leaving the animal soft and exposed while its new carapace grows.

Molting usually takes place in the summer, and in decapods the female's molt is usually followed immediately by mating. In lobsters the period from mating until the hatching of the eggs can be as long as two years. The sperm is stored by the female for about a year. Then the eggs are extruded, fertilized by the sperm, and cemented to hairs on the swimmerets on her abdomen. Between 5000 and 12,000 eggs are carried by the female for between 10 and 12 months, after which they hatch out in spring or early summer.

Toward the end of the incubation period the eggs

Above: a lobster molting. The new shell, which forms under the old one, is too soft to give the lobster protection, so the animal hides from its enemies until it hardens. A young lobster molts four times in the first month of life.

can clearly be seen in clusters that look like a bunch of berries on the lobster's abdomen. In some parts of the world, conservation laws protect berried female lobsters. They are not allowed to be landed, and when they are found in a lobster pot, they must be carefully returned to the sea. In the United Kingdom, for example, government fisheries inspectors keep a sharp eye on lobster catches in the early spring, not only for berried females but for those that have been scrubbed of their eggs by the unscrupulous.

After hatching, the lobster larvae resembling small shrimps molt almost immediately. They live as plank-

Left: the common or American lobster. This species is the most valuable commercially. The lobsters are marketed alive, and only turn red after being boiled. They live in Atlantic waters from Labrador to North Carolina.

ton in the surface waters and after more molts take on a more lobsterlike look, with well-developed claws. They then sink to the seabed. This period in the young lobster's life lasts about a month and the casualty rate is very high: probably only one in a thousand of the larvae survives on the seabed. Lobsters molt frequently and grow fast while they are young, but by the time they become sexually mature at about seven years, molting takes place only once a year. Lobsters can reach a weight of 40 pounds and a length of 4 feet from tail to claws. They often live to a ripe old age – the American species up to 100 years.

Crabs can also reach great sizes. Aside from the monstrous Japanese spider crab already referred to, the British edible crab can attain a weight of 14 pounds. There is a bewilderingly wide range of species

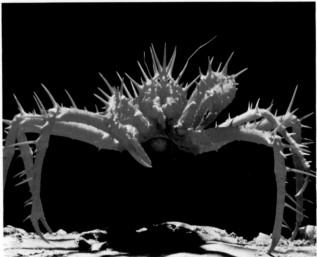

Left: the underside of a female devil crab showing the mass of eggs it carries. This is known as being in berry.

of crab, from the giants to the tiny pea crab that lives as a parasite within a mussel. The female pea crab often grows to a size that prevents her escape from the mollusk, and during the mating season the much smaller male has to move from one mussel to another to mate. The larvae that result are shot into the outside world by the mussel's exhalant system.

Not all crustaceans spend most of their lives on the seabed in relatively shallow waters. Some, particularly decapods, are to be found at depths of nearly 20,000 feet, and these are often grotesquely adapted for survival in the abyss. Deep ocean spider crabs, for example, have long spindly legs that enable them to move swiftly over the mushy sediments without becoming bogged down.

Crustaceans among the invertebrates are to be found throughout the depth and breadth of the world ocean. The vertebrates – the fishes in their myriad variety – also have worldwide distribution.

Chapter 6

The Fish Population

Fish, which dominate the seas and therefore two-thirds of the Earth, are the ancestors of human beings after going through reptile and bird stages. From slow-moving, heavily armored creatures that grubbed along the seabed for food, fish have developed into the most successful marine animal. There are thousands of diverse species that today colonize every part of the world ocean from tropical rock pools to the dark, silent, near-freezing depths of the ocean trenches. This chapter looks at the development of the fishes, from sea squirts and blood-sucking lampreys to the sleek, efficient, 60-miles-per-hour tuna. It describes the little known habits of some of the species, including sharks that must keep swimming constantly, codfish that shelter in the deadly tentacles of jellyfish, and herring that chase the moonlight.
Marine mammals, especially the highly intelligent whales and dolphins, also have their part in this account.

Opposite: there are thousands of species of fish in the waters of the world, from little to large, from colorfully beautiful to drab, from preyed-upon to predator.

Ancestors of the Fish

The flabby, lumpy organisms called sea squirts do not look to be very interesting creatures, but in fact are fascinating. Shaped like bottles or vases and often beautifully colored, adult sea squirts are immobile. They anchor themselves to the seabed – to a rock or to piers – and there they stay, taking in water through one opening and, after filtering plankton and other organic matter, pumping out the water and waste material through another opening. Touch a sea squirt and it shrinks to become a lifeless looking jelly bag. As it contracts it squirts out a stream of water, the action from which it gets its name.

The sea squirt is most fascinating in its larval state when it looks like a tadpole. Its bulbous forepart contains the rudiments of its inhalant tubes as well as its gut, and its body is stiffened along the back by a thin rod connected to a bundle of nerves in the head. This rod or notochord is the early stage of the backbone found in the embryos of all vertebrates, including human beings.

The sea squirt keeps its spinal rod for only a few days. As soon as the larva cements itself to a rock and loses its tail, it becomes an anchored adult. But those few larval days are of great significance on the evolutionary scale: the humble sea squirt represents the great leap from the invertebrates to the vertebrates. They are the link with the fishes which, when they crawled painfully from the sea, became reptiles and then birds. From these developed the familiar land-based mammals – and eventually men and women.

The larval sea squirt with its simple backbone is called a protochordate. The lancelet is another of these organisms. An inch or so in length, it looks like a fish but has no head and feeds by filtering plankton through gill-like clefts. It spends its day in the sand or mud of shallow water areas and at night ascends into the plankton at the surface.

The next stage in the development of the fish is represented by an animal of rather repulsive appearance and unsavory habits. This is the lamprey, which

Top: the sea squirt is encased in a jellylike tube called a tunic, one end of it fixed permanently to something. Sea squirts live either as individuals, in small clusters, or in large colonies.

Left: a lancelet sifting the water for food. Lancelets look somewhat like a fish, and also have segmented muscle blocks similar to fish. But they are invertebrates and have no distinct head or tail.

in its larval form closely resembles the adult lancelet although it is bigger. The larval lamprey has no eyes or fins, feeds by filtering, and lives in mud. After growing to the size of an eel, it develops eyes, a flexible backbone, and long fins. It is not, however, a true fish. Its head ends not in a set of jaws but in a circular disk in which a spine covered tongue is set. The lamprey clamps this slimy, rubbery disk onto the side of a fish and rasps away scales and skin with its spiny tongue until it can begin to suck blood. It releases an anticlotting substance to keep the blood flowing. Although a healthy fish may be able to rid itself of its unwelcome visitor, it may often be so weakened by the attack that it leaves itself open to infection or to easy capture by other predators.

Lampreys may be found in fresh and salt water. Their relative in the class of Agnatha (jawless fish) is the hagfish, which spends all its time in the sea. It lives in a mud burrow with just its head poking out. Hagfish have a keenly developed sense of smell and leave their burrow when they sniff out a dead or sick fish. Then like the lamprey, they attach their head to

Above: a hagfish at rest. This primitive marine animal secretes a vast amount of slime, seen in the background.

weight of the armor kept them close to the seabed where they nuzzled through the mud in search of food. Certainly nobody would have given these species much of a chance. If a prediction had had to be made at that time as to which of the ocean's creatures

Left: a lampern, also known as a river lamprey because it spends its time in freshwater before adulthood. There are about 30 species of lampreys, which are related to the hagfish. The lamprey has some teeth even on its tongue to aid it in puncturing a fish's skin to begin sucking blood.

the animal and suck out the flesh and blood, leaving a skin-covered bag of bones behind.

The lamprey and hagfish are the living remnants of a number of jawless fish – the very first invertebrates – that swam the world ocean over 500 million years ago. It would have been difficult for anyone to predict that these crude creatures would develop into fish, the most sophisticated and successful of all marine creatures. Jawless fish were cumbersome beasts with a muscular, fin-fringed tail projecting from an armor suit made up of bony plates. A certain amount of movement was gained by thrashing the tail, but the

were going to dominate the world, the vote would almost certainly have gone to the cephalopods – the agile, speedy, and intelligent squids and octopuses – or to the mighty sea scorpions that dominated the seabed and which were the ancestors of today's land insects.

The jawless fishes did more than survive, however. They developed jaws, paired fins, streamlined bodies, keen senses of sight and smell, a remarkable ability to communicate with each other – and became fish, which rule the oceans. That means that fish rule more than two-thirds of the earth.

Primitive but Successful Shark

The transition from slow-moving, heavily armored, and bottom-dwelling jawless fish to the fast moving, sleek shark might seem somewhat abrupt, but it is logical. When fish as they are today developed from their lamprey-like ancestors, there was an evolutionary division. Some developed bony skeletons, but sharks and rays developed cartilage which is softer, lighter, and more flexible than bone. Their make-up

has not changed for some 250 million years. Therefore, for all their fearsome look and the legends that have grown up about their cunning and skill, sharks are fairly low on the evolutionary scale. Indeed, if it were not for their fast swimming speed and their ferocity, many species would probably have died out long ago.

When they first developed, sharks must have seemed indomitable. By having bodies made up of light gristly cartilage, they would have had a far greater turn of speed than the armored fish. But because the shark lacks a swim bladder, it has to move almost constantly to stay afloat. Its buoyancy is increased by an extremely oily liver, but if it stops flexing its body for long, it will eventually sink. Moreover, once it has burst into top speed it cannot suddenly stop, as can many of the more developed fish, nor can it go into reverse. All it can do is swerve and hope it does not bump into something. The shark swims in the same way as its primitive jawless ancestors did: it flips the

Top: a white shark, one of the large species that can reach a length of 36 feet. It is considered dangerous.
Above: a Port Jackson shark, the best known of the 10 species of primitive horn sharks. The female lays eggs of a peculiar corkscrew shape and strange horny material.

rear half of its body and thrashes its tail. But if it had this equipment only, the shark's progress would be ever downward in a nosedive. So it has a pair of pectoral fins for stability.

The surface of the shark's body differs from that of the more developed fish. Instead of being scaly, the skin is covered with tiny teeth, each with a cavity and each coated with enamel. These are enlarged around the jaw to present the onlooker with the fearsome sight of rows of fangs. In many species, however, the teeth are there not to bite or rip flesh but merely to prevent the escape of prey that has been trapped in the jaw cavity. Most sharks swallow their food whole. One of the strangest looking but harmless species is the hammerhead, whose wide-spaced nostrils enhance its acute sense of smell to make it a very effective hunter in the murky waters it favors.

All sharks are capable of biting. Catch one – even one of the small varieties – and it will snap. But only a few species, perhaps 35 out of the 300 known, are killers. The great white shark, widespread in tropical waters, is highly dangerous. It is a large, surface-living creature and comes from the same family as the mako and the harmless porbeagle. The latter is the one most frequently found in northern European waters, where it is a favored food fish in Scandinavia. The basking shark is also found in temperate zones and, although frightening in appearance with its length of up to 25 feet, it feeds almost exclusively on plankton. So does its even bigger tropical cousin, the whale shark, which can reach up to 30 feet long. Some sharks hunt in packs and seem to acquire a corporate ferocity that outweighs that of the individual. The tiger shark, reaching 20 feet in length and 1500 pounds in weight, is a particularly deadly specimen. Packs of tiger sharks have been known to herd shoals of other large fish into shallow water before going in for the kill.

It has to be emphasized that people's fear of sharks is based more on an irrational phobia than on hard statistics about the incidence of shark attacks. It is the sort of fear that makes parents lock up their children when they hear that a tiger has escaped from a zoo some 20 miles away – and then let them ride through hazardous traffic on bicycles when the animal is recaptured. It is the sort of fear that makes films starring a plastic, radio controlled monster shark one of the biggest box office successes of all time. In fact, there are less than 100 attacks on people by sharks in a year. About half of these are fatal. To put

Above: a hammerhead shark, easily recognized because of its head though the body is just like all other sharks.

these figures in perspective, about three times as many people die each year from bee stings in the United States. About three times as many die from being struck by lightning. But such is the fear that authorities at tourists resorts will go to elaborate and expensive lengths to prevent that chilling dorsal fin appearing anywhere near their beaches: vast expanses of offshore netting to trap the fish or "curtains" of bubbles from undersea hoses and "walls" of sound to frighten them off. Reflective screens, chemicals, and even electric shocks have also been used to keep the beasts at bay.

Below: a shark gliding through the water. The shark's mouth is on the underside of its head and the gills on the sides of the head, with five to seven openings. It is constantly looking for food to satisfy its continual hunger.

Some Relatives of the Shark Family

Not all sharks have a shape that brings fear to people. Some are welcomed. The dogfish, which grows to a length of four feet, is a shark, but is appreciated by anglers and commercial fishers alike – and by fish-and-chip lovers in Britain where it is sold as rock eel, rock salmon, or huss. Anglers would be justifiably terrified of the tiger shark, but they happily handle the tope, its smaller relative from the Carcharinidae family, which is common in European waters. The monkfish is also a shark, but its flattened body often means that it is thought of as a ray.

The monkfish is not the only shark with a flattened body. Some sharks, tired of so much moving to stay

Right: the dogfish is a small shark, having the toothed skin, the mouth on the underside, and the big appetite of the shark family. Dogfish eat mostly crustaceans, worms, sea cucumbers and hermit crabs, which they seek on the sea bottom by smell rather than sight.

Below: a monkfish. Although it is a true shark, it looks entirely different from other sharks. The tail is more slender, the mouth is on the front of the head, and the broad flattened head and front fins resemble a ray in shape.

afloat, take rests by lying on the seabed. Others have given up the perpetual struggle and lead an entirely sedentary existence on the seabed. The carpet shark, found in the western Pacific and off Australia, is a good example. It has adapted to a seabed existence by becoming flattened from top to bottom and growing flaps of skin and long bristles. Its color is mottled so that when it lies motionless on the ocean floor it closely resembles a rock or a patch of sand. To mistake it as such is fatal for any passing fish or crustacean, however.

Skates and rays, which also have gristly rather than bony skeletons are likewise flattened. Their pectoral fins are greatly enlarged, and by a birdlike flapping of these winglike fins, they propel themselves gracefully. The tail, no longer essential for movement, has consequently become thin and whiplike.

The biggest of the rays, the manta, is an impressive and beautiful creature that can reach a breadth of 25 feet. It cruises the surface layers of the oceans, its huge "wings" conforming to our modern ideas of an aircraft – smooth, streamlined, and sleek. The manta

does not have the same turn of speed as a shark but glides slowly, each flip of its great pectoral fins giving it forward motion, while the very area of these "wings" keeps the animal afloat. The manta has a vast slit of a mouth and, as it makes its stately progress, it sucks in small fish and copepods.

Most rays, however, live on or near the seabed, sailing close to the bottom where they grub for crustaceans and mollusks, crushing the creatures in their powerful jaws. For this grubbing process the ray has its mouth on the underside of its head, but this means that it has had to adapt its breathing equipment. Its shark cousins suck water in through their mouths and filter it out through their gills.

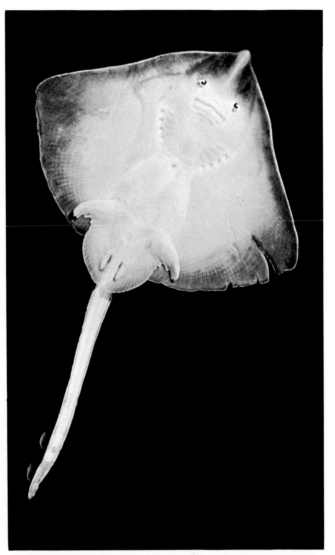

Above: the bat ray in motion. Like all skates and rays, the bat ray spends much of its time lying on the seabed. When it moves, it looks like it is flying through the water because of the way it flaps its winglike fins.

Above right: a hornback ray, *Raia clavata*. There are certain low-density substances in the liver of these fish that help make them buoyant, though slow moving.

Below: an electric ray lying on the seabed – its usual habit except when hungry. There are about 36 species of electric rays, the largest reaching about 5 feet long and the smallest being about 17 inches in length.

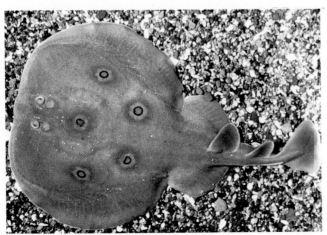

The bottom-feeding ray would quickly find its breathing apparatus clogged up with sand and rubbish if it adopted this approach, so it sucks in water through slits on the top of its head. The water feeds through to gills and is pumped out on the underside of the fish.

From these descriptions the rays may seem to be harmless browsers of the oceans, but many of them have a sting in the tail. The smaller cousins of the giant manta, for example, have tails loaded with poisonous spines. The poison from the Australian ray, which reaches 14 feet long and 7 feet across the wings, and the American stingray, which is 10 feet long, can be fatal.

Some species of electric rays, which are distributed throughout the world, can deliver up to 220 volts – enough to knock a diver off his feet. The electric organs are distributed around the wing-fins of the fish and are used as a weapon to make up for the slow speed of the animal. Faster swimming prey are stunned by the electric shock so that they slow down enough for the the ray to scoop them up in its vast mouth.

Sharks, rays, and skates, with their gristly skeletons, are a successful marine group, but they are geographically and physiologically limited by comparison with the bony fishes.

The Swim Bladder in Fish Evolution

There are over 16,000 known species of bony fish in the world ocean, so the best a brief survey can do is to look at the characteristics that have made them so successful and to examine a few important groups and species.

What liberated so many species of bony fish from the problem of having to thrash their tails ceaselessly merely to stay afloat was the development of the swim bladder. Some early amored fish migrated to coastal areas having swamps and lagoons. The warmer water there contained less dissolved oxygen so that the fish needed to lift its head above the surface and take regular gulps of air. This air went into a pouch which was rich in oxygen-absorbing blood vessels, which allowed the fish to keep up its oxygen quotient.

In the next stage of development the fish dispensed with the laborious process of coming to the surface for oxygen, a process that was also potentially dangerous, since it made fish vulnerable to predators. The fish filled its air bag by diffusing air from the blood. This had an additional, and important, advantage. The gas-filled bag acted as a float so that the fish no longer

Above: the well-developed pectoral fins of the flying fish are used like glider wings to keep the herringlike fish airborne. Some flying fish have both the pectoral and pelvic fins enlarged to enable them to fly.

had to expend energy merely to stay in one place. The bag, known as a swim bladder, secretes oxygen from the blood on demand. As the fish dives it increases the supply of compressible air to its swim bladder. As it ascends some of the gas has to be expelled; in most fish this is achieved by reabsorption into the bloodstream.

The body of the fish developed to take advantage of this benefit. The pectoral fins no longer had to project sideways, like the hydrofoils in the shark, to give lift; the tail did not need to be upturned to give maximum push; and the head no longer had to be flattened. Instead, the whole body of the fish is flattened from side to side and the pectoral fins are used to give balance and maneuverability. They can be used as brakes but they also allow the fish to hover or go into reverse. Combined with the dorsal and pelvic fins, the pectorals give the fish a fairly sophisticated system of stabilization, steering, and braking.

This new-found ability led to an explosion of diversity in species of fish. Some achieved the ultimate in pectoral utilization: the flying fish, in a desperate escape from a predator, tucks its large webbed pectorals flat against its body and pushes toward the surface at a speed of up to 20 miles an hour. It breaks the surface, spreads its pectorals, flips its tail rapidly in the water to gain even more speed, and achieves lift-off at around 40 miles an hour. Usually it will glide at a height of four or five feet for a distance of a few hundred yards before submerging again or, if it wishes to fly further, will beat its tail against the water to maintain flying speed. If the flying fish takes off into the wind, an up-current may lift it to 45 feet,

Below: a diagram showing the changes in the volume of a fish's swim bladder at varying depths. Volume changes are less in deeper waters, so deepsea fish can go up and down over a greater distance without having to make abrupt corrections to the volume of gas present in its swim bladder.

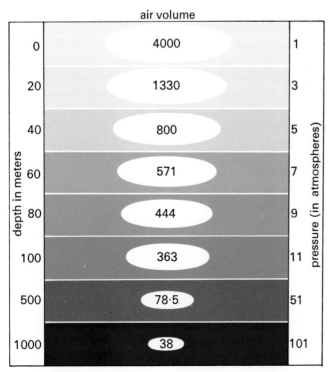

air volume

depth in meters	air volume	pressure (in atmospheres)
0	4000	1
20	1330	3
40	800	5
60	571	7
80	444	9
100	363	11
500	78·5	51
1000	38	101

pectoral, dorsal, and pelvic fins only for steering and braking. When it is at full speed the fins tuck into grooves, the gills fit flush with the head, and even the eyes hardly protrude. Scales are small and thin as paper, and the body is lubricated by mucus to cut drag still more. The tuna can produce a turn of speed to match any land-based animal, but it is not merely a short-burst sprinter: a tagged bluefin was recaptured some 4000 miles away 50 days after being tagged. This meant that it had swum at a rate of at least 80 miles a day.

To achieve such performance the tuna also has a mechanism that, unlike almost any other fish, enables it to maintain its body temperature within the range 77–86°F, even when swimming in a water temperature of 50°F. By staying warm the tuna is able to maintain its swimming speed and to migrate from warm to cold waters without flagging.

The tuna group represents the pinnacle of development in speed and power, although other fish developed equally remarkable characteristics to adapt to their chosen environment.

Left: a shoal of tuna, also called tunny. This sleek fast fish moves in shoals in which the individuals are all about the same size – smaller fish in big shoals and bigger fish in small ones. They migrate according to the movements of the fish they feed on and to find warmer waters.

Below: a seahorse swims upright, using rapid movements of the dorsal fin to propel itself. This fin may oscillate 35 times a second when the fish goes full speed.

at which height it will make good use of crosswinds to mislead oceanbound pursuers, even to the extent of doubling back upon itself.

To survive, the flying fish needs its ability to break the surface of the sea. For although its subsurface speed can be about 20 miles an hour, it is still a plodder compared with one of its main predators – the tuna. Seeing one of these streamlined fish streaking toward it at a speed of up to 60 miles an hour with a lunchtime glint in its eye should be enough to encourage any small fish to learn how to fly!

The species of the scombroid group, which includes tuna fish, are the fastest of the ocean swimmers and their adaptation for speed has given them a sleek beauty. The tuna powers through the surface layers of the world oceans by beating its tail, and it uses its

Flatfish: Adapted for Seabed Life

Although the swim bladder was the key factor in the better adaptation of the fishes, some species do not need it for their niche in the underwater world. The flatfish that are prized catches of fishermen – dabs, flounders, plaice, sole, turbot, and halibut – are good examples of marine animals which can dispense with the swim bladder.

The plaice is typical of the flatfish in having a development that is as peculiar as its appearance. In European waters the female plaice releases thousands and thousands of eggs between January and April. After being fertilized by the male, the eggs join the plankton in the upper layers of the sea. They float with currents and tides for up to 16 days and then the young larvae hatch partly, swimming attached to their eggs and feeding on the yolk sac. When this food source is finished, the young fish feed on plankton and other larvae.

At this stage of its life the plaice is a perfectly ordinary looking round fish, complete with swim bladder and well-developed pectoral, dorsal, and anal fins. But adult plaice live on their sides on the seabed, venturing only short distances off the bottom. This sort of life hardly needs a swim bladder to regulate buoyancy over the water column; nor is there much to be gained from having one eye permanently stuck in the mud. So after about six weeks of life the plaice undergoes a strange metamorphosis.

The left eye of the place moves upward and over its head, finishing up just above and in front of the right eye. The head twists and the mouth moves sideways. The swim bladder disappears, the swimming position changes, and the fish moves along on its side with what used to be the left side pointing downward. It retains the pectoral fins but these are little used: the pronounced dorsal and anal fins do all the work.

The plaice also changes color during its metamorphosis. Its left, or downward-facing, side loses its dark coloring to become silvery-gray, while its right, or upward-facing, side stays dark but develops orange spots. This type of pigmentation is to be found in many fish, from slow-moving flatfish to voracious sharks.

Below: a close-up of the plaice to show its eyes, which are both on one side of the fish's head. Living in waters from Iceland east to the White Sea and south to the coasts of France and the western Mediterranean, it is one of the most commercially important of all the flatfish.

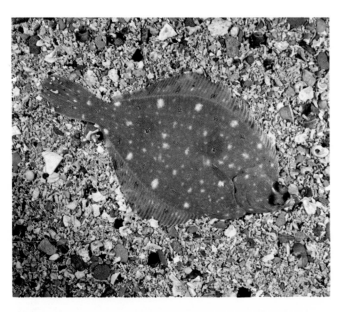

Above: a plaice resting on shellgravel bottom, showing its excellent camouflage. The upper surface is light mottled gray to blend with the color of the fish's habitat. Even the red spots turn pale if a plaice rests on whitish pebbles, and coloring varies widely according to need.

The white underside means that when a predator looks up toward the surface for prey, the silvery belly will be difficult to pick out against the light coming down. When the fish is viewed from above, its dark back tends to blend in with the seabed or the dark waters beneath it. The mottled back of the plaice blends in well with the sandy and pebbly bottoms along which it moves while looking for small mollusks and crustaceans. When preparing to rest, it further conceals itself by nestling into the seabed, flicking sand and gravel over its back with its tail.

An interesting sidelight has been thrown on this ability of the plaice to change its pigmentation to suit its metamorphosed body. During experiments in fish farming in Scotland, fertilized eggs were hatched out in tanks and the larval fish were reared in a carefully controlled environment with a regular supply of food and without the presence of predators. The larval

roundfish duly metamorphosed but scientists conducting the experiments were surprised to observe that large numbers of the little flatfish had not fully developed the typical protective pigmentation. Some had dark patches on their undersides, some had bright silvery sections in their orange-spotted top sides. These animals stood out clearly in their tanks: in the open sea they would quickly have been spotted by predators. What had happened? Could the fish, reared in an artificial, totally secure environment, have "forgotten" the need to protect themselves? This might have been expected after many generations of artificially reared fish had been produced. But should it happen in the very first generation? Or could it be that plaice which metamorphosed with this piebald pattern in their natural habitat were swiftly snapped up by other fish so that only the properly developed ones survived long enough to be caught by

Above: a plaice concealed beneath the sandy gravel. The fish flicks sand with its fins to create a cover for itself.

Below left: a flounder on the move. Like other related flatfish, flounders swim by undulating their flattened body vertically and then gliding down with the body held rigid. Though a marine fish, it can also live in fresh water.

fishing boats? The mystery was never solved.

Other European members of the plaice family (Pleuronectidae) are the dab, lemon sole, flounder, and halibut. The latter is the giant of the family, reaching a weight of 700 pounds and a length of 14 feet to fully justify its delightful Latin name tag of *Hippoglossus hippoglossus*.

There are three other main families of flatfish – the Bothidae, Soleidae, and Cynoglossidae. The Bothidae are distinguished by having their eyes on the left side, opposite from the plaice. The best-known species in the Bothidae family is the turbot. The soles make up the Soleidae family and are easily recognizable by their streamlined shapes and peculiarly twisted mouths. The Cynoglossidae family's best-known member is the tongue sole which grows to a length of 15 inches in tropical and warm temperate waters.

141

The Valuable Cod

Mystery can be found in the lives of even the seemingly most mundane and well-known marine animals. The cod is of vital economic importance to commercial fisheries. It is part of the great order of fishes called Anacanthini which includes other important Northern Hemisphere food fish like haddock, whiting, coalfish, and hake. Because of their commercial importance these fish have been closely studied, but one aspect of their lives defies explanation. No one knows why the young are afforded protection by an otherwise predatory jellyfish.

As with most demersal or bottom-feeding fish, the cod lays eggs on or near the seabed. The female does her best to ensure the furtherance of the species by laying between four and nine million of the transparent spheres, each of which is about one twentieth of an inch in diameter. These are fertilized by the male after which they rise to the surface to join the plankton and the young of other demersal fish, including the flatfish. Like the flatfish, the young cod feeds on its own egg for a few days until it has developed jaws and a working digestive system. Then it is on its own, competing for the tiny plankton which are its only food source. It is powerless to resist the swarms of fish that plunder the rich surface layers of the northern seas in spring.

Countless young cod perish so that perhaps only a few of the millions of fertilized eggs survive this planktonic stage. Of course, if all the eggs of all the cod grew into adults there would be no room for water in the sea, in much the same way that if all the potential progeny of a single pair of houseflies reached adulthood the land would be covered with the insects

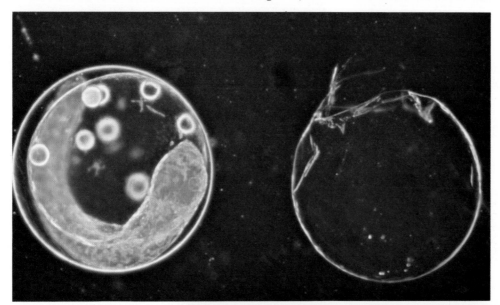

Opposite top: the whiting differs from most of the other members of the cod family by living mainly in shallow waters. Silvery sides make it easily distinguishable.

Left: an enlargement of an unhatched fish egg and a fish just hatching. All fish lay large numbers of small eggs, and pelagic fish usually leave them suspended in the open water. The cod family lays their eggs in the millions.

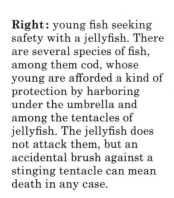

Right: young fish seeking safety with a jellyfish. There are several species of fish, among them cod, whose young are afforded a kind of protection by harboring under the umbrella and among the tentacles of jellyfish. The jellyfish does not attack them, but an accidental brush against a stinging tentacle can mean death in any case.

142

to a height of several miles. Nature imposes checks and balances or, put another way, nature's checks are what gives it its balance. The cod has to lay millions of eggs because there is only a remote chance that any will survive to adulthood. Once the eggs are laid, the female has no control over their destiny. The female lobster produces only a few thousand eggs and keeps them on her body until they hatch, which increases the chances of survival. The advanced mammals produce just a few or often just one infant, but the chances of survival are vastly improved because the mother carries the baby in her body until a fairly advanced stage of development and looks after it after birth, often for years.

The unprotected cod makes an effort to improve the odds for survival and it does this, together with the young of the haddock and whiting, in this strange way: it heads for protection to the tentacles of the huge jellyfish called *Cyanea*, also known as sluthers or lion's mane. The *Cyanea* has tentacles of up to 200 feet long and a body of up to 8 feet in diameter. It can stun a fish as large as a mackerel in order to draw it into its digestive tract.

Cod, whiting, and haddock – just an inch or so long – live within the great protective mantle afforded by the trailing tentacles of the *Cyanea*, safe from attack by predators. Yet these small fish are not themselves stung or consumed by their protector. There has been no satisfactory explanation of this mystery.

Cod stay with their jellyfish host for perhaps two weeks and then leave, settle to the bottom, and begin their adult lives. Often they will have drifted far from the spawning grounds during the planktonic stage and will usually have finished their long hitch-hike provided by the jellyfish in a bay or inlet in a water depth of around 60 feet. There they start to hunt, preferring hard rocky seabed where they feed on small crustaceans and mollusks. The growth rate of the

fish depends on the waters in which it lives: those in the North Sea will be about 8 inches long at the end of their first year; those in more northern, colder waters will be only about 3 inches. Sexual maturity is reached at about four years when the fish is some 3 to 4 feet long.

Fish caught before this stage are often referred to as codling and are equally often thought to be a completely different species. This is probably because they are caught in bays and estuaries, while the adult

Below: coalfish or saithe, known as pollack in North America. Although a member of the cod family, coalfish feed on their cousins the Atlantic cod. Food is swallowed whole and takes five or six days before it is digested.

fish is found in far deeper waters. For with maturity comes the urge to merge, and the cod begin to migrate to the spawning grounds in the winter – to the central and northern North Sea, to the northern coast of Norway, to the waters off south and southwest Iceland, and to the Grand Banks of Newfoundland. Other migrations are made in the summer to water depth usually of around 250 feet, although cod have been found at depths as great as 1500 feet. Apart from spawning, the chief factors influencing the distances over which cod migrate are water temperature, of which the fish prefers a range of 32–41°F, and food availability. The adult cod eats herring, sprats, haddock, and even its own young. By 10 years of age, if it has escaped the trawl, a cod can be up to 5 feet long. The oldest caught have been 20 years old and have weighed over 50 pounds.

The cod is a handsome fish, olive green or brown, with dark spots. It is distinguished by its three dorsal fins, protruding upper jaw, and most particularly by the long barbel on its chin.

Its close relative, the haddock, is smaller, more rounded, with a less pronounced barbel and a smudgy mark above the pectoral fin on each side. This mark is often referred to as St Peter's thumbprint. The haddock is not found as far north as the cod and appears to be a more sociable animal: while cod move in loose-knit shoals, haddock group more densely and concentrate their gastronomic attentions on shellfish on sandy or muddy bottoms rather than on fish.

Another close relative is the coalfish or saithe, called pollack in the United States, which is nearly black and has no chin barbel. It was once served in British fish-and-chip shops under a wide variety of

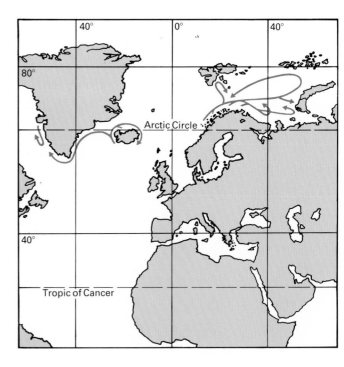

Below: the cod's long chin barbel is clearly seen on this photograph of four dead cod.

names, including the unlikely one of rock salmon. Another frequently undervalued food fish is whiting which is silvery, streamlined, and found over wide areas, mostly in coastal waters at medium depths. The young whiting, however, is like the cod in that it matures in shallow waters where it feeds mainly on shrimp.

It is a frightening thought that people's swiftly changing fads and fancies for particular types of food can often mean the difference between the healthy flourishing or the swift near-extinction of an animal species. The hake is a good example. At the beginning of this century hake was regarded by the British as a nuisance when hundreds of the fish spilled from the trawl onto the decks of fishing boats. The crew would pick the cod and haddock from the catch and shovel the hake back into the sea. Hake was simply not popular with the fish-eating public. But tastes changed and hake became one of the most popular for fish-and-chips, particularly in the north of England. No longer did fishermen groan when hake poured from the trawl – they went fishing in areas where the fish was known to be plentiful. Today it is a high-priced fish, and over-fishing has made it scarce.

Hake, a relative of the cod, is found in the Irish Sea and down to the Bay of Biscay. It is one of the deepest water fish regularly hunted, living on the bottom in water depths between 600 and 2500 feet. They are big fish – typically up to 4 feet long – with pronounced dorsal and anal fins. The European hake, *Merluccius merluccius*, has a southern hemisphere cousin, *Merluccius capensis*, found mainly off South Africa.

Other deepwater members of the Anacanthini order are relatively unknown compared to their commercially sought-after relatives. Rat-tails or macrourids have, as the name suggests, long whiplike scaly tails and are usually about 2 feet long, although specimens up to 4 feet long have been captured.

Opposite top: a map showing cod migration throughout the waters of the North Atlantic.

Above: a Norwegian fishing boat, showing the crew sorting and gutting cod. The cod ranks among the top as one of the world's most valuable food fish.

Right: a large mixed catch of fish on board a British vessel. At the beginning of this century, fishing boats would discard all but the cod and haddock from such a catch. Today, the growing scarcity of overfished species is bringing about a change in popular taste.

Herring: Popular in Fable and Fact

No other fish has enriched folk history as much as the herring. Since the Middle Ages it has been immortalized in etchings, paintings, prose, poems, and song. Perhaps this is because the herring has played an essential part in the economic well-being of whole European communities. It is probably the best-known, if no longer the most economically important, of all the pelagic fish. Human knowledge of the herring, nonetheless, has been acquired not just for curiosity's sake but in order to better exploit the long popular food fish.

Herring are pelagic fish, which means they rove the full extent of the water column from surface to seabed, but do not rest on the seabed or use it as their primary hunting ground like demersal fish such as cod, haddock, and flatfish. The herring differs from other pelagic fish in that it is one of the very few bony fishes whose eggs are hatched on the seabed. Paradoxically, it lives mostly in midwater and lays its eggs near the surface, but the eggs sink to the seabed. There they attach themselves to rocks and seaweed in great clumps, for they are laid in the billions and billions, each female producing between 50,000 and 100,000.

After a week or two, the incubation period depend-

lives. Billions perish. The young fish can only feed in daylight and it has to dash after plant and animal plankton. At first it darts about, snapping at everything, but usually succeeding in making a catch only once in every 20 attempts. If it cannot achieve a better

ing on water temperature, the eggs hatch and the larvae feed on the yolk sac for a few days. Then they are on their own in the most critical period of their

Above: a scene at Scarborough, England in the early part of the 20th century, showing workers gutting herring in the harbor. It has been said that the British Royal Navy was founded because of fights over herring fishing.

146

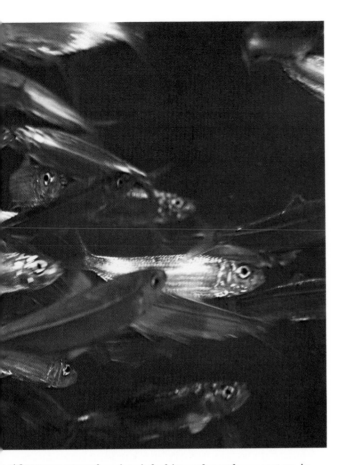

which the most important species is the tiny copepod crustacean *Calanus finmarchicus*. There are probably about six million tons of copepods of various species in the North Sea alone. The copepods make daily migrations to the surface, moving up at night and down during the day over distances as great as 1000 feet. In following the animal plankton to the surface of a calm sea at night, the herrings look like they are chasing the moonlight. Herrings use their swim bladder to maintain buoyancy, so when a huge shoal moves up to feed, the water seems to boil because of the countless billions of bubbles that are released by the herring as the pressure decreases.

Herrings are distributed all over the North Atlantic from Europe to Canada. But they form distinct races, each with its own spawning, migration, and feeding habits; some spawn in spring, some in autumn.

All herring races have suffered badly from over-fishing, and the populations of some of them have been decimated. With fishing bans and quota systems introduced to protect some races and a reduced number inhibiting exploitation of others, fishermen's attention have turned to a close relative, the sprat. The adult sprat is not as big as the herring, reaching a length of about 6 inches compared to the herring's 1 foot. Nor does it live as long. A sprat's life span is five years, while herring have been known to reach 25 years or more. In the North Sea the sprat population has grown fast and the young are now much exploited. Other relatives of the herring include the sardine or pilchard and the anchovy. All differ from the herring in that their eggs float and hatch on the surface.

Above: mature herring inhabit cool northern waters in enormous shoals. They feed mainly on plankton.
Below: the larva of a herring, seen attached to the yolk sac on which it feeds for a few days just after hatching.
Below right: a shoal of sardines. The tiny young are packaged as food in vast numbers.

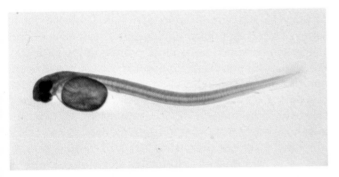

success rate than one in 20, it starves. If it snaps at something too big to be swallowed whole, the food sticks in the gullet and the larva dies. To add to its danger, there is the continual threat of being preyed on by cod, haddock, and other bottom-dwelling fish.

Millions of herring do survive and these increase their catching efficiency from one in 20 to as great as eight in 10. The amount of food consumed by herring provides another insight into the sheer proliferation of food sources in the oceans: a million-strong shoal of herring, all snapping for food perhaps 10 times a minute; that means eight million tiny organisms consumed every minute, and 48 million every hour – just by one shoal.

These rapidly gobbled organisms are plankton, of

Great Migrators

Above: sockeye, or blueback, salmon on their migration.

There would seem to be little similarity between the Atlantic salmon and the European eel. The first cannot be mistaken for anything but a fish and its pink flesh is almost universally popular. The second looks much like a snake and has a narrow appeal as a food fish. Both, however, are the great migrators of the northern seas, and the unerring navigational sense of each still has not been entirely explained.

The Atlantic salmon, prized by anglers, spawns in the rivers of northern Europe after which it leads a freshwater existence for about two years. It then moves out into the open sea, adopting a silvery pigmentation to enable it to blend into its surroundings.

Until recent years not much was known about the Atlantic salmon's life at sea. But then tagging of young fish as they left the rivers revealed that most of them make their way clear across the North Atlantic to Greenland, where they live on herring and other smaller fish. Why do they make this long journey when there are other waters nearer home that are perfectly capable of sustaining the fish? This is a mystery. One theory says that the fish has inherited a behavior pattern from the times when the North Sea was mostly dry land and European rivers flowed

Below: salmon fry only four days after hatching. Like other species, they feed for a time on the egg's yolk sac.

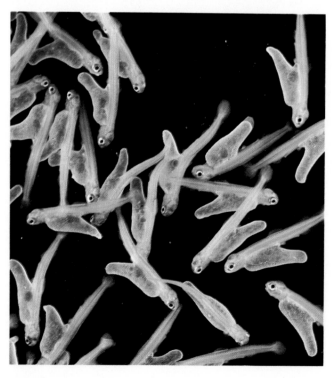

straight into the Atlantic. Even if true for the Atlantic salmon, this would not explain the long migrations – sometimes up to 2500 miles – made by the Pacific salmon.

Mystery also still surrounds the salmon's almost unerring accuracy in returning to the river in which it was born in order to spawn. Navigation over the final part of the journey – from bay to particular estuary to particular tributary – is achieved by a sense of smell and taste: each salmon remembers over the years the particular smell, made up of animals, minerals, and plants, of the river in which it was born and can detect this odor even if diluted to as much as one part in a million. This has been proved by the fact that when the nasal sacs of a salmon are blocked, it gets hopelessly lost. Also, if eggs spawned in one river are moved to another, the fish will return to the waters in which they hatched rather than to where they were moved.

How do the young fish learn to migrate to greatly distant feeding grounds? And, having matured on those grounds, how are they able to find their way back to where they were spawned, even to picking up the scent of their home river? Are salmon able to navigate by plotting the position of sun and stars, or do they have a hypersensitive reaction to ocean currents? One thing that is known is that the adult fish swim against the prevailing current flows to get back to their spawning grounds. This gives their young an advantage in migrating to the feeding grounds from the rivers of their birth because they can then drift with the prevailing currents. The young fish heading for the sea will glide swiftly over rapids and weirs and be swept effortlessly along by

Above: male and female sockeye salmon in their nesting and spawning grounds, the Adams River in Canada.

Below: this leaping Atlantic salmon is fighting to get to its traditional spawning place in the Laxá River, Iceland. The Atlantic salmon breeds in European rivers from Spain to Russia and along northeastern North America.

fast currents.

Perhaps the most interesting question about the salmon's migration is simply, "Why?" Why do they do it at all? Why do salmon battle their way across hundreds of miles of ocean against strong currents and then fight their way upstream against still stronger river currents to reach the headwaters? They could easily choose a calm, quiet river. Does it matter if they spawn in exactly the same spot as preceding generations?

Often the return to the spawning ground is a mighty battle. It starts by the young fish having to make the complex chemical adjustment to living in fresh water rather than salt. Then, frequently, as an adult it will have to fight every inch of the way to reach the place it was spawned, this time adjusting from salt water to fresh. Sometimes it has to make scores of attempts to leap a waterfall in fighting its way upstream. In all, its migration is one of the great battles between animal and environment.

The salmon has to have a formidable armory with which to overcome the difficult natural obstacles it faces. One of its strong points is a keen sense of smell; another is good eyesight to pick out the quietest section of a weir or the lowest lip of a waterfall. It also uses the sixth sense that is unique to fish: the lateral line organ, a fine line that runs down the body of the fish from head to tail. The line is made up of pores connected to a canal beneath the surface of the skin. Sense organs in this canal are nerve-connected to the fish's brain so that it is able to pick out with great accuracy changes in movement in the water through which it swims. By the very act of swimming a fish creates a pressure wave that moves out in front of it. When that pressure wave meets something ahead the fish can "feel" the object through the lateral line. This is a great benefit in murky waters. The lateral line also explains the way in which a shoal of fish moves as if it were one linked organism: each fish is aware of its neighbor and of its neighbor's movements.

In its long migration back to the spawning ground and in its long upstream fight, the salmon uses its lateral line to pick out points at which currents are marginally weaker and exploits this to speed its journey.

Eventually the salmon reaches home – the place in

which its parents and grandparents spawned and hatched. There the female digs what might be called a nest in the gravel and silt while the males fight for the privilege of joining her for the simultaneous release of eggs and sperm. Each group of eggs is covered with gravel and the female guards them for as long as she has the strength. That time is limited, for salmon make their long and arduous migration just once. After they have spawned, they die.

Fish which, like salmon, migrate from sea to fresh water to spawn are called anadromous species. Those that go the other way – from fresh water to salt – are called catadromous. The common eel is a catadromous fish. The adult male eel migrates to sea when about seven years old, the female when about 12. Until the 1920s nobody knew where they went nor how the young eels got back into their chosen estuarine environments in order to grow and develop, but in the 1920s the Danish scientist Johannes Schmidt made a series of nettings in an eastward direction across the North Atlantic. At each position as he progressed eastward he caught bigger eel larvae. This led him to develop a theory about the eel's life history which is generally accepted but which still leaves some interesting questions unanswered.

Schmidt's theory postulates that when freshwater eels reach sexual maturity they migrate to the Sargasso Sea – that calm area in the center of the great North Atlantic current gyre. There they spawn. The eggs hatch deep in the water and the larvae, perhaps just two fifths of an inch long, rise toward the surface where they drift on the major ocean currents. At this

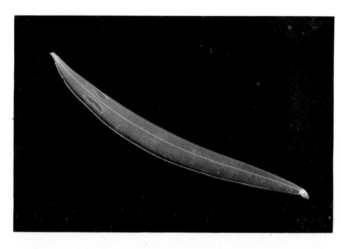

stage they are delicate leaflike creatures. During the course of their journies they mature, taking on their familiar elongated shape. The currents sweep them to widely separated destinations: some to the eastern coast of the United States, some to Iceland, some to the British Isles, some to the Mediterranean and North Africa. These great migrations take a long time. For example, eels that finally settle in European rivers may have spent three years getting there. They will have tolerated extremes of temperature and made the hard adjustment from salt to fresh water. With their specially designed gills, they will even have been capable of crossing stretches of dry land to reach their freshwater home, which some have to do.

That is the theory and it holds good, but it produces some mysteries. For example, it requires acceptance that all the common eels found in the rivers around

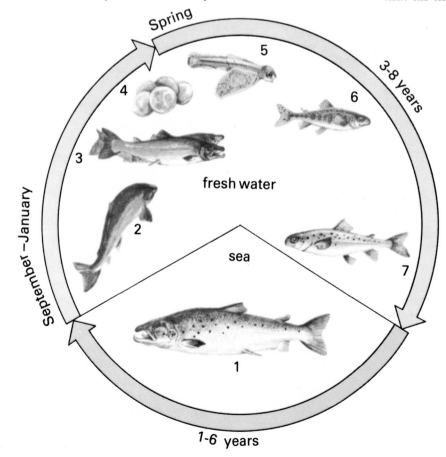

fresh water

sea

Spring

3-8 years

September – January

1-6 years

Left: the life cycle of the Atlantic salmon. (1) Mature salmon migrate from the sea to the rivers of their birth. (2) They leap rapids where they must. (3) The males fertilize the females' eggs. (4) and (5) Eggs hatch into fry. (6) The young spend the next three to eight years in fresh water, passing through the parr stage. (7) As smolt, they are ready to travel to the sea, spending from one to six years there before returning to breed and die themselves.

Below: sockeye salmon eggs. The dark spots are the eyes of the unborn embryo.

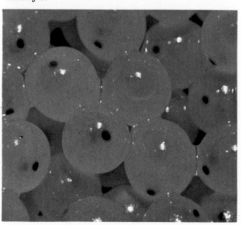

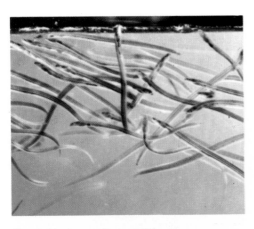

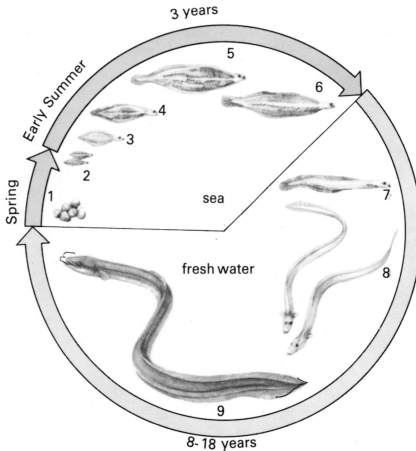

5

4

3

2

1

Early Summer

Spring

sea

6

7

8

fresh water

9

8-18 years

Opposite top: a larva of the European eel – leafy thin and glassy clear.
Above: the elver stage of the European eel – transparent and more snakelike.
Right: the life cycle of the European eel. (1) Eggs are laid in spring in the Sargasso Sea. (2) The eggs hatch in summer. (3) to (6) The larva grow and change shape, meantime drifting in the Gulf Stream to European coastal waters. (7) They become transparent elvers. (8) They are nearly mature when they reach the rivers of their aim. (9) The adults remain in fresh water before returning to the sea to spawn.

the North Atlantic originate in the Sargasso Sea and that they are driven from there by currents to their ultimate freshwater homes. But there are physical differences between the American and European eels. The latter generally have more vertebrae than their American counterparts. Are they distinct races? If so, how does each race manage to separate in the Sargasso Sea, each to go in its own direction? Some authorities suggest that the number of vertebrae in a species of fish can vary according to the temperature of the water in which they hatch and breed. If this is so, is there one common race whose members adopt different characteristics as they drift along a particular current? Another difficult question is raised by the fact that no adult eels have been caught beyond the limits of the European and African continental shelves during the time that millions of them must be heading to the Sargasso Sea to spawn. This has raised yet another theory: that there is but one common stock of eels and that these are American. They head for the Sargasso Sea and spawn. Then their offspring drift in all directions, taking up their freshwater existence before heading toward the Sargasso Sea again. But only the American eels make it. The rest perish on the way. It is a reasonable theory, but it does not provide an explanation for the drastic physical changes that the mature European eel makes in preparation for its long sea journey: its skin, glands, eyes, and digestive organs all adapt for a seawater existence. Does it undergo all this merely to perish soon after it reaches the sea?

It is also a mystery how the eel manages to navigate to the Sargasso Sea spawning area, swimming against prevailing currents as it must do. It is difficult to believe that the Sargasso Sea, in the middle of the Atlantic, has a distinction of the sort that draws a salmon back to its native tributary. Does electricity provide the answer? Recent research shows that eels are able to detect the extremely weak electric currents set up in the oceans by the combination of the earth's magnetic lines of force and moving salt water. Eels have been shown to be highly sensitive to the tiniest electric fields, and it could be that they are able to navigate by them.

All these theories set up still more questions. They are but another example of the great amount of work that needs to be done before the ocean's many remaining mysteries are revealed.

Below: European eels have only very small pectoral fins.

Migration Patterns of the Atlantic Salmon and the Common Eel

© Geographical Projects

East Kamchatka Humpback

Bristol Bay Sockeye

Columbia River Chinook

Atlantic Salmon

Above: this map shows the migration routes of three different species of Pacific salmon (see key) and the single species of Atlantic salmon. The information was obtained scientifically by putting tags on salmon and following the tagged specimens through their migratory cycle.

152

Larvae 10 mm

Larvae 30 mm

Larvae 45 mm

Limit of unmetamorphosed larvae

Larvae 25 mm

Limit of unmetamorphosed larvae

Larvae 15 mm

Larvae 10 mm

© Geographical Projects

American Eel

European Eel

Above: this map shows the migration of American (blue) and European (pink) eels. The circular lines, marked with a size, indicate the different stages of development between birth and adulthood and show how much faster American eels mature. The barbed lines hugging the coastlines (blue along North America, pink along Europe) show where the adult eels live in freshwater coastal streams.

Adaptations to Life in Deepest Waters

Sunlight quickly dims as it passes through the surface layers of the oceans. The water filters out the primary colors of the spectrum as depth increases. Red goes first and, depending on the turbidity of the water, a diver can be quite shocked even at depths of 100 feet to see that a scratch on his hand will ooze green blood. Go deeper and the yellows and greens disappear from the spectrum. There is only blue. Eventually, at around 2000 feet, there is no trace of sunlight at all. It is a world of darkness.

Yet in this darkness there is life. Even in the very deepest parts of the oceans there are fish, crustaceans, and mollusks. This raises a seeming contradiction. It is known that life in the ocean depends, as it does on land, on photosynthesis: the process in which the energy provided by sunlight is converted to produce the plant life which is vital to animal life. So how can life exist and perpetuate itself in regions where there is no sunlight at all?

Whatever the answer, there is a richness and variety of life in this world of darkness. It was only 100 years ago, however, that we became aware of such life: that was when William Carpenter and Wyville Thomson made the expedition disproving the dogma that no life existed below 1800 feet.

There is a source of food for deep-sea animals in the form of dead organic matter, which reaches even the deepest trenches of the ocean. It would be difficult to see this descending matter. Certainly the corpse of a fish would only sink a few hundred feet at most before it was devoured by other fish. But dissolved organic matter and the tiny excreta of plankton find their way to the deep ocean floor. Microscopic dinoflagellates, the halfway organism between plant and animal, live on this oceanic debris. So, too, do bacteria. They exist

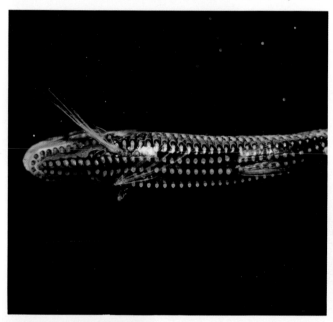

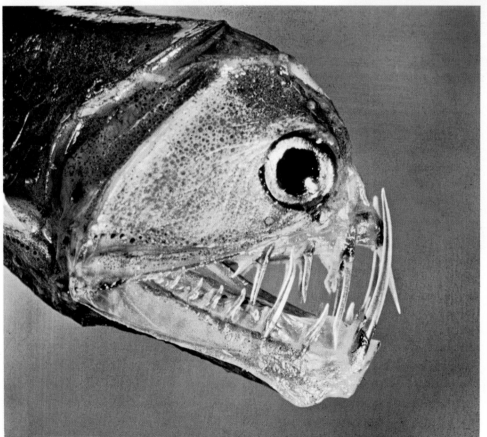

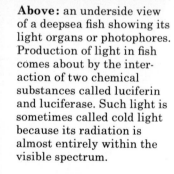

Above: an underside view of a deepsea fish showing its light organs or photophores. Production of light in fish comes about by the interaction of two chemical substances called luciferin and luciferase. Such light is sometimes called cold light because its radiation is almost entirely within the visible spectrum.

Left: the head of a viperfish showing its amazing movable jaw as it looks in repose.

154

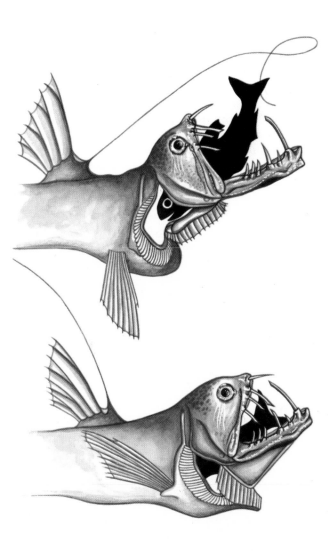

Left: the movable jaw of a viperfish in action. When swallowing large prey, the fish first throws back its head. This pulls down the lower jar so that the bones supporting the gills are pulled out of the way, leaving clear passage.

on the compounds of shells and skeletons, converting these into organic substances. Mollusks and crustaceans, in turn, feed on the microscopic dinoflagellates and on the seafloor substances produced by bacteria. Fish feed on the mollusks, the crustaceans, and each other.

Deep-sea food sources are nonetheless sparse compared with the almost daily explosion of plant and animal life in the surface layers of the ocean. This scarcity is one reason why deep ocean fish have such grotesque shapes. They often have huge mouths so that they have a better chance of sucking in the few meals that pass their way. They may have hugely distensible guts so that even if a dead or injured animal three times the size of the predator is found, it may be packed into the large mouth and squeezed into the stretchable stomach to provide a lasting meal. They may have wildly extended pelvic fins and chin barbels which they use to probe the dark sediments and disturb burrowing crustaceans. The crustaceans may have extraordinarily long spindly legs so that they can support themselves on the silty sediments of the ocean floor.

It might be thought that in a world of eternal darkness the pattern of evolution would make eyes unnecessary, but this is not so. Deep-sea dwellers have eyes and sight.

It is true that sunlight does not penetrate into the oceans. However, there is another source of light. This is bioluminescence – the light produced by certain animals themselves. As many as half the deep

Below: the body of a viperfish, showing the double row of light organs along the lower side.

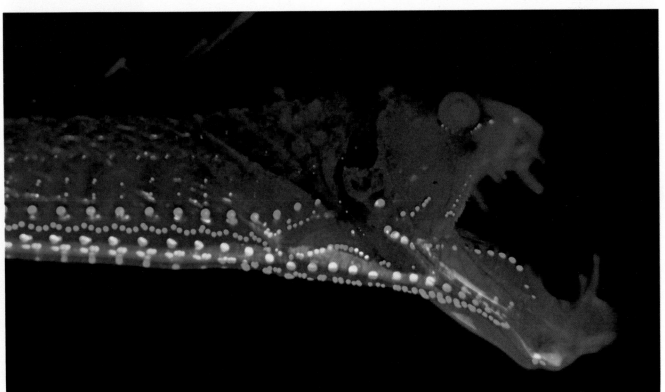

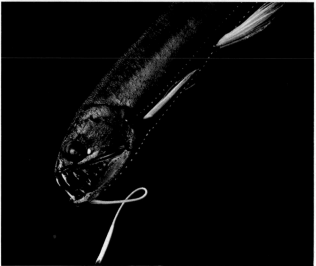

ocean creatures – from zooplankton to crustaceans to fish – may have the light-emitting organs called photophores.

Most photophores do not give a constant light but flash on and off intermittently, some for perhaps a second or so, some for as long as five minutes. This irregular light display presents an underwater kaleidoscope of golds, reds, blues, strong greens, and soft pinks. Some fish have all the colors, with perhaps a row of red and blue lights along the tops of their bodies and a row of green and red along their lower halves.

Some fish flash their intermittent lights spontaneously. Others flash only in response to another light, or when touched, or in response to another external stimulus. It all depends on the purpose the light system of any creature has been designed to fulfill. Some, it must be supposed, have no purpose at all, but most do. The lights are used to convey messages. They say: "Come here, I want to eat you." Or "Go away." Or "Come and be my lover." In the case of mating, at least four species of the anglerfish have a rather grisly solution to the problem not only of finding a mate in the eternal gloom but also of ensuring that they stay together for life.

The female has on top of her snout a fin-ray which ends in a bulb. In each species a different type of light is displayed to enable the male to pick out the female of his particular species. To make doubly sure, the female anglerfish is thought to emit a characteristic scent to which the male is attracted. This has been surmised by scientists who have noticed that the female anglerfish does not have highly developed organs of smell while the male has very large ones.

Having been attracted to the much larger female of up to 3 feet long, the small male of about 6 inches in length grips her body with his jaws and continues to cling on until he becomes fused with her flesh; only tiny openings at the sides of his mouth remain to enable him to breathe. The very tissues of the male and female grow into each other so that the male becomes a parasite, completely dependent on the female's blood to stay alive. But it is a two-way relationship, for the female depends on the male to fertilize her eggs. In some species the female takes more than one mate: some have been captured with three males fused to their bodies. In other species the male attaches himself to the female only during the reproductive season.

Not only is the deep ocean a place of flashing lights but also it can get noisy. The oceans ring with clicks, grunts, and strange drumming noises. Fish do not have lungs or vocal chords with which to produce sounds, so many of them use their swim bladders as drums instead. The males of the deep-dwelling rat-tails, for example, have a pair of large muscles at the forward end of the swim bladder and these drum on the gas-filled bladder to set up a low-frequency booming sound that travels considerable distances (remember that

Top left: a close-up of an anglerfish's head, showing its lure. In this species, the lure is fleshy rather than rodlike. After attracting prey, the anglerfish snaps it into its huge mouth almost faster than the eye can follow the action.

Above left: one of the deepsea fishes that has a light organ on its cheek along with others on its underside.

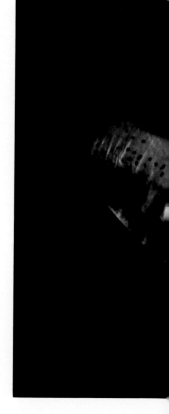

Right: viperfish have fang-like teeth tipped with slight barbs. They live in depths of 1500 to 9000 feet.

water is an excellent sound carrier). Other fish grind their teeth; some rap their fins on the seabed; some crustaceans flick their claws to produce a distinct clicking. These sounds are produced to attract mates, to establish territory, to frighten off predators, and as alarm calls to other fish in a shoal.

Above: an anglerfish of a species in which the lighted lure looks more like the "fishing rod" it is often called. This lure attracts the male mate as well as prey. A parasitic copepod can be seen near the anglerfish's tail.

Sound in many species is picked up by the swim bladder, amplified by it, and transmitted by small bones to the inner ear. As in human beings, this inner ear also controls balance. In a deep-dwelling fish this sense of balance is of vital importance. Human beings know which way they are oriented in relation to the earth – gravity sees to that. But a fish moves in three dimensions, so how does it know which way up it is swimming in the eternal darkness of the ocean depths? In cavities behind the brain the fish has calcareous plates called otoliths. These rest on hairs that line the cavities and that are connected to the brain by nerves. As the attitude of the fish changes in the water, the otolith moves to touch hairs at the top, bottom, or either side of the cavity. Nerve impulses reach the brain, which then signals to the fins and tail so that the fish automatically rights itself.

The otolith is of particular interest to fisheries scientists, whose study of it has enabled them to determine the age of many species of fish with great precision. Each year the otolith grows to produce two concentric rings. By counting these, the age of the fish may be determined in exactly the same way as the age of a tree may be calculated by counting the rings in its trunk. The technique, however, only holds good for fishes of temperate waters. In the tropics, fish do not display ring growth on their otoliths; it is therefore almost impossible to tell their age.

157

Mammal Species of the World Sea

Whales are mammals of the order Cetacea, whose members are among the most advanced animals in the world. The smallest cetaceans are the porpoises and dolphins. The largest is the blue whale, which weighs as much as 25 elephants and is the biggest living mammal. Cetaceans have also been around the longest, with a 50-million-year history. They were once land dwellers and almost certainly are descended from tiny mouselike animals. These returned to the sea and adapted to a marine existence that encouraged their growth. Their forelegs became paddles and their back legs disappeared, although beneath the blubber at the back of a whale's body are the rudiments of the hind

Dolphins have revealed such outstanding intellectual gifts that there is a tendency to attribute greater powers to them than they in fact possess. But even those gifts that have been scientifically observed and documented reveal an intellect that puts the animal very high on the evolutionary scale. Some species have a brain the same size as that of a person in terms of comparative body size, although this does not necessarily indicate the same intellectual development. It often seems that dolphins have many of the more pleasant attributes of the human race without the evil of which people are capable.

Dolphins have the power to conceptualize: to originate an idea, to put the idea into action, and to change the course of that action in the interests of greater efficiency – or sheer fun. Dolphins play for play's sake, and very few animals do that. A kitten does not skid a ball of paper across a carpet just for fun. It is part of a self-imposed, instinctive training for hunting and trapping real prey. But there are many stories of dolphins playing silly games purely for the enjoyment they get from them. This is why

Left: the bottlenose dolphin, known in the United States both for its performing skills in aquariums and for its role in a series of navy research programs.

Below: a dolphin swimming close enough to the surface to make its blowhole visible. The bottlenose lives in Atlantic waters from Florida to Maine, in the Bay of Biscay and the Mediterranean, and off West Africa as far south as Dakar.

limbs. Because fur is not an effective insulating material in water, that also disappeared, to be replaced by thick layers of fat that maintain body heat even in the freezing waters of the Antarctic Ocean.

Porpoises and dolphins are the best-known of the whales. They are the most numerous and their natural sociability makes them an easy subject for study by humans both in the wild or in captivity. There is plenty of evidence to suggest that their larger cousins, which have not been studied as much, share the same remarkable characteristics – one is tempted to say "personality."

Above: this dolphin is not just surfacing for a quick breath of air. It is making a signal that it has sighted a sonar target as part of a research program conducted by the United States Navy. Dolphins show high learning ability.

they are so easily trained to perform in dolphinariums. The natural sense of fun that will lead a dolphin in the wild to play for hours with a car tire can quickly be channeled into more formal tricks by the animal's human keeper. Dolphins learn faster than any other animal with the exception of people, and is also capable of displaying boredom if asked to perform the same task too often.

Dolphins are solicitous of each other. If one of their shoal is injured or in difficulties, they will race to its aid, combining to push it to the surface if it is unable to come up to take air. They are fiercely protective of their young. The mother suckles her calf every half hour for the first two weeks or so and after that insists that her baby stays close to her for several more weeks. She takes swift disciplinary action if the youngster strays from her side.

Their natural gregariousness extends to the dolphin's long-chronicled relationship with people. Plutarch marveled at the dolphin's gift of "disinterested friendship," observing that the animal "has no need of man, yet it is the friend of all men and has often given them great aid." Stories abound of the ways in which dolphins have not only befriended people – returning year after year to renew old acquaintanceships – but have saved bathers and shipwrecked sailors by pushing dinghies toward shore, allowing people to ride on their backs, and steering those in danger of drowning away from treacherous currents to safe shallow waters.

Dolphins are part of the suborder of cetaceans called Odontoceti – the toothed whales – of which there are some 80 species. Most species are less than 20 feet long but the group includes the 60-foot long sperm whale as well as the killer whale that reaches 30 feet long. The toothed whales live on fish, squid, and crustaceans, although the killer whale will prey on seals and penguins and is one of the very few predators of the dolphin. In the wild the killer whale can be a frightening aggressor but, like its smaller cousins, it is easily tamed in captivity.

The second suborder of cetaceans, the Mysticeti, includes the biggest mammals on earth. Yet they do not prey on fish. These are the baleen whales, called so because they have a sieve of fine plates, or baleen, hanging from the roofs of their mouths instead of teeth. Each whale swims slowly, takes a great gulp

Right: a dolphin getting a treat for a job well done during an experiment by the US Navy.

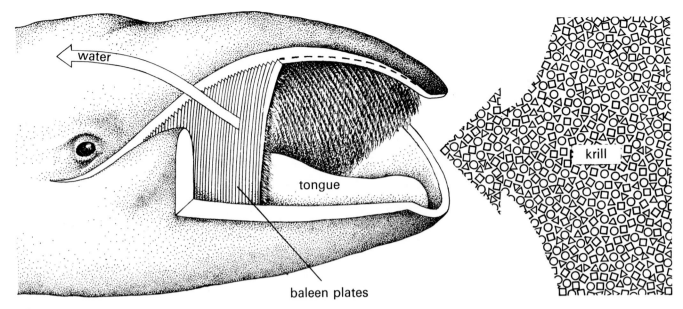

water

krill

tongue

baleen plates

of water, then pushes it out through the sieve of baleen. Trapped behind the plates are thousands of krill, the tiny copepods that throng the southern ocean, which the whale swallows.

One family of Mysticeti is known as the right whale. They have thick layers of blubber, a high yield of oil, and move comparatively slowly. This gave them their name simply because they were the "right" whale to catch. The other main family includes the fin, sei, Bryde's, minke, and blue whales.

The blue whale is probably the biggest mammal that has ever lived, reaching a length of over 100 feet and weighing 120 tons. No land animal has ever, or will ever, reach this size because its limbs would simply break under the weight. The sea, however, supports the weight of the blue whale, which is four times heavier than the biggest known dinosaur. There is also a positive advantage for a sea-living, warm-blooded mammal in being large: the bigger the animal,

the lower the ratio between volume and surface area, and the easier to keep warm. A small fish has a very large surface area in proportion to its body volume and therefore loses heat quickly. Even the very smallest of the cetaceans, the porpoise, is still over six feet long, with 45 percent of its body weight consisting of heat-retaining blubber. A big whale has a blubber content of 25 percent.

Mammals breathe air, so whales have to come to the surface to fill and empty their lungs. They are very efficient when it comes to breathing. One big breath allows a blue whale to stay submerged for 50 minutes; a bottle-nosed whale can survive on one breath for two hours. They all breathe swiftly. The animal comes to the surface, exhales through the blowhole on the top of its head, and inhales again – all in two or three seconds. When people are breathing normally, they usually get rid of about 15 percent of the air in their lungs in a single exhalation. A whale, with a

Top: how the baleen whale eats. Instead of teeth, this whale has long thin strips of bone called baleen plates. When the whale gulps a mouthful of water, its tongue lifts and forces the water out through the sievelike plates, leaving the krill to be eaten.

Right: a close-up of the blowhole of a pilot whale. Some whales have two blow-holes instead of one, but all use them for breathing. Exhalation produces the familiar spout of water when whales are on the surface.

160

great roaring spout of water, clears about 90 percent. This is one reason why it is able to exist for so long on one breath. Another reason is that the whale's body chemistry is such that it can function for much longer without taking in oxygen. Its muscles contain very high concentrations of a substance called myoglobin which stores oxygen and gives whalemeat its characteristic dark color. The whale's blood is also rich in oxygen-carrying hemoglobin. It all adds up to the whale being able to utilize 90 percent of inhaled oxygen compared to 20 percent in people.

Whales are able to dive to great depths and at great speed. Dolphins have been known to dive to over 500 feet and back in less than three minutes. Sperm whales regularly go still further in pursuit of squid, which is one of their favorite foods. Their bodies have been found entangled in submarine cables on the seabed as deep as 3700 feet.

turn of speed. The blue whale, for example, can sprint over short distances at 25 miles an hour; dolphins can reach over 30 miles an hour; and a killer whale over 45 miles an hour. An examination of the bodies of dolphins shows that their muscle power is not enough, possibly by a factor of three, to produce such speeds. What does it is that they somehow manage to reduce the friction on their bodies. How they do this is not fully understood, but it is thought that ripples on their skin, combined with natural secretions of oil, set up nonturbulent flows of water along their bodies, so reducing the drag of the water. If their secret could be fully understood and utilized by humans in boat building, the world shipping industry would save millions and millions a year.

The giant whales are impressive as much for their endurance as their speed. A big blue whale generates up to 500 horsepower when it flexes its mighty tail

Left: the baleen plate of a southern or black right whale can be seen in this picture. Right whales do not gulp water in to feed, but rather swim through a mass of krill with their mouth open so that water constantly flows into the mouth and out of the baleen, sieving the krill out.

There is still some mystery surrounding the ability of a whale to dive to such depths and return so quickly without damaging its body tissues. If people ascend from a deep dive too quickly, the nitrogen in their blood comes out of solution to form bubbles. These increase in volume as the person ascends, causing the painful, often crippling, and sometimes fatal bends. Whales do not suffer from the bends even after their lightning rise from the depths. The most widely accepted explanation is that, because of the whale's breathing efficiency, it dives with the maximum of oxygen and the minimum of nitrogen in its lungs.

When it makes its sensational dives for squid, the sperm whale snaps up the animal and swallows it whole. The bonelike beaks and sucker pads of countless squid often form themselves into an indigestible, dark, and sticky lump, perhaps weighing as much as half a ton, in the stomach of the whale. This substance is ambergris, much prized as a fixative for perfumes and cosmetics.

Despite their size, whales are able to produce a good

flukes, and it can maintain a steady speed of around 15 miles per hour for days on end. The baleen whales use these great powers of endurance every autumn when they make their long migration from the Antarctic to warmer subtropical waters, perhaps as far north as 20° S, where they breed. The exact location of most of these breeding grounds is not known. During this stay in warmer waters the animals eat very little, subsisting mainly on the stores of blubber they have built up during the Antarctic summer. A blue whale calf is 25 feet long at birth, and it grows at a faster rate than any other mammal. Suckled by its mother, it doubles its weight in its first seven days. It more than doubles its length after six months to reach 55 feet. By the age of two it reaches sexual maturity, and by then it is over 70 feet long and weighs 60 tons.

When they are moving at high speed, dolphins need very good navigational capabilities. These are based on an echo-location system that makes the most advanced sonar devised by people seem crude by comparison. The animals emit sound and listen to the

return echoes in a frequency range of between 150 and 120,000 cycles per second. (The human ear can detect sounds in the range of 150 and 20,000 cycles per second.) The dolphin has a vast repertory of sounds – clicks, squeaks, creaks, bleeps, and whistles – as well as a range of sounds beyond the human hearing range.

Low frequency sound travels farthest. High frequency sound is soon absorbed but has greater resolution – it enables small objects to be identified clearly. The dolphin transmits low frequency sounds over a wide area just to detect an object, whether it is prey, another dolphin, a predator, or merely an obstacle. Having found an object or objects, the animal will switch frequencies, using higher-pitched sound to gain an accurate picture of the object. It will zoom onto its target, its head swinging from side to side as it saturates the object with narrow sonar beams. From the type of echo it receives, it can tell the difference

Below: a killer whale. Found in all seas, this species has distinctive coloring in its black and white markings.

Above: a southern right whale breaching. Right whales have huge heads, sometimes being as much as one third of the total length of the body. Their baleen is so long that it folds up on the floor of the mouth when the mouth is shut.

between one species of fish and another. A blindfolded dolphin can thread its way through a maze, detect and avoid a single strand of fine wire, or distinguish real food from objects of exactly the same shape and size. In the latter case there is no question of the dolphin merely sniffing out its meal – all whales have little or no sense of smell.

Dolphins also use their sonar systems for communication with others of their own kind, and the degree of sophistication of the messages that are passed back and forth remains the subject of a lively scientific controversy. Some students of the animals are convinced that the dolphin has a highly developed language and that if only the "code" could be cracked, people and dolphins would be able to have meaningful conversations that would go far beyond the issuing and understanding of commands to animals in captivity. Other scientists say that, although there is a complex pattern of calls for navigation, distress, recognition, and so on, these noises do not in themselves constitute a language any more complex than that of some birds and primates. Those who support the language theory respond that, until we can fully understand and analyze the complex stream of noises that pass between dolphins, the theory cannot be discounted. So the fascinating debate continues.

There is evidence to suggest that the larger whales share the dolphin's great intelligence as well as a sense of loyalty and fun. But examples are harder to come by since the big whales are much more shy. However, big whales also communicate and some even sing. Humpback whales have been recorded calling to each other in long repeated themes that are

Above: a sperm whale on the deck of a whaling ship off Durban, South Africa. This whale prefers the warmer seas, although it lives in all waters. It is one of the few species in which the male is much bigger than the female.

ocean layers over distances of hundreds of miles.

Nearly all the species of whales, from the high-speed dolphin to the gently browsing mammoth blue whale, are not merely harmless to people but seem often to welcome their company. Human response as far as the larger animals are concerned has been to wage ruthless war upon them. The war has been so mindless that, even if whale products had been absolutely essential to survival, we would still have continued hounding some species to extinction with no thought for the future. In fact, whale products are no longer essential: whalebone corsets are no longer used; carriages with baleen for springs are rare; flexible steel has replaced baleen in fishing rods and umbrella stays; synthetic materials have replaced whale oil in lamps, soaps, and as lubricants.

There are encouraging signs at last that the whale will be spared. It would be our loss not to have this gentle, shy, giant of the oceans.

musical enough to have been issued on a long-playing record. The sound is like songs of eerie beauty. Other fin whales use the sound-carrying properties of the deep sea to bounce their messages through the

Below: the sea mammals, drawn to scale to show their comparative sizes with each other and with a human.

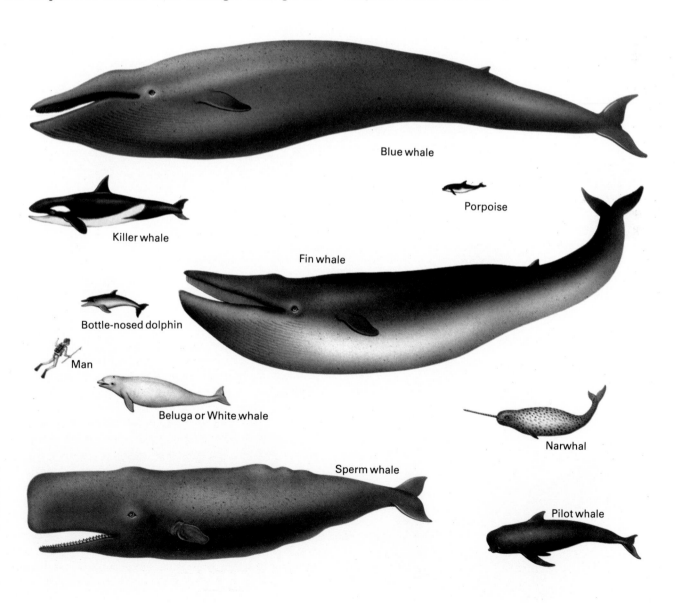

Blue whale

Porpoise

Killer whale

Fin whale

Bottle-nosed dolphin

Man

Beluga or White whale

Narwhal

Sperm whale

Pilot whale

Chapter 7

Exploiting the Ocean

The sea could provide far greater amounts of protein food than it does. Yet, paradoxically, many species of fish have come close to being fished to extinction. Why is this? Part of the answer is that only six areas of the world's major fishing areas yield 80 percent of the world's supply of fish. Consumer preference is largely responsible for this uneven exploitation, which concentrates on the traditionally popular fish. Technology also has affected the way fisheries are exploited, and this has proved a curse as well as a blessing.

Other marine resources could increase the quantity of food available, especially if new tastes are developed, as well as provide many new medicines. The sea's rich source of dissolved chemicals might even save the world from exhausting certain essential minerals. Deep ocean mining holds a bright promise for keeping industry supplied with vital metals for many years, and great efforts to harvest this deepest of all ocean resources are being made today.

Opposite: still the most heavily exploited of the sea's riches are fish in their great variety. Other resources – such as minerals and drugs – await fuller utilization.

World Fish Catch: Present and Future

The living resources of the sea are huge, but they are being exploited in a haphazard way that has more to do with power politics and greed than with the need to feed a world in which two-thirds of the population does not have enough to eat.

Since 1970 the world fish catch has been steady at between 65 and 70 million tons a year. It reached this plateau rather quickly: the catch was 29 million tons in 1955 and 53 million tons in 1965. The fact that this plateau has been reached does not mean that we are taking the maximum possible amount of fish from the sea. There is a potential world catch of over 100 million tons a year that could be had without damaging the total resource.

This seems to indicate that all is well with the world fishing industry and that all we need to do in order to add another 30 million tons of protein to the world diet each year is to build more ships and catch more fish. But can this comfortable prediction be reconciled with some unpalatable ecological facts? For example, fisheries biologists warn of the total extinction of some races of herring; international catch quota systems have had to be set for certain species in the North Atlantic so that they will have a chance to survive; gunboats of one nation chase the fishing boats of other nations away from self-established fisheries boundaries.

Surely in a world where another 30 million tons of fish a year could be caught, shouldn't there be plenty to go around? The answer is to be had from a deeper dig into the statistics. We know that 90 percent of the world fish catch comes from the continental shelves where plant and animal plankton thrive in their relatively shallow waters and so provide the main food for fish. The other 10 percent of the world catch is made up of oceanic animals living offshore, primarily tuna and whales.

Opposite: a fishing trawler at Wick, Scotland. One of the world's richest fishing areas lies in the North Sea between Great Britain and the mainland of Europe.

Left: a fish market in Japan. The Japanese depend on marine food for daily sustenance rather than for augmenting the protein diet.

Below: this map illustrates how the Japanese longline fishing fleet went farther and farther afield in pursuit of tuna in the period from 1948 to 1962. By then, not one tuna fishery had not been fished by the Japanese.

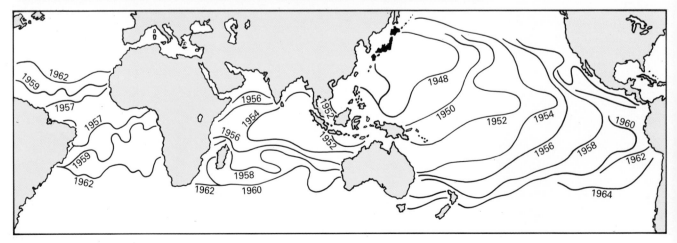

The distribution of the present harvest is far from being globally uniform, however. The Food and Agriculture Organization (FAO) of the United Nations divides the world ocean into 15 fishery areas. Nearly 80 percent of the total 70 million tons of fish a year comes from just six of these 15 areas. They are the northwest, northeast, east-central, and southeast Atlantic and the northwest and southeast Pacific. The reasons for this concentration of fisheries are complex. They involve the development of fishing technology, consumer preferences, and the economics of food production.

The world's top two fishing nations are Japan and the Soviet Union. The reasons why these two nations

have reached their position and the ways in which they have achieved it go a long way to answering some of the questions already posed.

Japan cannot support its vast population on the limited land space available, and so exploits the sea for anything that is remotely edible. A Japanese eats six times as much marine food as an American, and half of Japan's edible protein comes from the sea to only three percent in the United States. The Soviet Union has had to turn to the sea to boost the protein requirements of a population that has suffered from the failure of many harvests and also to achieve independence in protein supplies. Both nations have long exploited the water close to their shores and

Right: shooting the net – in this case a purse seine – from a Korean training boat. UN aid helped the Korean government set up a four-year training program to improve the country's fishing industry and fish catch.

continue to do so, but the near-shore protein yield proved inadequate and Japanese and Soviet fishermen looked farther afield. They colonized the world ocean. There is not a single tuna fishery in the world where a Japanese boat cannot be seen, from New Zealand to the Canary Islands. Japanese trawlers fish off Alaska for king crab, off South America for shrimp, and off Canada for salmon. Soviet trawlers prowl the English Channel for mackerel, search the waters off South Africa for hake and pilchards, and roam the North Atlantic for many marine animals.

Japan and the Soviet are not alone in the vast oceanic exploitation: some three quarters of the world fish catch is landed by only 14 nations. Some of these keep close to their shores, but most have gone farther away. In doing so, they have concentrated their efforts on the six areas singled out by the FAO as yielding 80 percent of the world catch while the other nine

Right: a weighted trawl on deck, ready for lowering into the sea. Trawling can be done by a few individuals on small non-mechanized boats as well as by full crews on large mechanized vessels.

Left: haddock in a Scottish fish market. Related to the cod, though smaller, haddock is widely fished on both sides of the North Atlantic, where it is most plentiful.

areas remain less well utilized.

As a result, vast fleets converge on the prime fishing grounds of the world ocean. They go there because the types of fish preferred by the consumers back home abound in those areas – or used to abound. The international onslaught has taken a vicious toll of favorite food fish such as cod, haddock, and herring. The world fishing industry now has to examine closely all its operational methods in the face of the crisis of too many ships chasing too few species of fish in too crowded parts of the ocean. It is a problem that is difficult of resolution – and it is a problem that must be resolved.

Below: this map divides the world into the 15 main fishing areas as defined by the Food and Agriculture Organization (FAO) of the United Nations. The fisheries of about half the areas – including almost all the North Atlantic – are now exploited to the limit of their yield.

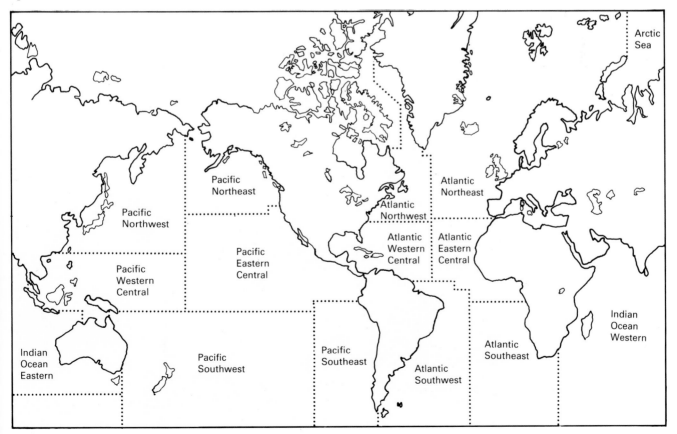

Fishing Methods

Anyone who is patient, keen of eye, and quick of hand can catch a fish. You stand near a quiet shore up to your knees in water and wait for a fish to swim by. Then you pounce. You swiftly scoop your hands beneath the wriggling creature and whisk it out of the water and onto the beach. It takes a lot of skill and practice. Alternatively and more easily, you can pick shellfish from intertidal rocks.

That is how fishing started. Then came the use of hooks and lines, of static traps, and of nets pulled or pushed by one person along the shoreline. It was a short step to the use of boats to deploy the hooks and lines, to set the traps, and to tow the nets. Fishing moved from the shoreline out into the open sea and more specialized methods were adopted to catch specific varieties of fish. These methods may be broadly categorized as demersal and pelagic fishing as related to demersal fish, which live and feed mainly on the seabed, and pelagic fish, which swim throughout the water column.

Below: a 19th-century painting of a Dutch scene showing a sailing vessel with beam trawls for fishing. Historically, the beam trawl gave way to the otter trawl when the steam engine enabled ships to tow bigger nets.

Demersal fishing methods are based mainly on the trawl. When all fishing boats were operated by sail, the vessels towed a beam trawl. This was a conically shaped bag of netting pulled along the seabed with its mouth held open horizontally by a wooden beam. At each end of the beam were skids which kept the net open vertically and which prevented it from ploughing into the seabed. It was towed from the sailing ship at the end of two long ropes – the warps – and the net was hauled in laboriously by hand-powered winches and capstans.

Beam trawls are still used in some specialized fishing operations but their universal use diminished with the introduction of the steam engine to the fishing fleets toward the end of the 19th century. The advent of the steam engine, and later the diesel, meant that bigger and bigger nets could be towed, their size limited only by the difficulties of handling the heavy beam when the trawl was brought aboard. So the otter trawl was developed. This dispensed with the beam and relied for horizontal spread on otter boards, which are rectangular steel or wooden boards up to six feet long and three feet deep. The conically shaped net has its sides stretched out to form "wings" so that the mouth of the net can be 200 feet across. One other board on each side of the net is attached to the wings. Each board has a towing brack set in such a way that when the net is towed along the seabed at a speed of up to three miles an hour, the boards tend to pull away from each other and thus keep the mouth of the net open horizontally.

The boards also tend to scrape up sand from the bottom so that fish in the path of the trawl are fright-

Left: an otter trawl, the bottom fishing net largely in use today. Otter trawl design has evolved over three quarters of a century. The horizontal spread of the mouth of the net is achieved by using otter boards, one on each side of the net connected by wire ropes called bridles. The top of the net's mouth is kept open by floats while the bottom of the mouth is kept down by iron or wooden weights.

ened and get herded into the net. The bottom leading edge of the net – the ground rope – is weighted by iron or rubber bobbins, great beads up to 2 feet in diameter through which the ground rope is threaded and which enables the trawl to roll along the seabed. The top edge of the net is kept high above the seabed by plastic or alloy floats.

Because the length of the wooden beam might be up to 70 feet, sailing vessels towing trawls had to bring the net over the side. When the big otter trawls were introduced, they hd to fit in with existing designs of ships and were also handled over the side. So for many years the biggest vessels in the fishing fleets were the sleek and graceful "sidewinders" with their low freeboards to aid hauling the net.

Hauling the net is arduous and dangerous. The fishermen line the low rail of the ship and, as it rolls downward, they reach out to grab the net. As the ship rolls upward, they hold on tight to haul a section of net aboard. The procedure is repeated, with the crew frequently having to stand up to their waists in water, until the main body of the net is on board and the cod end – the bag of netting at the end of the trawl into which the fish are channeled and trapped – floats at the side of the ship. The crew on deck and the skipper on the bridge look anxiously for this first sight of the cod end, which gives them a good idea of whether or not the past two or three hours spent in towing the net along the bottom have been worthwhile. Sometimes the cod end will float sadly empty on the surface because at some stage during its bumpy journey along the seabed it will have been ripped and all the fish will have escaped. The sight that makes the crew feel that the hard work has been worthwhile is that of a bulging cod end surrounded by hundreds of squabbling, shrieking seabirds attacking the fish trapped inside.

Above: the crew of a fishing boat waiting to haul in the trawl. Even on modern boats, this operation is still a dangerous and laborious task.

Right: a full cod end floating on the surface at the side of a boat. The cod end is the narrowest part of the otter trawl – the final trap of the fish caught within it.

The cod end is winched aboard the vessel and hangs for a few moments, suspended from the hook of a derrick. Then one of the crew stoops beneath it, gives a sharp tug to the quick-release knot at the bottom of the bag, and steps smartly back as the fish cascade to the deck. The net is then prepared for another tow and the fishermen settle to the task of gutting, cleaning, and sorting the fish, working on the open deck with almost no shelter from the elements. The fish are stowed below deck, sometimes in bulk and sometimes in boxes. When the fishing is good and a bulging cod end appears at the end of every tow, there is little time between one deck load of fish being stowed away and the next load arriving. In such cases the crew may work up to 18 hours at a time with very little respite.

There is an obvious illogicality in towing a net from a vessel and then having to go through lengthy and complicated maneuvers to bring it on board, so in the 1950s British and German trawling companies adopted some of the techniques of whaling ships. They built vessels with sloping stern ramps. The net was hauled up this ramp and onto the deck with the handling crew protected by the bridge superstructure from wind, spray, and the more than occasional wave. Hand in hand with this development came quick freezing at sea. Before this development, fishing vessels were limited in their range of operations by the amount of ice they could carry to last them from the time it took to get to the fishing grounds, fish for a few days, and

Above: gutting, cleaning, and sorting fish. Such work has to be carried out between the time that one catch is unloaded and the next hits the deck for unloading. On good fishing days, this can mean a work period of 18 hours.

Below: quick-frozen cod on a modern stern trawler. The whole fish has been packed into blocks of a size that can be managed by hand, and the blocks have been arranged in layers. The fish can be seen clearly through the ice.

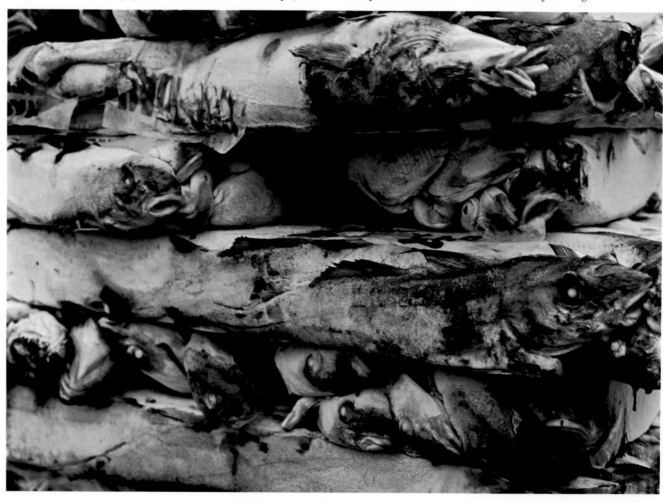

Right: checking the fishing nets on a trawler. There has been a swing from natural fibers to the use of synthetic fibers for the manufacture of netting.

Below: the stern end of a present-day freezer trawler. Fish are hauled up the ramp in the nets, which can be seen stowed on each side of the ramp.

get back to port before the first-caught fish spoiled. With quick freezing, there was no limit to the size of ship or how far it could go to fish. On the modern stern trawler the net is hauled onto an open deck and the fish are immediately poured through a hatch onto a completely sheltered working deck below where they are gutted, often filleted, and quick frozen. Today, much of this work is mechanized with machines to gut, fillet, and even turn waste fish products into fishmeal and oil.

Technology removed the natural checks and balances in commercial fishing. Ships of many nations could roam the oceans at will with no need to return to port until their holds were full. Some nations –

notably the Soviet Union, East Germany, and Japan – went still further. They sent out flotillas of fishing vessels which supplied a mother ship – a vast quick-freezing factory – with raw material. Uninhibited by the traditional need to get to market before the first-caught fish spoiled, these fleets were able to stay at sea for months. Most of their effort was concentrated on the North Atlantic grounds, and stocks of demersal fish soon became depleted. But the world ocean was – and still is to a large extent – wide open to this efficient harvesting method, and international fleets began to move freely around the world in search of the comparatively few species that were in greatest demand by the consumers back home.

Right: three kinds of trawls in
current use (top to bottom) – a
fleet of drift nets, a one-boat
pelagic trawl, an otter trawl.

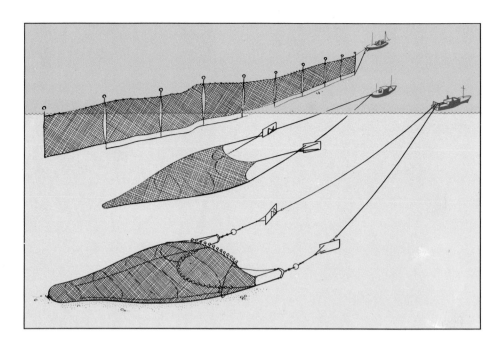

Another method of catching demersal fish such as cod or haddock is by longlining. Hundreds of baited hooks are used, each fastened on short lines of about 5 feet long called snoods. These are tied to the longline at six-foot intervals, and sections of longline are joined to make one great line perhaps two miles long. This is an excellent way of catching fish that will be of guaranteed high quality because they are not damaged by being crammed into a trawl. However, the hooks have to be baited and the snoods attached to the longline by hand, so unless a satisfactory method for mechanizing these tasks can be perfected, it seems destined to die out – at least in highly developed fisheries.

The Japanese are masters of the art of longlining, which they use for catching not demersal fish but pelagic, ocean-swimming tuna. In this method the lengths of line are buoyed so that they float in loops from the surface. And such loops! Each section of line is about 150 feet long. From these hang shorter lines with hooks at the end. Sections of main lines are joined to form a single line that stretches away from the boat to a total length of perhaps 60 miles. A single cycle of casting and hauling the line may take up to 24 hours. This fishing method is practiced by the Japanese in every area of the world where tuna occur, from South America across the Atlantic, Indian, and Pacific Oceans in tropical latitudes.

Another method of catching pelagic fish, which is used primarily for open sea salmon fishing, is driftnetting. This was the method on which the great European herring fisheries were built. Each net was 100 feet long by 45 feet deep, and anything up to 100 of them were joined to form a "fleet." This was paid out from the bow of the ship with floats on the top rope and weights on the footrope so that a wall of netting perhaps two miles long hung vertically in the water with the top of the net between six and 75 feet beneath the surface, depending on the fishing conditions. Her-

ring rising to the surface at night would swim into this wall of netting and become trapped by their gills. This passive form of fishing gave way to active methods of which the main ones are midwater trawling and purse seining.

Midwater trawling uses the same type of conically shaped bag of netting used in bottom trawling but the

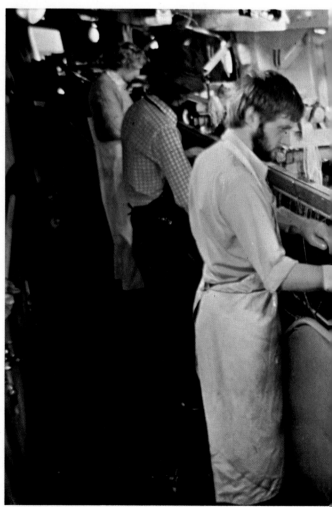

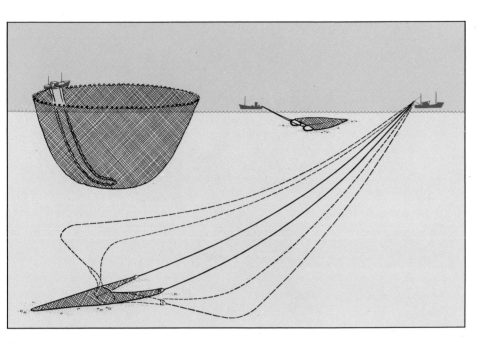

net is towed at a particular position in the water column – from near the seabed to near the surface – depending on where the shoals of pelagic fish are swimming. The technique was developed in the late 1940s and utilized two boats towing a single trawl between them, with one warp towed by each boat. This gave horizontal spread to the net. Depth was regulated by floats, weights, and the speed at which the net was towed. The next development was the one-boat midwater trawl which had light otter boards to give horizontal spread. The midwater trawl is used to catch all species of pelagic fish but, because it can be towed directly through a shoal, it is particularly suited to species that live in dense shoals, such as herring.

Next came the purse seine which goes beyond the midwater trawl to catch an entire shoal in a single operation lasting only about 20 minutes. It is the most ruthlessly efficient of all fishing methods. The largest purse seines are nearly a mile long and over 500 feet deep – a vast curtain of netting. When a shoal of fish is detected, this wall of netting is run out by the ship in a great circle so that it surrounds the shoal. Floats support the top edge and weights keep the lower edge down in the water. When the shoal has been surrounded the bottom of the net is closed – or pursed – and a basin is formed inside which the shoal is trapped. The whole net is then hauled on board, forcing the fish to the surface. The mass of panic-stricken fish is scooped out of the artificial pond alongside the boat either by a mechanically handled scoop of netting – a vast ladle called a brailer – or is simply sucked into tanks on the vessel by a hydraulically powered pump like a giant vacuum cleaner. The method used depends on the type of fish being caught and the purpose for which they are to be used. Purse seining was developed by the United States to catch salmon and tuna off California. These fish were destined for canneries and therefore had to be in reasonable condition when landed, so calling for a brailer to be used. In the Peruvian anchoveta and Scandinavian herring and capelin fisheries, on the other hand, the fish were only

175

needed for conversion into fishmeal. Because their condition did not matter, they were unceremoniously sucked from the ocean to be delivered into the holds of the vessel in a way that often reduced them almost to pulp.

A single invention was responsible for the explosive growth in purse seining for pelagic fish throughout the whole world. This was the hydraulic power block. The power block is simply a large hydraulically powered V-shaped sheave through which the entire net – weights, floats, and all – is hauled. Its use meant that bigger and bigger nets could be handled by the same number of crew members, with the result that otherwise uneconomical fisheries could be exploited for the first time and traditional pelagic fisheries could be exploited more efficiently. The power block started life in the California salmon fishery, was

Above: a harbor view of fishing vessels equipped with hydraulic power blocks (foreground). A simple enough device in itself – being merely a large V-shaped sheave hydraulically powered – it had far-reaching results. Because it enables huge nets to be handled easily without an increase in the size of the crew, it makes the exploitation of previously low-yield fisheries more economical.

quickly adopted by the tuna fleets, and then was taken up by Peru. Its use by that nation saw it rise from nothing to develop the world's biggest fishing industry in only 10 years. In Europe as well, catches of pelagic fish – particularly herring – soared when the power block/purse seine combination was introduced. In the 1960s, for example, Norway built over 600 purse seiners fitted with power blocks over a period of seven years. The biggest of these ships, about 160 feet long, were able to make two trips a week during the herring

Right: the bridge of a modern trawler showing its electronic equipment. Included is the ship's radar display and monitoring and control devices.

season, catching up to 400 tons of fish with only one set of the net and frequently landing up to 600 tons of herring per week.

Underwater sound detection has played an equal part in the enormous success of modern fishing methods of all types. We know that the echo sounder bounces sound back from the seabed so that the depth of water beneath the keel of a ship may be accurately determined. In addition, fish swimming above the seabed also show up on the recorder, so the conventional echo sounder has been adapted to make it a powerful fish-finding tool. Because different species of fish have different shapes and sizes of swim bladder, the echo returned and displayed from these air-filled sacs is individualistic enough to enable skilled fishermen to identify particular species merely from the smudge that shows on the recorder in the wheelhouse. Today's trawlers carry more electronic equipment than passenger liners. There are two echo sounders and ancillary devices to expand the scale of the bottom few feet of the water column so that a single fish in 2000 feet of water may be detected; there is a gyro compass, autopilot, radar, rudder angle indicator, a device which measures trawl warp tension, and winch control and monitoring equipment; there is engine monitoring and control equipment, and a room full of radio equipment to enable the skipper to talk not only to other ships in the vicinity but to the head office thousands of miles away.

On a bottom-fishing trawler the echo sounder searches a small area of sea and seabed immediately below the keel of the vessel. Sonar is an acoustic searchlight: its beam may be trained from straight ahead to directly beneath the ship and in a 360 degree arc in any of these positions. A typical fish-finding sonar can scan an area of 1000 million cubic yards of seawater in about two minutes.

For the purse seiner the sonar is an indispensable aid. As the ship approaches the fishing area a narrow beam of sound is projected through the water to the front and sides of the vessel over a distance of a mile or more. When it makes contact with a shoal, the ship homes in, keeping the fish on target while the net is prepared for launching. As the ship gets close, the sonar is switched to a wide angle mode to keep the fish in view. The shoal will show up on the graphic recorder in the wheelhouse in a characteristic shape that will enable the skipper to judge its size; the ping from the headphones on the set will also help him keep in contact. The net is cast and the boat makes a wide circular sweep to trap the shoal, which is being monitored the whole time by the sonar so that last-minute adjustments may be made to the set of the net.

Midwater trawlers also use sonar and echo sounders, and in even more sophisticated combinations. Again, sonar is used first to pick out a shoal of fish ahead of the vessel, the net is lowered, and the ship moves in. But to be able to tow the net right through the middle of the shoal, the skipper must know the depth both at which the fish are swimming and at which the net is being towed. So he uses the technique of "aimed" trawling. A transducer mounted to the top rope of the trawl bounces sound off the seabed to give an exact reading of the net's depth. The sonar tells him the depth of the shoal. By heaving in wire on the net or by increasing or decreasing speed, the skipper can aim the trawl straight at the thickest part of the shoal. Another acoustic device on the bottom rope of the trawl looks upward and bounces sound off the fish entering the net. In this way the skipper can judge when the trawl is full and may be hauled in.

Other equipment has been developed to take a large amount of the guesswork – some would call it intuition – out of fishing. Today's vessels have satellite navigation aids linked to autopilots so that the trawler can steer a course along a particular seabed contour on which fish are known to gather. More and more devices are being developed to monitor the performance of the "invisible" underwater net and computers take in statistics on weather, water temperature, salinity and other vessels' catches to calculate "best-bet" fishing areas for whole fleets of vessels.

The fishing industry is now dominated by technology. In many respects this technology is the key to increasing supplies of protein to a hungry world. In other respects it has spelled the ruin of many fisheries and threatened the extinction of whole species of fish.

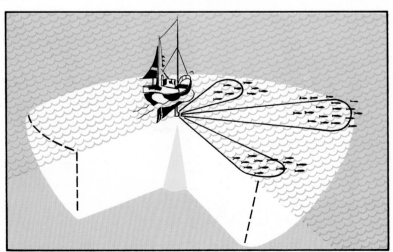

Left: this diagram compares the search area covered by an echo sounder (yellow) and sonar (white). Sonar coverage is wider because of a movable transducer that acts like a searchlight.

The Dangers of Overfishing

When European fishing vessels were able once again to venture into their traditional North Atlantic hunting grounds after World War II, the crews were amazed at the huge catches they made. Like arable land, the grounds had been "rested" for five years and so yielded bumper harvests. Since then there has been no rest for these grounds. Consequently, fisheries in the northern hemisphere, particularly in the northern Atlantic, are in a sorry plight. Catch quotas have been imposed as well as total bans on fishing for certain species in some areas, a good example being the North Sea herring fishery.

When the main pelagic fishing methods depended on nets and patience, fishing crews used their skill and experience to be in the right place at the right time. Advancing technology changed all that. Midwater trawling and purse seining, being active methods, enable a ship to search out and capture entire herring shoals at one time – perhaps 400 tons for half an hour's work.

Right: a purse seine full of herring being winched aboard a Norwegian vessel. Modern fishing techniques and technology have so increased catches of such very popular fish that these species are threatened with extinction.

Formerly, too, demersal fishing trawlers were limited as to how long they could stay on a fishing ground because preservation with ice gave fish a limited "shelf life." The longest a side trawler could stay away from Hull, England to operate in the Newfoundland Banks was about a month.

Quick freezing on the ships removed this constraint, allowing vessels to remain at sea until their holds were full. Fleet fishing, practiced mainly by East European nations, took this technique one stage further. A Soviet fleet arriving on the North Atlantic fishing grounds can consist of one or more mother ships of up to 40,000 tons. Each is equipped with a helicopter, club, library, theater, concert hall, hospital, and swimming pool as well as vast fish processing facilities: truly a floating factory. Mother ships are served by flotillas of catching and processing vessels: freezer trawlers which both catch and freeze fish; smaller catchers which take fresh fish to the freezer trawlers or mother ships for processing; and transport vessels which take frozen, canned, and smoked fish back to port. Some Soviet fleets stay away for up to four months, during which they can produce 10,000 tons of fish products, 1000 tons of fishmeal, 10 million cans of fish, and 100 tons of fish oil. A fleet like this can saturate an area. It can conduct a blitzkrieg on the fish stocks of a particular ground sweeping away everything – demersal, pelagic, and even shellfish – in its path.

The technological onslaught on the North Atlantic grounds from Spitzbergen to Newfoundland began in earnest in the 1960s by conventional side trawlers, individual stern freezer trawlers, midwater trawlers, purse seiners, and factory fleets. The North Atlantic grounds could not take it.

Each species of fish in a given area has what fisheries scientists call a maximum sustainable yield (MSY). This means that there is a finite amount of any species that can be taken from any area. The MSY point is reached when the total catch of the species from the area does not increase no matter how efficient the fishing methods are. In other words, more and more vessels uneconomically chase fewer and fewer fish. The fishing fleets waste money and the species is driven to near extinction.

Fisheries scientists have long spelled out this simple message but they were largely unheeded until the past few years. When the economic effects of overfishing became dramatically obvious to fishermen the world over, they then started paying attention to the scientists. They learned first that fish stocks could be preserved and a decent living made from the principal species if a minimum mesh size was established for trawls. This would insure that only older, mature fish that had spawned a few times would be caught, allowing younger, immature fish to slip through the trawl and live to propagate the species. Improved forecasting techniques also enabled the scientists to define the total number of fish of a particular species that could be taken from an area year after year with no adverse effect on the stock.

International fisheries commissions took the scientists' advice and laid down rules for minimum mesh sizes. In some countries these rules were strictly enforced: in Great Britain, for example, government inspectors are equipped with gauges to measure mesh sizes in trawls, and if these are undersize the skipper of the trawler is prosecuted and his catch and gear confiscated. Other nations have paid lip service to this sound conservation principle, applauding and adopting it in theory but proving remarkably lax in practice. In other words, they cheat, and with the advent of factory trawling it is difficult to furnish proof of this cheating. Undersized meshes can be incorporated into trawls and the immature fish that are caught can quickly be converted into fishmeal on board so that there is no evidence of illegality.

That there is cheating is evident from the declining catches made in the world's major fishing areas and from the ever-decreasing sizes of the most popular species such as cod and haddock. As it became obvious that control by mesh size was not working, scientists recommended the adoption of quota systems for nations fishing certain species in highly fished areas under self-regulatory schemes. Quota systems, however, can only work when all the nations fishing an area not only agree to the quotas but also follow the

Left: the Soviet factory ship *Vostock*. It carries 14 fishing boats, one of which is seen being lowered over the side. The boats bring their catch back to the factory ship, which can process as much as 300,000 tons of fish in one day. The *Vostock* can stay at sea for four months without any port calls, during which time it produces canned and frozen fish, fishmeal, and oil.

Right: herring – one of the most popular food fish of all times. Overfishing, especially in the North Sea, resulted in near extinction, and the reestablishment of the species is still an uncertainty.

guidelines set down under the agreement. It depends on international trust and goodwill since quotas are unenforceable in practical and political terms. Nobody is going to engage in all-out war over a haddock – although it has come close to that. If a factory fleet arrives on a multi-species fishing ground, who is to say that the fishermen are conscientiously fishing their quota of cod and ignoring the haddock quota they fulfilled on previous expeditions? Besides, what is the point of the United Nations organizing a Committee for East Atlantic Fisheries that includes 20 nations if the Soviet Union, which takes a quarter of the fish from that area, is not a signatory to any of the mesh size and quota agreements that the committee reaches?

We have to hope that in the end, economic and plain common sense, based on the MSY theory, must prevail. Even extremes of individual greed and the expediencies of politics ought to bow to the simple fact that it is senseless to continue to pour money and resources into the effort to catch overfished species.

There are now enough examples of the follies of overfishing to impress everyone with the need for common sense. The following set of statistics, from J Moller Christensen's book *Fishes of the British and*

Northern European Seas (1978), gives one alarming example.

The North Sea herring fishery was stable for centuries. The distinguished British fisheries scientist F D Ommanney, while sounding warnings about overfishing of demersal fish, said in 1950 that, "There seem to be such vast stocks of herring that there is no danger of any shortage arising from any man-made cause – at least in the near future." Dr Ommanney could not have been expected to foresee the introduction of the purse seine, power block, and sonar to the fishing industry just a few years later. At the time he made that statement in his book *The Ocean* (1949), the North Sea herring fishery was dependent on the passive driftnet methods of fishing. In the 1960s midwater trawling and purse seining eclipsed the drifters. The purse seiners sought herring not for the traditional purposes of selling the fish in fresh, smoked, or canned form, but in order to supply raw material for factories that would convert them to fishmeal for livestock feed. Technology paid off initially and the herring catch soared: in the Norwegian Sea it reached two million tons by the mid-1960s. Then it slumped. Within a few years it was down to a hundredth of this figure. There were, simply, no herring left.

This did not particularly matter to the fishing fleets. They were not fishing for one species over another to satisfy the palates of consumers. They were fishing for any species of a similar protein content to supply the fishmeal manufacturers, and they readily switched to mackerel and capelin. The annual North Sea mackerel catch rose from 100,000 tons to over one million tons in a few years. Then it dropped and the population was down to about 20 percent of its original size. Meantime the herring stocks had had a chance to revive and attention switched back to that resource. The stock was literally decimated: by 1975 the adult herring population was less than 10 percent of what it had been in the early 1960s. Again there was a switch to other species including capelin, Norway pout, and sprat. Stocks of these species had increased as herring and mackerel had decreased because there was more food to go around. However, if capelin, Norway pout, and sprat are in turn overfished, we cannot be certain that the herring and mackerel they have replaced will reestablish themselves. North Sea herring may have been exploited to the point of no return.

When we talk of switching the fisheries effort from one species to another, even to the extent of a total ban on catching a particular species, there is still no guarantee that the ban can be enforced. How could it? When a giant purse seine snakes around a school of, say, capelin, and the catch is pulverized into fishmeal, who is to say – or prove – that a proportion of that catch did not consist of small, immature herring?

If pelagic fish populations in the North Sea area continue to be decimated, stocks of bigger demersal fish will be affected in turn since these feed on smaller pelagic fish. Because the animal and plant plankton on which the pelagic fish and the larvae of demersal fish depend will then have fewer predators, they will multiply and survive in their billions. They will also compete with the ever-increasing amount of organic matter released from the land for the sea's supply of oxygen. The result will be the death of marine life. It would be too simplistic and sensational an argument to say that the introduction of the purse seine and sonar to the North Sea is going to reduce that area to a stinking desert, but we have many warning signs of such a catastrophe. There is probably plenty of time to take remedial action – if the warnings are heeded.

Above: a gauge for measuring mesh size. Fish conservation depends largely on regulating the size of the mesh of trawls. The mesh has to be large enough to allow the escape of undersize fish so that they can reach full maturity.

Left: loading mackerel in boxes. This fish is also an endangered species.

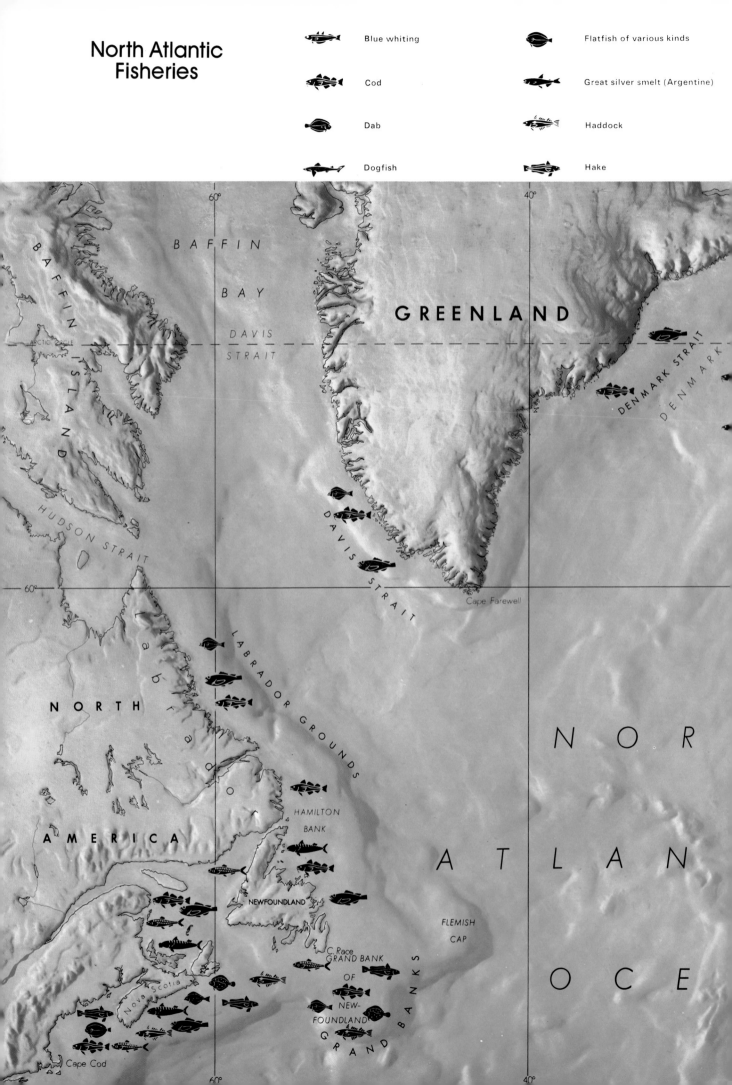

North Atlantic Fisheries

Blue whiting		Flatfish of various kinds	
Cod		Great silver smelt (Argentine)	
Dab		Haddock	
Dogfish		Hake	

BAFFIN

BAY

DAVIS

STRAIT

BAFFIN ISLAND

Arctic Circle

60°   40°

GREENLAND

DENMARK STRAIT

DENMARK

HUDSON STRAIT

DAVIS STRAIT

Cape Farewell

60°

NORTH

L a b r a d o r

LABRADOR GROUNDS

N O R

AMERICA

HAMILTON BANK

NEWFOUNDLAND

FLEMISH CAP

A T L A N

C. Race

GRAND BANK

OF

NEW-FOUNDLAND

GRAND BANKS

O C E

Scotia

Nova Scotia

Cape Cod

60°   40°

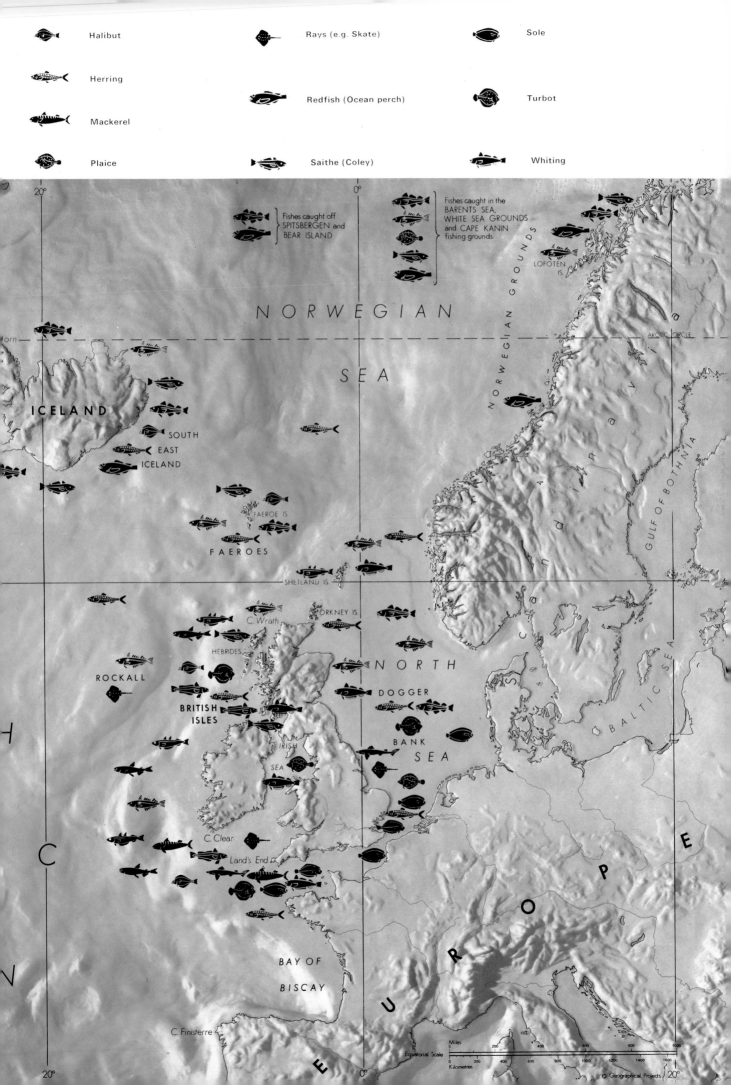

Halibut	Rays (e.g. Skate)	Sole
Herring	Redfish (Ocean perch)	Turbot
Mackerel		
Plaice	Saithe (Coley)	Whiting

Fishes caught off SPITSBERGEN and BEAR ISLAND

Fishes caught in the BARENTS SEA, WHITE SEA GROUNDS and CAPE KANIN fishing grounds.

NORWEGIAN GROUNDS

LOFOTEN IS.

NORWEGIAN

SEA

ARCTIC CIRCLE

Horn

ICELAND

SOUTH
EAST
ICELAND

FAEROE IS

FAEROES

SHETLAND IS.

ORKNEY IS.

C. Wrath

HEBRIDES

ROCKALL

BRITISH
ISLES

IRISH

SEA

N O R T H

DOGGER

BANK

SEA

SCANDINAVIA

GULF OF BOTHNIA

BALTIC SEA

EUROPE

C. Clear

Land's End

BAY OF

BISCAY

C. Finisterre

Miles
Equatorial Scale
Kilometres

© Geographical Projects

20°

The Way Ahead for Fisheries

How can the crisis of too many ships chasing too few fish be resolved? How can we reconcile the fact that some areas of the world ocean are being swiftly rendered into marine deserts with the fact that there is another 30 million tons of fish a year to be had without damaging the overall marine resource?

The answer lies mainly in changing the preferences of the consumer. In other words, we must learn to eat different fish. The term "consumer preference" is being used in a broad sense. In underdeveloped nations with sadly deficient supplies of protein, for example, the populations must be convinced to put to one side

Above and left: the octopus and the grenadier are two examples of edible marine life that could be exploited if the average consumer were educated to accept them. In fact, octopus is already part of the diet in some countries, including Spain, Italy, Japan, and Hawaii.

traditional dietary habits and customs in order to take advantage of the rich supplies of marine protein. In the developed nations the reluctance to change eating habits will also have to be overcome: witness the face of a Briton when first presented with a plate of lightly grilled octopus in Italy, or that of an American asked to try the East London delicacy of jellied eels from a roadside stall.

Such consumer reluctance is the chief reason for the gap between the present 70-million-ton world fish catch and the potential of 100 million. Even in the pillaged North Atlantic there are unexploited stocks. Trawler crews from Britain curse the redfish. In fact, on conventional side trawlers it used to be thrown over the side while on stern freezers it is ground into fishmeal. In Germany, on the other hand, an intensive advertising campaign persuaded consumers that this species is a delicacy. The campaign was such a success that many German trawlers went out to fish exclusively for redfish.

Fish like the blue whiting, the grenadier, the rabbit-fish, and the anglerfish are all present in quantity in the North Atlantic fishing grounds. Some of them, the grenadier for one, are difficult to process because they are so bony or have such strange shapes; but technology can solve this problem by the development of far more precise filleting machinery. Some, like the anglerfish, are particularly unappetizing to look at; but fish can be processed and served in many ways – in fish cake, paste, paté, and soup.

Most of these little-exploited species live in deep water, which is an additional incentive to their greater exploitation. Many nations, seeing the onslaught that has been made close to their shores by the ships of other nations, have sought to protect what they consider their resources by claiming exclusive fishing rights to distances as great as 200 miles from their coasts. They have done this to protect spawning grounds near their shores and to give their own small-boat fisheries a chance to survive. The legality of some of these claims has been challenged. Iceland's justification for its unilateral declaration of exclusive fishing areas was its massive economic dependence on the fishing industry. This action provoked two ugly "cod wars" with the United Kingdom, with gunboats from both sides playing highly dangerous cat and mouse games – Iceland seeking to arrest and bring to port British trawlers and Britain seeking to protect its disputed right to fish.

The claims made by Iceland to back its action provide a good example of how a fish stock can swiftly suffer under a sustained trawling attack. One of the danger signs of overfishing is a reduction in the average age of fish being caught. Fisheries scientists in Iceland showed that in 1938 cod older than 10 years formed 35 percent of the total Icelandic catch of that species. In 1948, after the grounds had been rested during World War II, that percentage leaped to 62 percent. Then came the big trawlers. By 1968 the proportion of cod older than 10 years had dropped to four percent, by 1971 to three percent, and by 1976 to two percent.

The imposition of exclusive fishing areas means that international fishing fleets have to work farther from shore. That means that they have to work deeper and to look for new species in new areas. Already fleets from Japan, the Soviet Union, Eastern Europe, and Spain roam the fishing banks off Argentina and South Africa and penetrate deep into the Pacific Arctic areas. Increasingly they are also turning south to Antarctica to catch the three-inch long planktonic crustacean that is the main diet of the great baleen whales – krill. With the slaughter of the whales it is estimated that, theoretically, there are now over 100 million tons of krill a year waiting to be taken. Already some 125,000 tons a year are being caught. Krill has a strong taste and flavor that would make it unacceptable in its natural state, but it is packed with protein and could be processed as a protein concentrate or turned into animal feed. Some nations, particularly the Soviet and Japan, are actively researching the harvesting of krill. There is some irony in the fact that it took the near extinction of the great whales to make this vast potential harvest available.

Left: an Icelandic gunboat warns off a British trawler during the 1958 "cod war" when Iceland imposed a 12-mile limit on fishing in the North Atlantic. Confrontations recurred in the 1970s after the imposition of a 100-mile limitation.

Peru's Anchoveta: an Object Lesson

If only one example has to be chosen to illustrate the way in which the too-rapid adoption of new technology can cause the rapid collapse of a thriving fishery, it is to be found in the extraordinary case of Peruvian anchoveta.

The waters off Peru teem with life. There are over 500 known species of fish living partly on each other but mainly on the fantastically rich concentrations of plankton along the Peruvian coast. An oceanographic phenomenon known as upwelling is responsible for this richness of marine life. The Humboldt, or Peru, Current flows northward along the coast of the country, but the southeast trade winds blowing along the northern and central parts of the coast cause a transport of surface water away from the coast. It is replaced by cool deep water which rises to the surface and which is rich in nutrient salts such as nitrogen, nitrate, phosphate, and silicate. The concentrations of some of these minerals off Peru are twice the average of those found in similar latitudes.

The combination of sun from above and nutrient salts from below is one on which phytoplankton – the plants of the ocean – thrive. Some 99 percent of the diet of the herringlike anchoveta is made up of phytoplankton, so these small fish also thrive in their billions. Until the mid-1950s the single main predator of the anchoveta was the seabird population. They consumed an amazing three million tons of the fish a year. Other fish, squid, and mammals accounted for another 3.5 million tons. The local fishing fleet in 1955 took a mere 59,000 tons although this does not represent the total catch, because before 1955 there had been far easier ways to reap the marine harvest without even going to sea. There are three main species of birds feeding almost exclusively on anchoveta, the best-known being the guanay. The 40 million or so of these birds produce an incredible amount of excrement, which piles high on nesting sites on the coast. Known as guano, it is formed from the nutrient-rich anchoveta on which the birds feed. It is harvested in the hundreds of thousands of tons for use as fertilizer.

In the mid-1950s came the introduction of the power block to California fisheries. The new invention was quickly taken up by the Peruvian anchoveta fishermen because it enabled them to use big purse seines.

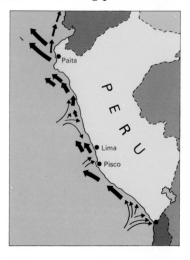

Above: a map of the currents off Peru. The thick arrows indicate the cold upwelling water of the Peru coastal current, and the thin arrows indicate the warmer water from the open ocean.

Left: anchoveta, once the pride of Peru's fishing industry. Overfishing brought a collapse of the formerly teeming fishery.

Above: a fishmeal factory near Pisco, Peru. The whole anchoveta catch was processed into fishmeal for export.

The industry's growth, thanks to the power block, was nothing short of explosive. From a 59,000 ton catch in 1955 the exploitation grew to nearly nine million tons by 1964, to 10 million tons by 1967, and to nearly 12 million tons in the early 1970s. Almost over-night the world's biggest fishery in terms of weight of catch of a single species had sprung up. It made Peru the world's leading fishing nation in terms of weight of catch of all species, outstripping even Japan which had for many years been the principal nation. At one stage, over 2000 boats of between 65 and 95 feet in length were engaged in anchoveta fishing, nearly all of them equipped with power blocks and sonar.

The whole catch was processed into fishmeal, nearly all of it for export to Europe and the United States. Peru is in an extremely favorable position to manu-facture fishmeal, its main manufacturing centers being in areas of little rainfall so that the plants can stand in the open air and the millions of sacks piled up without the need for permanent roofing. Costs of processing are therefore low.

Heavy dependence on a single species of fish carries many risks, and the Peruvian anchoveta fishery is no exception to this rule. The whole ecological system off Peru depends on the upwelling effect caused by the Peru Current being blown offshore, but this pheno-menon is not completely reliable. In some years a warm tropical current, which usually flows south only as far as Ecuador, pushes farther south to Peru where it

tends to blanket the cold water of the Peru Current. When this movement coincides with abnormally weak winds along the Peruvian coast, the Peru Current is not pushed out to sea and little upwelling occurs in a phenomenon known as El Niño. Starved of nutrient salts, the phytoplankton does not reproduce in the usual numbers. Countless millions of anchoveta starve to death as a result, as do the fledgling birds that depend on them. El Niños have occurred in-frequently – in 1891, 1941, 1953, 1958, and 1972. Up until 1972 the anchoveta stood a good chance of sur-viving these occasional years of misfortune and the stock quickly reestablished itself. But the additional toll exacted on the stocks of anchoveta by having 10 million tons removed by fishing vessels in addition to the natural mortality inflicted by birds and other marine animals proved too much for them to bear. The 1972 El Niño struck a heavy blow at the heart of the anchoveta fishery and today the annual catch has been reduced to around two million tons a year. Many of the ships equipped to fish for anchoveta have been converted to fish for other species and there is every indication that Peru is seeking to establish a more well-rounded fishing effort, less dependent on a single species.

An important lesson to be learned from the Peruvian experience – aside from the perils of overfishing one popular species – is that the upwelling phenomenon is almost certainly not unique. Research in other tropi-cal areas, notably the Indian Ocean, could reveal huge stocks of marine protein in the upwelling – and perhaps near the nations of greatest need.

187

Prospects for Fish Farming

Despite the mass of technology that is now available to the average skipper, fishing is still a remarkably crude way of obtaining protein when compared to other food harvesting techniques. Electronic and mechanical aids may make the job easier, but fishing as such is like obtaining meat by going out with a gun to track and shoot wild animals over whose movements and breeding habits we have no control at all.

It would therefore seem to be a logical step for the fishing industry to move in the same direction as agriculture in developing a high degree of control over stock and environment, with the added ability of adapting the end product by selective breeding and genetic control to meet market demands. Fish farming has a very long way to go before it can approach the level of sophistication represented by factory farming of poultry and cattle. It is unlikely that even then fish farming will ever reach the stage, as some have suggested, in which vast amounts of marine protein can be intensively farmed to feed a hungry world. The problems of production on the scale required to reach that goal are just too great. At best, fish farming will

iron out some of the peaks and troughs of naturally fluctuating supplies. It could never be a universal solution to protein deficiency.

Freshwater fish farming, which is easy, has been practiced for centuries. Fish such as trout, carp, and catfish are simply kept in fresh water, their basically natural environment, protected from predators, and given an adequate supply of food. Marine fish farming, however, presents a more complex set of problems. To rear marine fish it is necessary to create a replica of the animal's natural environment – open sea conditions – and to maintain this environment against pollution, excessive rainfall, wind, and heat. The creation of this environment is costly, which is why most marine fish farming so far has concentrated on luxury foods such as crabs, oysters, and tuna. Most marine farming today is only a holding operation.

Above: a rare example of successful commercial farming of algae for use as feed at an oyster farm. The water is artificially circulated to remove waste gases and to distribute the nutrients.

Left: a Japanese shrimp (or prawn) farm. During this stage of farming, the shrimp are kept in open enclosures protected from predatory eels by a surrounding net.

Animals such as crab and lobster are caught in the normal way and merely held captive until market conditions are most favorable for their sale. A slightly more advanced method involves taking the young hatched from the eggs of captured adult fish, putting them in the sea in cages, and letting them grow to maturity under reasonably controlled conditions. But there is no real control of the life cycle of the fish. It is not possible to select individuals with a fast growth rate, or those that eat little, or those that have a resistance to disease. Nor is there any way in which fish can be selectively bred so that they eventually inherit these favorable characteristics.

Nevertheless, much success has been obtained in the farming of shellfish, particularly in the Far East. For example, female prawns caught at sea shortly before they spawn are placed in tanks where each releases over half a million eggs. The larval prawns are fed phytoplankton and larval brine shrimps until they reach the post-larval stage. Then they are placed in open ponds where they reach market size within about nine months after being fed on a diet of minced shrimp, fish, and clams. Survival rates of up to 60 percent have been achieved by this method.

Mollusks also lend themselves well to cultivation since they are nonmigratory. Mussels, for example, are grown by suspending them on strings from a surface raft so that they are able to take advantage of the higher density of plankton near the surface. Oysters are hatched and planted in a similar way and left to mature on slats of wood.

The great challenge is to produce marine fish in quantity under controlled conditions. Experiments have shown that it is extremely difficult to obtain this degree of control when the fish are kept in vast ponds: they are too prone to the effects of pollution and changes in heat and salinity. The most successful method has been to rear the fish from eggs in nursery cages and move them, once past the larval stage, to floating or submerged net cages where they can be reared to market size. These techniques have been used successfully, on an experimental basis, in Britain, with sole, plaice, and turbot. The experiments have raised as many problems as they have answered, however. The food required by the young fish is so expensive that attempts have had to be made to wean them onto cheap products of the offal type. Greater numbers of fish have to be packed into each enclosure to reduce capital and maintenance costs, but by packing more into the cubic foot, they are far more liable to be wiped out by disease.

Despite all these problems, the work is still going on. There may never be hundreds of acres of coastline devoted to producing millions of fish a year, but the next few years could see the development of an industry something like factory poultry farming. Fish farmers will probably buy basic stock – the young larval or post-larval fish that are the equivalent to the day-old chick – and equipment from a central supplier. They will then rear the fish while being supplied with food, antibiotics, and farming advice from the same centralized organization.

Left: trout that are a product of farming. Fish farming of freshwater fish such as trout has a longer, and so far, more successful history than saltwater fish farming.

Below: floating nursery cages used in farming saltwater fish in Scotland, a technique developed around 1965. Plaice can be grown from the egg to about 10 inches in just over two years.

189

The Ocean as a
New Drug Source

The living resources of the sea that are of use to mankind are not limited to the flesh of such animals as fish and shellfish. The sea's plant life is another rich resource. In the Far East, particularly in Japan where anything the sea produces is looked on as a potential resource to make up for the severe lack of agricultural land, seaweed is an important part of the average diet and certain species are regarded as delicacies. In addition to extensive harvesting of the naturally growing crop, the Japanese cultivate some species of red seaweed. They collect the spores in special nets and grow them on buoyed lines in the sea.

Red seaweed yields the important substance known as agar. This is a bleached extract practically insoluble in cold water but easily soluble in boiling water; it sets to form a clear gel. The gel is used to give texture to a wide range of manufactured foods including canned meat, cake icing, sweets, and pet food. Another use is as a coating for pills because of its property of being able to pass through the stomach without being digested. Perhaps the most important use of agar is in biological and medical laboratories where it is used in culture media on which bacteria, molds, and isolated tissues are grown.

Brown seaweed yields alginic acid from which a range of alginates is derived. The uses of alginates are many: as a stabilizer in ice cream; to add body to jellies and pie fillings; to give longer shelf life to salad dressings; to stabilize the foam in beer; to thicken

shampoos; to provide surface coating for paper; and to thicken fabric dyes, latex rubber, and paints. Carrageenin is another seaweed substance used in a large range of foods, drinks, and manufacturing processes. It is from a red seaweed known as Irish Moss, although it is harvested throughout the North Atlantic and from as far away as the Philippines.

The flora and fauna of the sea are rich not only in the protein resources they offer. They also represent a huge resource of natural drugs, the potential of which is now beginning to be tapped. There are many hundreds of marine organisms from microscopic plankton to giant rays capable of harming people with their stings, by their venom-laden bites, or by the poisons they can deliver when they are eaten. The notorious puffer or blowfish is an example. Certain parts of the flesh of this fish contains the poison tetrodotoxin, and it can kill anyone who eats it in only 30 minutes.

Although poisonous itself, tetrodotoxin and many

Above: partially exposed kelp, or brown seaweed, at low tide on the southwest coast of England. This kelp yields alginic acid, which has wide commercial applications in such industries as pharmaceuticals, and cosmetics.

Above: a tropical sea cucumber, one of some 1000 species. Research has shown that several species produce a substance which has successfully brought about regression of tumours in mice.

Left: cultivation of laver, a marine red algae, in Japan. Seaweed is a common component of the Japanese diet and its cultivation is a major food industry. Red seaweed is also eaten by the Welsh, who call it "laver bread."

of the other toxins found in marine organisms have potential benefits. Tetrodotoxin has local anesthetic properties many thousands of times more powerful than cocaine. It cannot be used directly as a local anesthetic because it spreads through the tissues, but studies of its molecular structure and other properties are giving chemists and pharmacologists new insights into biological processes that could lead to the synthesizing of new drugs.

The search also continues for new "natural" drugs from marine organisms. In the late 1960s, in the general burst of interest in marine work, there was vast enthusiasm at the prospect of producing many new, life-saving drugs from the sea. Researchers pointed eagerly to the properties of such substances as holothurin, which when isolated from the toxins produced by certain species of sea cucumber caused the regression of tumors in mice. Could this, they asked, be a cancer cure? Among other marine biotoxins that were identified are ones blocking conduction in nerves; causing a sustained contraction of muscles or alternatively relaxing them; and affecting the behavior of the cardiovascular system. Hundreds of marine bacteria, plankton, sponges, plants, shellfish, and fish contain substances with marked antibiotic properties. Marine organisms, it was suggested, could yield more drugs than the land-based plants and animals that currently yield half our pharmaceuticals, the rest being synthetic.

The enthusiasm was understandable but somewhat misplaced. The drugs industry prefers to work with synthetic materials because they are cheaper to mass produce and easier to patent. The research, development, and clinical trial program needed to develop a new drug is lengthy – typically 10 years – and costly – in the millions. Clearly the pharmaceutical industry is not going to rush out to sea to develop hundreds of new drugs given such costs and time scale. Development will probably be slow. But it still holds tremendous promise.

Drinking Water from Seawater?

Over the next 30 years people will use as much fresh water as the world has in all the previous years of human existence combined: for domestic use and drinking, for industrial processes, and for irrigation. Augmentation of the natural supplies that fall on the land will therefore be much needed, and it seems a logical step to turn to the vast resources offered by the sea, because the oceans contain 98.5 percent of all the water on Earth.

At first look it would seem to be an easy process to separate the minerals that make seawater unpalatable in order to produce fresh water. Just allow the seawater to evaporate and collect the condensed vapor. But desalination of this type, even when the sun's rays are directed through glass to give them more power, is extremely inefficient. Using the same analogy that described the ocean's affect on weather, it is like trying to boil a pan of water by applying heat from the top rather than the bottom. The amounts of water obtained by this form of desalination simply do not make the operation worthwhile, and methods to increase the amount are expensive.

So far, no process of desalination has been developed for areas of reasonable rainfall, such as northern Europe, that could produce fresh water more economically than if the same investment were put into more efficient utilization of existing natural supplies. Therefore, big desalination plants are used only under certain circumstances. One is in areas where there is a high seasonal demand, such as island tourist resorts, and where the cost of producing the water can in-

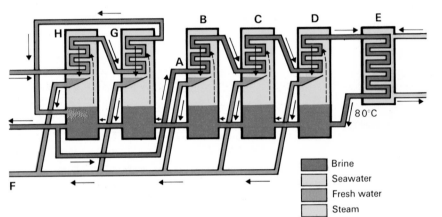

Brine
Seawater
Fresh water
Steam

Left: a diagram of the multistage flash distillation process of desalination. Brine at (A) passes under pressure in the condenser coils of flash chambers (B) (C) (D), to heat exchanger (E). As it flows in reverse, water vapor flashes off and is condensed on the cooler brine-filled coils above. This condensate forms part of the freshwater outflow at (F). The brine passes into flash chambers (G), (H), which contain condenser coils fed with raw seawater. This is recycled into the concentrated brine of the last flash chamber. The liquid that results is partly run off as waste and partly recycled to (A).

Right: a desalination plant in al-Khobar, Saudi Arabia. Desalinated water is still relatively costly to produce, but given the importance of drinking water and the aridity of the Middle East, it is not surprising to find substantial investment in desalinating the available seawater.

Above: an icebreaker near a huge iceberg in the Arctic. The frozen water that makes up an iceberg contains no salt. This has given rise to the idea of using icebergs to increase the water supply by towing them to the populated areas that need more water.

Left: the ocean provides an ample supply of bromine, which is often extracted by the chemical replacement method. Bromine, which is highly corrosive, is widely used for antiknock gasoline.

directly be added to the customers' bills. Another is when high initial outlay and running costs can be easily paid for, such as in the oil-rich nations of the Middle East.

There are several different methods for the desalination of water. The process selected depends on the amount of fresh water required, the location of the plant, and the use for which the water is intended. The most favored production for large scale production of drinking water is multistage flash distillation.

Of the 1.5 percent of the water on Earth existing in a fresh state, some 75 percent of it is locked up in the icecaps. Some 90 percent of it is Antarctic ice; as much fresh water drifts out into the oceans in the form of Antarctic icebergs each year as falls on the United States in rain.

Although it sounds like a joke, it has been suggested that the fresh water in the icebergs could be tapped by lassoing them, towing them to wherever the water is needed, and pumping the ice ashore in a slurry. In fact, serious studies are being conducted into the feasibility of towing icebergs to California and the Middle East, possibly with several two-mile-square table-top icebergs linked together.

All the known minerals exist in seawater. Each cubic mile of seawater contains 166 million tons of dissolved salts worth thousands of millions in money. Getting them out is a problem, however. Scoop up a bucketful of seawater anywhere and there will be gold in it, but only one ten-millionth part of that bucketful will consist of the precious metal. The costs of ex-

tracting it would be far greater than its value.

Chemicals extracted from seawater are far more mundane than gold but are present in far greater concentrations. Salt or sodium chloride is, next to water, the most abundant chemical in natural form on Earth. Around 30 percent of the world supply is extracted from seawater, either by direct evaporation in shallow ponds, by freezing seawater and processing the remaining brine, or by taking brine that has been partly evaporated by the sun and finishing it off in heated units.

The chemical element most confined to seawater is bromine, over 99 percent of it being in the oceans. It is extracted by a process of evaporation, precipitation, and chemical reaction, with the end product used as a gasoline additive and in dyes, medicine, and metallurgy. Over half the world's output of magnesium also comes from seawater. Magnesium is used for producing light alloys – especially useful in the aerospace industry – in pharmaceuticals, in fertilizers, and in synthetic fibers.

Other minerals will be extracted from seawater as more efficient processes are developed, particularly if lower-cost energy resources are developed to make the extraction process more economical. Research continues into the extraction of uranium, tin, and titanium. The increasing scarcity of these and other commodities and a corresponding improvement in the technology available to extract them will undoubtedly insure a steady development of this form of harvesting the ocean's wealth of resources.

Mining the Seabed

In 1964 John L Mero's book *The Mineral Resources of the Sea* was published. It was a slim, learned volume based on the research Mero had undertaken into the extraction of chemicals from seawater and the harvesting of beach minerals, continental shelf deposits, and the mineral resources of the deep ocean. This last subject occupied over half the book and was devoted to the most comprehensive review published up to that time of the distribution, quantities, mineral content, and likely harvesting technology and mining economics of a certain mineral that lies on the floors of the Atlantic, Pacific, and Indian Oceans in handy lumps. The lumps are called manganese nodules. Mero's dissertation was seized upon eagerly by academics, industrialists, and politicians alike. It was like an answer to a burning political question that had arisen the year before.

In 1963 the United States Navy nuclear submarine *Thresher* had made a test dive to about 1500 feet in the Atlantic following a refit. Something went wrong and the submarine sank to a depth of over 8000 feet with the loss of the whole crew. The disaster shook the American public, especially when it became apparent that even if the submarine had sunk to a mere 700 feet there would have been no system in operation capable of rescuing the crew.

A vast gap in the American technological armory had been revealed, and the shock was akin to that felt when the Soviet Union had launched Sputnik a few years before. Politicians wanted to know how this gap had been allowed to open up; militarists and industrialists set about finding ways to plug it. Public interest in the oceans had been awakened, and it was heightened more when Mero's book, popularized at conferences and by the media, showed that development of ocean technology could not only save lives but also make money.

Interest quickly grew in everything remotely connected with the sea's resources. For a while, everyone climbed on the bandwagon: aerospace companies built little yellow submarines, banks proposed vast fish farms, architects designed underwater cities.

For many enthusiasts, amateur and professional, it was their first encounter with the marine environment and many of the drawing board dreams faded under the harsh reality of one of the basic questions about making things work at sea: How do you stop it –

Above: manganese nodules on the deck of a mining ship. Modern oceanography has succeeded in surveying and plotting the biggest nodule fields, and modern technology is providing the means of extracting the minerals from the sea.

Left: a field of manganese nodules on the ocean floor, photographed by a deepsea camera. Although named for the main mineral found in them, manganese nodules also contain copper, nickel, and cobalt.

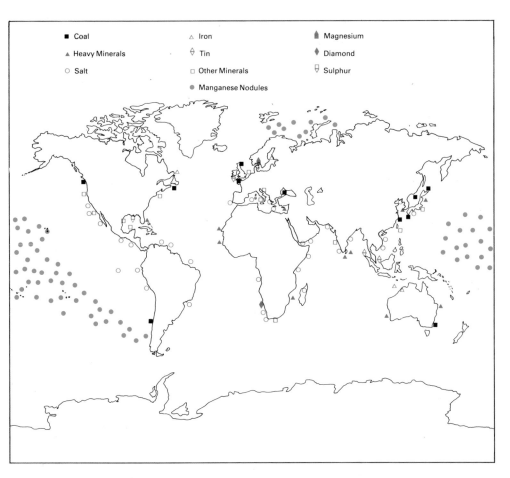

Left: this map shows how minerals are distributed in the world's oceans. The important question about marine minerals is how to get at them rather than where to find them. The technological problems in exploiting the vast potential lying on the deep-ocean floor have yet to be solved.

Map legend:
- ■ Coal
- ▲ Heavy Minerals
- ○ Salt
- △ Iron
- ⬡ Tin
- ▢ Other Minerals
- ⬢ Magnesium
- ◆ Diamond
- ⬗ Sulphur
- ● Manganese Nodules

whatever "it" happens to be – from being battered, crushed, buried, corroded, or lost? Some of the dreams survived. Deep sea mining of manganese nodules was one of them.

Manganese nodules are roundish objects varying in size from a small pebble to a huge rock weighing a ton. They are formed from substances precipitated from seawater gathering around a nucleus such as a shark's tooth, a small piece of fish bone, or volcanic lava on the seabed. Their presence on the deep ocean floor has been known for some time: HMS *Challenger* dredged them up in the Pacific during circumnavigation 100 years ago. Two things about Mero's book excited the business world: it showed that the mineral nodules are scattered in their thousands of millions over the deep ocean floor, and that they contain enough metals such as manganese, nickel, cobalt, and copper to keep industry supplied for hundreds of years. Moreover, the nodules are growing. They represent a mineral farm, with each pebble increasing in size a few thousandths of an inch every 100 years. Tiny though this growth may seem, when multiplied over the vast numbers of nodules, it represents a faster accumulation of certain metals than present world consumption.

Industry, in the shape of mining, oil exploration, and dredging companies, began first to test the feasibility of assessing the extent of the resource, then of developing methods to bring the nodules to the surface, and finally of finding ways of processing them. The companies involved have made quiet but remarkable progress in overcoming problems that are as great, if not greater, than those that confronted the administrators of manned spaceflight.

First came the assessment of a resource that lies under three miles of ocean as much as 2000 miles from shore. Under tremendous odds, surveys were made using television and still cameras at known nodule sites to identify deposits. Samples obtained by grabs showed that the mineral constituents of the nodules varied over short distances. This was unfortunate because it was important to any future harvesting program that metal content from one recovered batch to another be relatively similar in order not to complicate an already complex processing procedure. Therefore, areas of nodules with the right proportions of metals had to be pinpointed carefully. Even for these first reconnaissance studies, grabs and corers had to be developed as well as advanced underwater TV, movie, and still photographic techniques.

Having pinpointed mining sites and the area of ocean floor each covered, bulk recovery methods had to be developed. To put the scale of this problem in perspective, note that the most advanced seabed mineral recovery system in operation until the advent of nodule mining operated at a maximum depth of 250 feet. The nodules are at depths as great as 18,000 feet.

Work has now progressed to the stage where pilot harvesters have been developed and tried. Mostly these rely on hydraulic power or airlift principles to suck the nodules through a pipe to the surface – like giant vacuum cleaners. In their full-scale versions

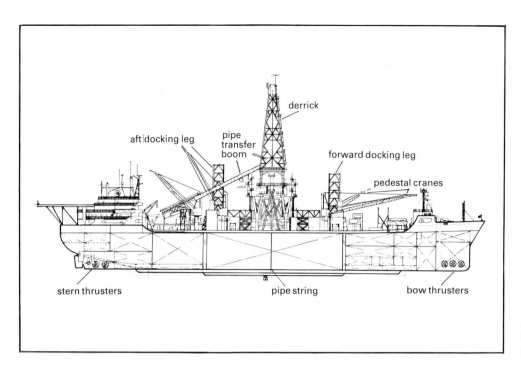

In the diagram, labels read: derrick; aft docking leg; pipe transfer boom; forward docking leg; pedestal cranes; stern thrusters; pipe string; bow thrusters.

these harvesters will be incredible machines. To justify the investment in ships, people, transportation, and processing they will have to pump perhaps 10,000 tons of material a day. The handling ship will have to be carefully maneuvered to avoid stressing a three-mile long pipe in adverse weather conditions, and it will have to work within fine navigational tolerances.

The problems will not end when the nodules are brought to the surface. They will have to be transferred continuously to giant barges which will ferry them to shore. Any ship-to-ship transfer system is a tricky operation. In such exposed locations and with the need for one of the vessels to drag tons of equipment behind it, all the usual problems will be considerably intensified. The nodules will have to be transported from barge to process plant in a slurry and then processed. New metallurgical processes have had to be developed to cope with the unique properties of the basic ore. A pilot plant that processes only 800 pounds of nodules a day has been built by the leading company of one of the United States' industrial consortia investigating nodule mining. This plant has demonstrated a breakthrough in nodule processing which, according to the management, has had a marked impact on the whole economic viability of the company's mining program.

There are four international consortia set up to investigate, conduct feasibility studies, and then engage in full-scale mining of manganese nodules. All now seem confident that the technical problems of finding, harvesting, and processing these strange lumps of ore into sheet metal can be overcome. But one major problem remains, and on this the fate of deep ocean mining hangs – for unless it is resolved the consortia will simply abandon all their expensive feasibility studies and call their pilot harvesters back to port. The problem is simply that of who owns the nodules. Internationally accepted legal controls for the ownership of the resources of the continental

shelf are already established, but the question of ownership of deep ocean floor resources was largely academic until the appearance of the would-be nodule miners. This has precipitated an angry debate at the level of international politics. It is a problem that the nodule miners, having scaled most of the technical heights, are largely powerless to overcome. The future of deep sea mining now rests in the hands of the politicians.

Managanese nodules are not the only deep ocean mineral resource that could soon be coming to the surface. Scientists on the United States oceanographic research ship *Atlantis II*, working in the Red Sea in 1963, expected to find that seabed temperatures at depths of 7000 feet would be close to those of the surface water which is 70°F. They were surprised when thermometers constantly gave readings of over 130°F. More research in this area of seabed volcanic activity revealed a hot salty brine on the seabed overlying thick layers of warm mud. These layers are extraordinarily rich in minerals. The Atlantis Deep, named after the ship that discovered it, has mud deposits between six and 90 feet thick extending over an area of 23 square miles. The mud contains 30 million tons of iron, 2.5 million tons of zinc, half a million tons of copper, and 9000 tons of silver. It is a mineral treasure trove, worth at least $2 billion.

Needless to say, the ocean miners have set their sights on this treasure. A West German consortium has built a pilot harvesting device, which is in effect a giant suction pump with a vibrating sieve. High pressure water jets loosen and fluidize the seabed soil, which is then sucked through the sieve and up a pipe to the surface. Much work needs to be done before a full-scale machine is built but there is every indication from the effort and money now being expended that the Red Sea treasure trove will soon be exploited.

Given these technologically advanced schemes to mine the very depths of the oceans, there is perhaps a tendency to overlook existing marine mining industries. There are a number of flourishing, profitable marine harvesting operations requiring considerable technical and operational expertise. Perhaps the ordinary nature of the product produced in most volume by these methods contributes to its lack of appeal, the most heavily extracted seabed deposit being a sand and gravel called aggregate. Aggregates are needed for making cement, so market forces have given this industry its greatest success in Europe, chiefly in Britain, and the United States. Building activity naturally flourishes in areas where land

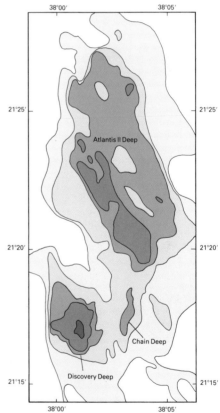

Above: a contour map of the Red Sea showing the areas in which brine has accumulated. Different depths are indicated by shaded colors: light green is 2200 yards deep; medium green, 2300 yards; and dark green, 2500 yards.

Left: one of the latest designs in a cutter suction dredger. Seen is the cutter head as viewed from the ladder gantry. A dredger of this type deploys a special cutting array to slice through seabed material.

197

values are high, making it pointless to use vast areas of that valuable land in digging for sand and gravel. Because the greatest single cost in the production of sand and gravel for cement making is transportation, the costs of transporting material by road from a distant land-based site to the center of demand have to be weighed carefully against those of dredging the material from the seabed close to the building areas. The quality of the deposit also plays a part in these calculations. These economic formulas have been carefully worked out, especially in Britain, and have led to the development of one of those oceanic rarities: a marine industry capable of producing a resource at less cost than its land-based equivalent.

Offshore sand and gravel deposits are exploited in fairly shallow water, 200 feet at most. Deposits are first delineated by surveying them with echo sounders and sub-bottom profilers. Detailed studies are then made to decide the exact length, breadth, and thickness of a deposit. Samples are obtained, usually by employing a seabed-mounted device called a vibrocorer. Contrarotating electrical motors powered from the surface drive a tube up to 20 feet into the seabed where it traps a core of the deposit and brings it to the surface for evaluation.

Once the deposit has been fully analyzed, the dredger moves in. A typical ship trails a pipe from one side to suck up sand and gravel, which passes through vibrating screens that separate the cargo into all-sand or all-gravel; some is left as a mixture. The ship carries up to 4000 tons of aggregate, sucking it aboard at a rate of 1800 tons an hour.

Dredging techniques of this type have developed from those used to provide new ports and harbors, to deepen and maintain shipping channels, and to provide material for land reclamation. Various types of dredgers are used. The bucket dredger has an endless chain of steel buckets, each of which scoops out underwater material and delivers it back to the vessel. The material is transferred to barges which dump it at sea or, in the case of land reclamation, take it ashore. The cutter suction dredger fixes itself to the seabed by piles and then deploys an array of steel knives on a rotating head which slices through seabed material; this is then sucked on board for pumping onto barges, or transported directly ashore by pipeline.

Bucket dredgers, cutter suction dredgers, and trailing suction dredgers have been instrumental in the success of the massive land reclamation projects undertaken in the Netherlands. They are also used to harvest a wide variety of offshore mineral deposits in addition to sand and gravel. These minerals are usually found close to shore and are mainly "placer" deposits – concentrations of heavy minerals formed where streams, waves, and currents have deposited them and washed away lighter materials. Typical heavy minerals are tin, gold, diamonds, and rutile. These have been exploited with varying degrees of success in several parts of the world.

The living and mineral resources of the sea, while not bottomless, exist in comfortable enough quantities to meet many modern needs. Sensible exploitation can guarantee essential supplies of food, drugs, and minerals for many years to come.

Opposite: sand that has been processed to be free of salt. Desalination of sea sand is necessary for some uses, as in land reclamation.

Right: a trailer suction dredger in operation. This type of dredger sucks up sand and gravel through a pipe trailed from one side.

Chapter 8

Energy from the Sea

Sudden and sharp rises in oil prices have shaken the world economy, but they have had at least one beneficial effect. They have made it worthwhile to reexamine many alternatives to fossil fuels for obtaining supplies of energy. This has led politicians, scientists, and engineers to investigate ideas that have been with us for some time but that now make more economic sense in the light of the spiraling costs of energy.

One of these ideas is to look to the enormous potential of the oceans, which are a source of self-renewing and environmentally clean energy. Many of the ocean's energy resources will take some time to tap efficiently, but in the end can provide a large part of our overall energy requirements. Such sources include the generation of power from waves, tides, winds, and water temperature differences.

Opposite: no matter how seemingly calm and peaceful, the ocean is in constant motion. Its movements – of currents, waves, and tides – are a potential source of energy.

Clean Power from the Ocean Waves

For thousands of years people have stood on beaches and clifftops and marveled at the power of the waves as they crash in from the ocean. Waves represent the stored energy of the winds. The energy contained in just three feet of the average Atlantic wave is around 70 kilowatts (kW) – the equivalent of the power needed by 70 electric heaters.

The first suggestions that wave power could be harnessed were made over a century ago. Now the need

wave power collection structures, anchoring, and transmission.

The first line of research – the gathering and analysis of data – shows just how far we are from the effective utilization of wave power. Only in the past 40 years has anyone looked at the physical properties of waves, and only very recently has anyone examined the total energy that waves carry and the peak power that they can deliver.

Gathering and analysis of wave data is, of course, essential for any wave power program. Waves do not move in a constant direction, nor do they have the constant height and timing that would guarantee a steady output of power. Computer programs have now been developed that can analyze the energy content of waves in a certain area over a certain period to predict the total power deliverable in that area over a year. Far more basic research of this type

Left: one of the big attractions of harnessing ocean waves to generate power is that it is clean energy. Much research is now going into ways of utilizing wave power.

for alternative energy sources has prompted the initiation of research programs in Japan, the Netherlands, Norway, Sweden, Britain, and the United States. Britain has the most intensive of these research programs, with a multimillion pound budget and 150 full-time employees working on various wave power projects.

Initial research has probably produced more problems to be investigated than it has solved to bring the prospect of harnessing wave power closer. But the intensiveness of the research now being conducted makes the successful use of wave power a question of "when" rather than "if."

Wave power research is being carried out on a number of fronts including the collection and analysis of wave data at likely harnessing sites, the design of

needs to be undertaken to decide the best sites for wave energy plants and to insure that they are of a design best suited to take maximum advantage of local conditions.

However, some sites have already been singled out as being suitable for the installation of power generating plants. One plant off northwest Scotland could generate 12kW for every yard of its length, another off southwest England some 8kW per yard, and a third in the North Sea around 5kW per yard. Other prime European sites that might be appropriate to development are the Bay of Biscay and the area off the coast of Ireland.

The attraction of these areas is that a significant amount of wave energy is generated nearly all year round, and that the waves are of a height most suited to machines now being developed. Wave height is important because a delicate balance has to be struck between the size of the waves that are harvested and the size, and therefore the strength, of the machine that will harvest them. For example, it is not necessarily most efficient to moor a machine in very deep water far from shore so that it can harness the enormous power generated by big sea storms. The structure and its moorings would have to be massive to withstand the continual pounding. This would mean that the costs of maintenance and of transmitting the power to shore would probably outweigh the value of the year-round wave power. The ideal sites are those in which there are steady and consistent waves. While the wave plant would have to be capable of withstanding violent storms, it would not depend on such storms to generate steady power. Instead it would rely on the day-to-day "average" waves.

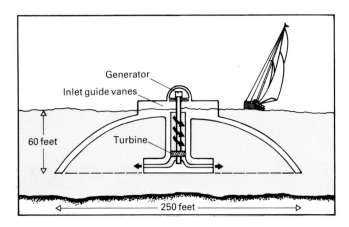

Above and below: a diagram and an artist's impression of a recently developed machine that generates power from ocean waves. It is designed to function like an atoll does in nature, bringing the waves into the center in spirals. A turbine wheel draws the energy continuously.

The plant would consist of a number of linked power-generating cells that could in some cases form a string of rafts up to half a mile long, moored in around 100 feet of water up to 10 miles from the coast. This distance from shore is necessary to get where the pattern of waves is undisturbed by headlands.

With these very basic guidelines at their disposal, designers have done their work. Hundreds of patents for wave power machines have been filed over the past 80 years or so, many of them amusingly impractical. Of the several practical designs on which active research is being undertaken, those in the most advanced state of development are the ones that float on the surface and simply bob up and down as waves pass. The energy represented by the bobbing movement is converted by the machine into electrical current, which is sent

ashore by seabed cable.

One of the best-known designs for a wave power generator is called Salter's Duck. Invented by Stephen Salter of Edinburgh University in Scotland, the device has been described as "a rocking boom," "an oscillating vane," "a cylinder with irregular cross-section" – or simply "a duck." The device has a pointed end that floats just above the surface facing the oncoming waves and a rounded bottom beneath the surface. The pointed end lifts under the passing wave and the bottom end stays beneath the surface. The energy generated is converted to hydraulic power and then to electrical power.

Scale models of Salter's Ducks have been tested at various sites and have worked. In full-scale form, each device would be up to 100 feet wide and 30 feet in

Left: a laboratory model of a Salter's duck, now being tested in Britain to produce electricity from waves. It is sometimes known as the "nodding duck" system.

Below: an artist's impresion of a string of Salter's Ducks in the water. The bobbing motion as the waves hit them drives pumps.

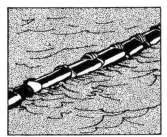

Below: waves rush into the submerged cylinder (**A**) of the Masuda light-buoy, pushing air up (**B**) to drive a turbine, creating a partial vacuum (**C**) to drive it again.

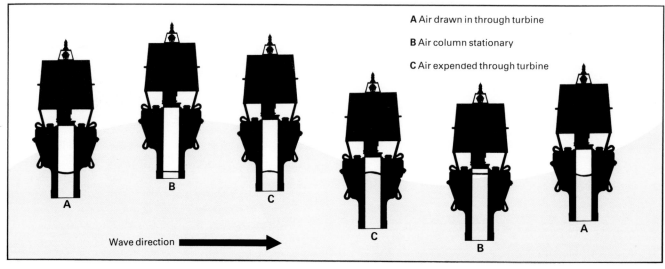

A Air drawn in through turbine

B Air column stationary

C Air expended through turbine

Wave direction ▶

diameter; up to 20 of them would be strung along a central, partly submerged spine. Each duck would have its own hydraulic pump connected to a common hydraulic line. The power would be fed to a power pack in the center of the spine for conversion into electricity and transmission to shore by a seabed cable.

A string of ducks this size would constitute a massive structure up to half a mile long. It should, however, be capable of extracting 25 percent of the available wave power, which would supply the electricity demands of a town of 85,000 people. Until that day, many more trials with strings of scale model ducks need to be carried out to iron out the problems that could arise with the installation, mooring, and maintenance of such a massive system in the open ocean.

Another wave power device on which high hopes

are currently pinned is based on an invention by Yoshio Masuda of the Japanese Marine Science and Technology Center. Some 15 years ago Masuda successfully marketed a navigation buoy that used waves to provide up to 100 watts for the buoy's light. The principle he used is now the subject of extensive trials to discover whether or not his invention can be adapted to provide electricity on a large scale.

The Masuda invention is basically a floating cylinder with a permanently submerged bottom whose end is open. As a wave passes, water rushes into the cylinder and pushes air up through a valve. The air drives a turbine that generates electricity. When the wave has passed, the water level in the cylinder falls and a partial vacuum is created. To fill this vacuum, air rushes in through a hole in the structure that is above the surface, again driving the turbine as it passes through.

Masuda's navigation buoys have a maximum diameter of 30 feet. Work is now in progress to apply the Masuda principle to buoys with diameters up to 500 feet, and to deploy a series in a string or to arrange a whole series on concentric chambers. Trials using the same oscillating water column principle have also been conducted on a 260-foot long ship in the Sea of Japan. The ship's hull was divided into 20 boxlike chambers open to the sea and its buoyancy was provided by longitudinal tanks. Waves creating pressure in the chambers powered 10 turbines.

Like the Salter's Duck, the Masuda buoy and ship systems present problems to be overcome before proving workable with full-scale models. They will have to be provided with completely reliable moorings capable of withstanding the millions of movements the structures would make each year.

All the wave power devices now being researched – and there are 11 such projects in Britain alone – have the common problems of mooring, huge size, and expense. A full-scale wave power machine could cost 10 to 20 times as much to build, install, and maintain as a conventional power station; and in addition, its output would not be constant and the power it generated at off-peak times could not be stored for future use. However, the rising costs of fossil fuel energy could make what is economically borderline today into what is economically sound in a few years. That is why the research continues. Researchers are also looking in another direction – to the example of the experimental project with tidal power in France. A few years ago it was written off as an economic failure, but it has turned into a success.

Above: an artist's impression of a very recent form of an oscillating water column converter. One of the desirable features of an oscillating column is that the turbine and other machine parts are easily accessible for repair and maintenance.

Right: how the above oscillating water column works. At the top left, rising water forces air up and is let into the turbine chamber when a louvered valve opens (shown opened). At the top right, air is shown moving out of a second valve that opens after the air has driven the turbine. The bottom two diagrams show the exact procedure in reverse, which means that the turbine is driven both when water rises and when it falls. The bottom diagram simply shows the water column in cross-section.

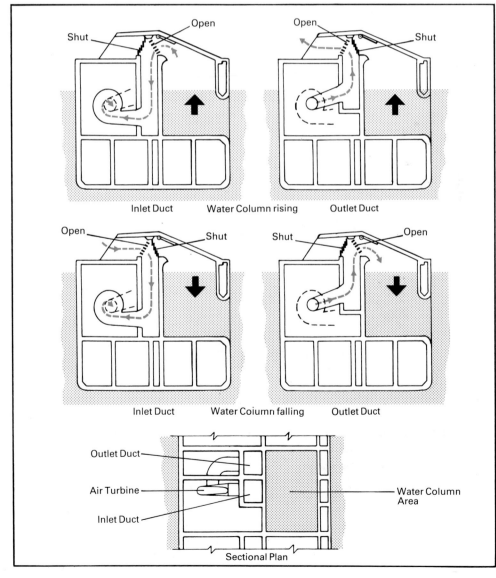

Harnessing Energy from the Tides

The French built a tidal power station in 1966 on the Rance Estuary. It supplies electricity to the towns of Dinard and St Malo. A few years after it had been commissioned, it was pronounced a failure. Its own administrators said that it would never recover the costs of building and maintenance. Then came the oil price hike. Energy prices went up and up. As a result, the Rance tidal power project is now an economic success.

In Britain, engineers solemnly conduct a feasibility study about once every five years to tap the enormous tidal energy of the Severn Estuary. Each time they shake their heads sadly and say that the capital costs could not be justified in terms of present-day energy prices. "But," they say, "if we had only gone ahead five years ago the construction costs could easily have been justified by now."

Above: a watermill now in use at Lower Slaughter, a village in Gloucestershire, England. It was the Romans who developed the method of driving a mill by making water stream down onto the wheel from above. Today, such ideas are gaining favor again.

In the energy business it is often necessary to play a hunch. France played a hunch with Rance, and today the country not only has a reasonably cheap supply of electricity to provide back-up energy to two towns, but also has amassed a fund of readily exportable expertise.

Above: a map of possible sites for tidal power plants.
1 Cook Inlet, Alaska **2** Lower California, Mexico **3** Passamoquoddy, Maine **4** Bay of Fundy and **5** Frobisher Bay, Canada **6** Maranho, Venezuela **7** San Jose Gulf, Argentina **8** Severn River, England **9** Rance River, France **10** Kislaya, Soviet Union **11** Cambay River, India **12** Seoul River, South Korea **13** Abidjan, Ivory Coast **14** Darwin and **15** Kimberleys, Australia. (**9** and **10** already exist.)

Tidal power has been exploited for centuries. There is reference to a tidal mill at Dover, England, in the Domesday Book of the 11th century, for example. But only in the past 50 years has this source of power been looked upon seriously as a means of providing energy on a large scale. With the enormous jump in energy prices in the very recent past, it is now being regarded even more seriously. The very engineers who have looked at the Severn Estuary from time to time since the 1930s will probably soon be given the green light to put their plans into action.

Of all the alternative energy sources at our disposal to replace fossil fuels, tidal power is a very good bet in terms of economics and reliability. The principle is simple enough. The tide comes into a bay or estuary. It flows through a damlike structure, the gates of which are closed at high tide. As the tide flows out again, the gates are opened and the pent-up waters rush through the dam to find their own level. On the way through the dam these waters drive generating turbines.

The attraction of tidal power lies in the predictability of the basic supply of energy. Because the height, speed, and direction of waves varies considerably from day to day and from season to season, the

accurate prediction of these variables is a major problem in insuring the success of wave power projects. But tides can be predicted with relative accuracy, and abnormally high or low waters do not greatly reduce the general degree of accuracy. Even if abnormal conditions are encountered, they have only a small effect on a tidal energy plant in terms of its overall annual output of electricity. Also, tidal power is clean energy, and with the growing public concern about the safety of nuclear power, far more attention will focus on the ways in which clean and safe energy can be exploited.

This does not mean that the world will soon be scattered with hundreds of tidal power stations. For a start, the natural conditions have to be right. Tidal power is only worth harnessing in areas where a funnel-shaped bay or an estuary is of just the right shape for construction of the dam, and the tidal range has to average about 15 feet to provide sufficient energy. The site must also be close to a center of heavy demand for electricity – certainly no more than 100 miles – so that transmission costs are kept down. Care must be taken to insure that the presence of the dam does not have any dramatically adverse effects on the environment. Finally, although running costs and

maintenance are low, the capital costs of the project are enormous. With such big capital outlays needed before a single watt of electricity has been generated, it takes many years to recover costs.

These restrictions greatly reduce the number of potential tidal power sites throughout the world. There are about 100 sites where it is technically feasible to build dams, but only a dozen of these completely fulfill the main economic and logistical criteria. Detailed studies are now being undertaken at sites in Canada, the United States, Australia, Argentina, India, China, and, as we have seen, in France and Britain. The Soviet Union already has a small experimental station working at Kislaya Bay. South Korea's first plant will be operational in 1986. It will produce 400,000kWh at a capital cost of over $300 million. All these studies draw heavily on the French experience at Rance.

Construction started on the Rance dam in 1961 and the plant came into full operation in 1967. The Rance Estuary is ideally suited to a tidal power project. It is narrow, so that the dam is only 2300 feet wide. Maximum tidal range at the spring equinox is over 44 feet; the average is 37.5 feet. Beneath the estuary bed there is granite bedrock, providing an ideal foundation for

Right: the Rance project system. **1** The difference in level between the rising tide and the basin forces water through turbines to generate electricity. **2** Some of this power pumps more water into the basin to raise the level for the next phase. **3** The outflow of water produces the most electricity as the tide falls. **4** More power is used to lower the basin to get the greatest height difference for the new cycle.

Below: the bridge over the Rance Dam, built as part of the Rance tidal energy project in France.

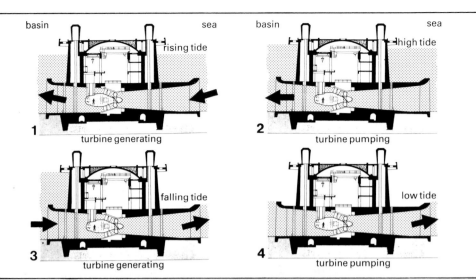

1 — turbine generating (basin / sea, rising tide)
2 — turbine pumping (basin / sea, high tide)
3 — turbine generating (basin / sea, falling tide)
4 — turbine pumping (basin / sea, low tide)

the dam.

Some 24 massive generating sets are fitted into the reinforced concrete of the dam. Each consists of a four-bladed turbine powering a 10-megawatt, 3500-volt generator.

These turbines operate on a two-way cycle. This is one of the variations possible on the basic principle of storing water at high tide and releasing it on the ebb to generate electricity, which would probably not be an economic proposition for most tidal stations. At Rance this drawback has been overcome by having the turbines operate when water flows in as well as when it flows out of the estuary. Another refinement is to have the turbines acting as pumps. This overcomes the basic problem presented by the fact that peak energy demand will not naturally coincide with tidal flow, so that the electricity generated cannot be

would have been 25 miles wide and, instead of the 24 turbines each producing 10 megawatts as installed at Rance, some 300 turbines each producing 40 megawatts would have been installed. The idea was abandoned in favor of a nuclear energy policy but could well be revived for economic and political reasons.

It is the way in which economic viability creeps up on potential tidal power projects, through outside factors, that will probably make the long talked-about Severn Barrage project in Britain come to fruition. Rance is designed to generate a peak of 540,000 units of electricity a year. The Severn plan could produce as much as 14,000 million units, representing about one eighth of total British demand. The dam would be 7.5 miles long, running from just south of Cardiff, Wales to Brean Down near Weston-super-Mare on the Somerset coast. It would enclose some

Two aspects of the Severn Barrage project in England. The Severn bore (**above**) is a high tidal wave that moves up the river at speeds up to 12 miles per hour. Low-tide on the Severn river (**right**) with a view of the bridge.

used to its best effect. At Rance, this off-peak power is used by the turbines to pump water into whichever of the basins on either side of the dam happens to be lowest. The head of water built up in this way can be released at peak demand times to generate electricity as and when it is required.

There have been some side benefits to the Rance project that will be closely studied by other nations. One is that the bridge across the dam has reduced the road distance from Dinard to St Malo by 20 percent. In addition, the dam itself attracts tourists to the area (some 500,000 vehicles a month use the bridge during the holiday season); boats increasingly use the lock through the dam (there were only 5287 transits in 1968 and 13,462 in 1976); and the take-up of power has reduced the tidal movement so that there is a broad expanse of calm water for recreational purposes.

The Rance plant was intended as a pilot project for an ambitiously large tidal power plant on the Chausey Islands in the Mont St Michel Bay. The dam there

150 square miles of water. Side benefits from the project would be the building of a new bridge across the Severn Estuary, and the provision of vast enclosed areas for water sports. Technically the Severn project is feasible. Politically and economically there are many problems to be overcome.

Tidal power is one of the first choices for an alternative energy source. As a footnote it is worth observing that exploitation of the energy contained in ocean currents comes a long way down the list. Studies have been made of the feasibility of exploiting the great ocean currents such as the Gulf Stream and the Kuroshio, as well as fast-moving tidal currents that in some areas can attain speeds of 10 knots at spring tides. Theoretically these currents could drive very light turbine blades, but the difficulties of installing them, of converting the tapped energy into electricity, and of transmitting it, means that there is little likelihood of any projects getting past the drawing board stage in the foreseeable future.

Using the Sea's Thermal Energy

Another alternative energy project to be picked up and dusted off in the wake of the 1973 oil supply crisis was that of ocean thermal energy conversion (OTEC). This idea, which was proposed as early as 1881, has had an up-and-down history. A small plant operated in Cuba in 1930 but failed after only two weeks. In the 1950s the French designed a station to be built off the Ivory Coast, but it never got past the planning stage. It was not until the 1960s that a company was set up in the United States merely to investigate the OTEC idea. Then came the oil crisis that spurred a full reappraisal of OTEC.

The reactions to this reappraisal were tremendously enthusiastic. A pilot OTEC plant is already operating off Hawaii and another is supposed to follow. Three more United States-sponsored plants are scheduled,

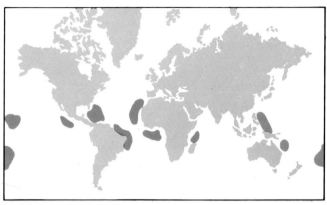

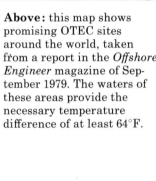

Above: this map shows promising OTEC sites around the world, taken from a report in the *Offshore Engineer* magazine of September 1979. The waters of these areas provide the necessary temperature difference of at least 64°F.

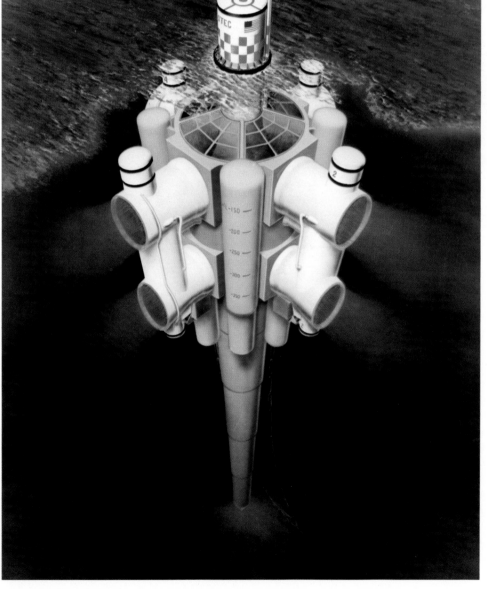

Left: an artist's impression of an OTEC (ocean thermal energy conversion) generator. It would supply enough electricity for a city of 200,000 people.

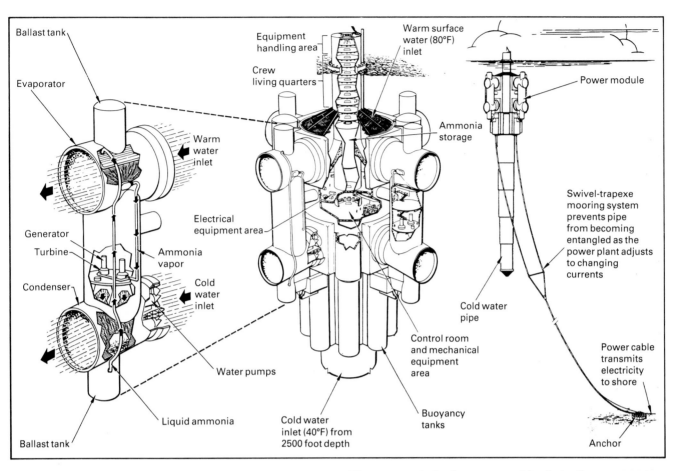

Ballast tank

Evaporator

Warm water inlet

Generator

Turbine

Condenser

Liquid ammonia

Ballast tank

Equipment handling area

Crew living quarters

Warm surface water (80°F) inlet

Ammonia storage

Ammonia vapor

Electrical equipment area

Cold water inlet

Water pumps

Cold water inlet (40°F) from 2500 foot depth

Control room and mechanical equipment area

Buoyancy tanks

Power module

Swivel-trapexe mooring system prevents pipe from becoming entangled as the power plant adjusts to changing currents

Cold water pipe

Power cable transmits electricity to shore

Anchor

Above: a drawing of the component parts of the OTEC with their contents labeled. The ballast tank contains the all-important generator, turbine, and condenser. On the far right is another view of the OTEC module showing the cold water pipe and mooring line extending from the bottom.

a European joint-venture prototype will be built soon, the French are considering building 10, and the Japanese will have a station operating by 1982.

What is OTEC, and what prompted all the enthusiasm about it? The principles and progress of the system were spelled out in the British magazine *Offshore Engineer* by Wilfred Griekspoor and Bart van der Pot, who are involved in the United States and European OTEC studies. Many of the following facts are drawn from their description.

We know that the oceans are the Earth's temperature regulator, absorbing vast amounts of solar radiation. In tropical areas the surface layers of the sea are heated to around 77°F. But deeper layers, from about 2000 feet, have a temperature of around 41°F. This temperature difference can be exploited on thermodynamic principles to run what is in effect a heat engine that produces electrical power.

The principle used is called the Rankine closed circuit cycle. Warm surface water is pumped through a heat exchanger (an evaporator) where it gives up its heat to a "working" liquid, usually ammonia. The liquid vaporizes and expands to drive a generator. The vapor then flows through a second heat exchanger (a condenser) and is liquidized, having given up its heat to cold water extracted from 2000 to 3000 feet depths.

The ammonia is then pumped back to the evaporator and the cycle is repeated.

One of the reasons for the high interest in OTEC is that there are many subtropical and equatorial regions in which the required minimum temperature difference of 64°F between surface and deep ocean layers is present night and day all the year round. This gives OTEC an immediate advantage over other marine energy conversion systems, such as waves and tides, in that the generation of electricity is constant. OTEC is also environmentally clean and can produce electricity at a cost competitive with coal, oil, and nuclear power. These systems also work on thermodynamic principles in the same way as OTEC, but they

Below: a diagrammatic illustration of the Rankine closed circuit cycle that enables the OTEC to convert the temperature differences of the ocean to energy.

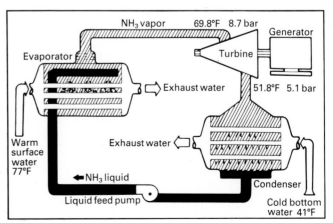

NH_3 vapor   69.8°F 8.7 bar   Generator

Evaporator   Turbine

Exhaust water   51.8°F 5.1 bar

Warm surface water 77°F

Exhaust water

NH_3 liquid

Liquid feed pump

Condenser

Cold bottom water 41°F

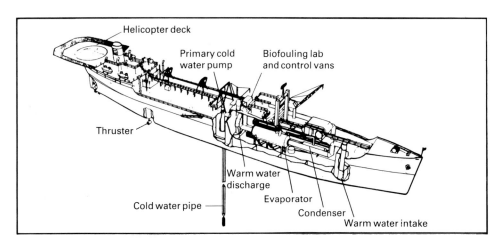

Helicopter deck

Primary cold
water pump

Biofouling lab
and control vans

Thruster

Warm water
discharge

Cold water pipe

Evaporator

Condenser

Warm water intake

Left: the United States Navy
ship that has been converted to
be a test platform for various
types of heat exchangers. The
kind that proves best will help
determine the final shape of
an OTEC power point.

can only exploit a temperature difference of hundreds
of degrees while OTEC requires a minimum of only
64°F. Although this means that the actual thermodynamic efficiency of the system is very low – in
practice it could be just over one percent – it does not
matter because the fuel for the power plant – the water
heat – is free.

It does, however, mean that enormous amounts of
water have to be processed. Griekspoor and Van der
Pot cite a hypothetical example of a plant working at
1.2 percent efficiency and utilizing a temperature
difference of 68°F. To produce one megawatt (MW) of
electricity, water has to flow through both the warm
and cold water parts of the system at 176 cubic feet per
second. To produce 100MW – which is the aim for full-
scale systems – flows of nearly 16,000 cubic feet per
second of both cold and warm water are required. The
machinery to handle such colossal flows will be enormous: for the 100MW station the surface area of the
heat exchangers alone would be between five million
and 16 million square feet. Research continues to find
the best type of heat exchanger for this work. The
United States Department of Energy has converted
a navy tanker, at a cost of $25 million, to test various
types of heat exchanger in a trial program off Hawaii,
using a 2100-foot long bundle of 4-foot diameter polyethylene pipes to bring in the cold deep water.

Given these principles of the process, the size of the
equipment required, and the water depth in which the
plant will have to operate, an idea of the shape of an
OTEC plant can be gained. It will probably be an
artificial island with, for a 100MW plant, a displace-
ment of up to 300,000 tons. That is a tempting figure
for many would-be OTEC plant designers because
there is a surplus of 300,000 ton tankers in the world
shipping fleet. But their use has some inherent
disadvantages. It would be difficult to install the huge
items of equipment required in anything but a purpose-built hull; moorings would present problems;
and a ship would tend to move more in a swell than a
purpose-built structure would.

The OTEC researchers are drawing heavily on the
experience of the offshore oil exploration and production industry, which has devised many designs for
large floating islands in recent years. A likely can-

didate is the spar-type platform, which is somewhat
like an angler's float, but a spar of the size required
to support an OTEC plant would be so huge that most
single-point moorings of the type now in use would
not be able to hold it in strong currents. New systems
would have to be devised. Other configurations proposed are for cylindrical and rectangular vessels that
float on the surface while being moored on the seabed.

Another taxing problem is that of the design of the
cold water pipe, which for a 100MW plant would
extend up to three-quarters of a mile from the bottom
of the surface platform. It would have a diameter of
up to 65 feet. (To get an idea of this size, think of six
buses one on top of the other.) This pipe, which would
be connected to the surface plant, has to be able to
withstand the forces exacted upon it both by the
movement of the surface vessel and by the often
counteracting forces of surface and subsea currents.
One scale test from a semisubmersible oil drilling
platform using a 5-foot diameter pipe 300 feet long
came to a quick and abrupt end when the pipe snapped

Above: a container ship unloading its cargo of oil at
Papeete, the main city of Tahiti. Such a tropical island
depends so heavily on imported oil to meet most of its
energy requirements that it is logically a prime target
place for the establishment of an OTEC project.

off in a storm. Engineers are now looking at various designs, including rigid pipes with flexible connections, flexible pipes, and bundles of small pipes.

Where will OTEC plants be installed and what will the energy be used for? Subtropical and equatorial islands relying on imported oil for most of their energy supplies are prime candidates, and feasibility studies have already been prepared for the Virgin Islands, Puerto Rico, the Polynesian islands, the Coral Sea islands, and the Canaries. The Pacific playground of Hawaii is a particularly suitable site for OTEC projects. One United States consortium has installed a mini-OTEC off the Hawaiian coast at Keahole Point to demonstrate the feasibility of a full-scale plant.

tween 10 and 100MW. Once plants have been established and have proved that they are capable of fulfilling this basic function, a wide range of other possibilities opens up. Slight modifications to the basic process would yield large amounts of desalinated water. Electrolysis of this water would produce liquid hydrogen, which in turn could be liquified. A further synthesization process would produce ammonia. There is also the possibility of transporting raw materials from the land to the OTEC plant where they could be transformed in traditionally energy-intensive processes. A good example would be the production of aluminum by the electrolysis of land-produced aluminum oxide off the West Indies. There is also the possi-

Above: the Mini-OTEC off the Kona Coast of Hawaii, the world's first plant of its kind, at its first site. It is being used as a demonstrator and test platform and could be the forerunner of huge units intended to serve medium size cities. OTEC units are effective in warm seas.

The mini-OTEC pontoon produces 50kW of electricity. It uses a 3000 foot polyethylene cold water pipe, the upper end of which is secured to a buoy that is in turn secured to the OTEC pontoon by cables. The cold water pipe in this way serves as the anchor line for the pontoon.

The prime aim of OTEC plants will be to transmit electricity ashore by seabed cable at powers of be-

bility of methanol production by combining gaseous hydrogen and carbon dioxide.

All the possible side benefits of the OTEC principle depend, of course, on the success of the many trials currently being undertaken by various nations and by individual companies. Certainly, OTEC offers the best prospects for continuous supplies of ocean energy as opposed to the inherently intermittent nature of wave and tide supplies. Success depends, however, on solving the problems presented by the sheer scale of the machinery required for the OTEC process, and therefore of the size of the platform on which that machinery would be mounted.

The Wind as a Power Source

They used to be called windmills. Now they are called aerogenerators. Whatever the name, they are devices to tap the power of the wind. Today in Canada, Denmark, the Netherlands, West Germany, New Zealand, Britain, and the United States, scientists and engineers are looking at bigger and better ways of exploiting the free energy source offered by the winds.

Wind power has been used for centuries – and still is – for milling corn and pumping water. The landscapes of Holland, and to a lesser extent eastern England, are still typified by the classic four, five, and sometimes six-sailed windmills, while some rural parts of the United States still use the tall multibladed turbine to pump water from the well.

These machines have very low efficiency, and so current research is being directed on two fronts. First, there is the need to improve the aerodynamics of "sails" and to introduce new materials so that greater efficiency can be obtained from small machines. These machines would be used onshore for their traditional pumping and milling activities, especially in lesser developed nations, and also to generate small amounts of electricity for domestic consumption by individual households or at most small communities.

The second area of research is concerned with huge wind machines for large-scale generation of electricity to be fed into an area or national grid along with that produced by conventional power plants.

Big wind machines are best installed offshore because the average wind speed is higher and the gusts are less disturbed than onshore where hills, buildings, and generally more complex wind patterns can affect the steady flow. The average wind power density offshore is something like two-thirds greater than that

Below: a 200kW experimental wind turbine located in Rhode Island on the east coast of the United States. It has been designed so that it can function even when there is very little wind. Its narrow blades swing freely in light winds and can withstand stronger air currents.

onshore. So the future could see windmills installed in shallow areas offshore – less than 70 feet deep and about 10 miles from shore. They would be in clusters carefully arranged so that each machine could take the fullest advantage from gusts of wind without canceling out another machine by absorbing all its power.

The windmills used for this scale of power generation – perhaps up to 1000MW per cluster – might be either of the horizontal or vertical axis type. Horizontal axis machines are of the traditional type. For use in offshore power development, each blade would have to be up to 100 feet long, weigh between five and 10 tons, and rotate at about 40 revolutions per minute. This would make the tip of the blade move at up to 330 feet per second. Much research will have to be done to assess the stresses and loadings to which such blades would be put.

Vertical axis blades, the second type for windmills, have two advantages. One is that the blades do not have to be turned to face the prevailing wind and the other is that the electrical power generated is led off to a subsea cable from the base rather than over halfway up the structure.

Big wind-gathering projects of this type now being

Above: a traditional windmill with canvas sails, in operation in Estremadura Province, Portugal. Today such power generators are often known by other names.

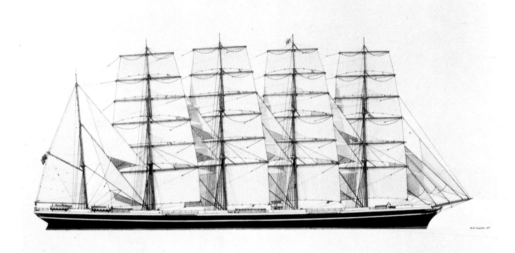

researched have the inherent disadvantage of an intermittent source of supply. The wind does not blow steadily and in the same direction all the time, and there are many problems involved with siting massive structures offshore. But if these can be solved, wind power could contribute a large part of a nation's electricity supplies. In Britain, for example, this could be as high as 25 percent.

It is not only for domestic electricity projects that people are again beginning to look at wind power. The rising cost of marine fuel has also led engineers in a diverse range of marine-based industries to look once more at the transport methods of their grandparents. Steel-hulled, four-masted sailing cargo vessels worked deep sea routes until the 1920s when they were superseded by coal and diesel-powered vessels. The tide, so to speak, has turned and there are now designs for a five-masted barque called *International Sailiner* which could be carrying bulk cargoes between Europe and the Caribbean within two years. Big crews will not be necessary to handle acres of canvas because halliards, steering gear, bracing yards, and windlasses will be hydraulically powered. The vessel of 450 feet long will have a crew of 51 and an auxiliary engine

of 3000 horsepower. Speed under power would be 11 knots, and maximum speed under sail would be 22 knots to give an average minimum speed of 16.5 knots.

Elsewhere, United States trawling fleets have been looking at the possibility of using hydraulically rigged sails to take fishing vessels to and from the fishing grounds. The time taken to tow a self-elevating rig from the United States to the Middle East was reduced greatly by rigging a sail from its elevated legs.

By the use of wave, wind, tidal, and thermal power, the world is preparing itself for the day when fossil fuels can no longer be relied upon as the primary energy source. What is encouraging is that so much research is being devoted to "clean" forms of "free" energy. Governments ought to look even more closely at these before irrevocably committing themselves, their funds, and their human resources to the environmental knife-edge of nuclear power.

It will be some time, however, before any of these energy-gathering projects can hope to usurp the position of the developed world's present primary fuel – oil. Today there is a massive effort being put into exploiting it from beneath the inhospitable oceans.

215

Chapter 9

Oil and Gas Resources

The world today depends on petroleum as its chief source of energy. But petroleum is a limited resource, and it will not be long before it becomes of minor importance in human energy consumption. Even so, the world is not yet ready to decrease its dependence on petroleum, and the next few years will see intensive efforts to find and exploit the remaining reserves of oil and gas that exist on this planet. The search will take place mainly offshore in some of the most inhospitable regions of the world: in the freezing Arctic, off iceberg-ridden Antarctica, in the hurricane-swept Atlantic. The huge costs of working in these regions will have to be met by the soaring price of energy throughout the world.

This chapter sketches the beginnings of the offshore petroleum industry and examines the way in which it has developed, from simple platforms to completely autonomous underwater oil-producing units. It can be said that the skills which the offshore oil industry is accumulating promises to be of value far beyond the petroleum era to all who seek to exploit the riches of the oceans.

Opposite: stark, massive, and dramatic – an oil rig pictured at night through the huge legs that keep it anchored. Such rigs are like small industrial cities.

The Last Years of
the Petroleum Era

It is difficult to imagine a world without oil, but soon – too soon for many people's liking – the petroleum era will be over. The world will then have to rely on other forms of energy.

We are so used to depending on oil and gas as the primary energy sources that their eventual depletion seems a frightening prospect. But that prospect has to be faced. Similar cases have happened throughout history. Wood was the primary heating and cooking material for thousands of years, and animals and wind-powered ships provided transport. Then came the use of coal, which fuelled the industrial revolution of the 19th century as the predominant energy source. Even as recently as the early 1940s coal provided half the energy used in the United States.

Petroleum and its associated products then took over from King Coal. In the United States, for example, petroleum now accounts for about three-quarters of the total energy consumption. Petroleum liquids – oil – account for 47 percent of this total and gas for 27 percent.

This level of dependence will have to change simply because the oil is running out. It will never be completely depleted, but sometime in the 21st century it will have to relinquish its position as the primary energy source of the world and be supplanted by a combination of other energy sources: solar radiation and nuclear, wind, tide and wave power, for instance. For the next 30 or 40 years, however, oil will remain the principal energy source. The non-Communist world currently produces around 50 million barrels of oil per day and this production level will level off at around 70 million barrels per day by 1985. There it will

Right: a European super-highway, jammed with cars. The dependence of the West-ern world on oil is increased as the number and use of private cars increases.

218

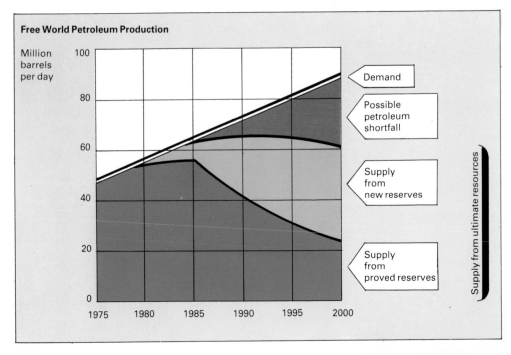

Free World Petroleum Production

Million barrels per day

Demand

Possible petroleum shortfall

Supply from new reserves

Supply from proved reserves

Supply from ultimate resources

(x-axis: 1975, 1980, 1985, 1990, 1995, 2000; y-axis: 0, 20, 40, 60, 80, 100)

Left: this diagram projects the supply and demand for petroleum by non-communist countries until the year 2000. The oil shortfall is expected to increase more and more sharply after 1990.

Below: oil flares light the sky at Dukhan, Qatar. The big oil producers of the Middle East are able to control prices and reserves of this much-wanted commodity.

stay for some years. It would probably be possible to produce more per day but the nations with the greatest reserves will almost certainly control the rate of depletion of reserves in order to eke out supplies and to maintain high price levels. The nations that will have the power to control supplies are those that themselves have relatively low consumption: over 70 percent of known oil reserves lie under Arabian sands. Saudi Arabia alone has 25 percent of the Earth's known petroleum reserves and by 1990 will probably be supplying over 20 percent of the non-Communist world's supplies.

Today a frantic worldwide scramble is in progress by the nations mainly dependent on oil for energy to find and discover their own reserves to bridge the gap between the next few oil-dependent years and the dawning of a new energy era. In the decade ending 1985 some $900 billion will have been spent on finding, developing, processing, and marketing oil. Over half of that will go into exploration and development, and a very high proportion will be spent offshore because more than a fifth of the world's total proven hydrocarbon reserves are offshore. This represents 200 billion barrels of oil (or gas equivalent). Only 18 percent of the oil reserves known to be offshore and capable of being recovered have yet been produced. So the world oil industry is increasingly looking to the sea.

The offshore search is very young even though there was some earlier exploitation of undersea reserves that were in shallow water known to be extensions of land-based deposits. Only in the late 1950s and 1960s did oil companies begin to look seriously for truly offshore deposits that had no geological link with the land. The search is in full cry today, from the frozen wastes of the Arctic to the equatorial seas of the Far East. Drilling is now being carried out in deeper and more hostile waters than ever before.

There is such a high price tag on the daunting technology required to exploit these undersea resources that everyone's attitude toward the use of energy will have to change. Far more attention will have to be paid to the way in which oil is used, and it seems likely that the remaining reserves will be used primarily as a feedstock for transportation fuels and for petroleum-based products like plastics. The day is swiftly approaching when the burning of oil to produce heat – either directly or by its inefficient conversion to electricity – will be supplanted by other energy sources. Until that day arrives, the search for oil is bound to continue at its present intensive rate, for all the formidable challenges this presents.

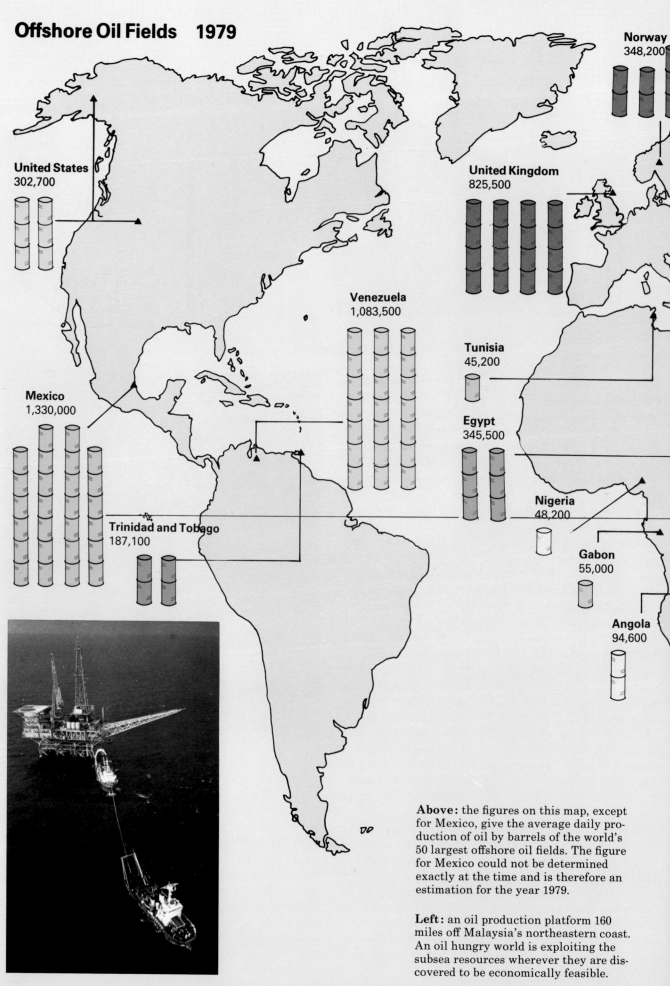

Offshore Oil Fields   1979

Norway
348,200

United States
302,700

United Kingdom
825,500

Venezuela
1,083,500

Tunisia
45,200

Mexico
1,330,000

Egypt
345,500

Nigeria
48,200

Trinidad and Tobago
187,100

Gabon
55,000

Angola
94,600

Above: the figures on this map, except for Mexico, give the average daily production of oil by barrels of the world's 50 largest offshore oil fields. The figure for Mexico could not be determined exactly at the time and is therefore an estimation for the year 1979.

Left: an oil production platform 160 miles off Malaysia's northeastern coast. An oil hungry world is exploiting the subsea resources wherever they are discovered to be economically feasible.

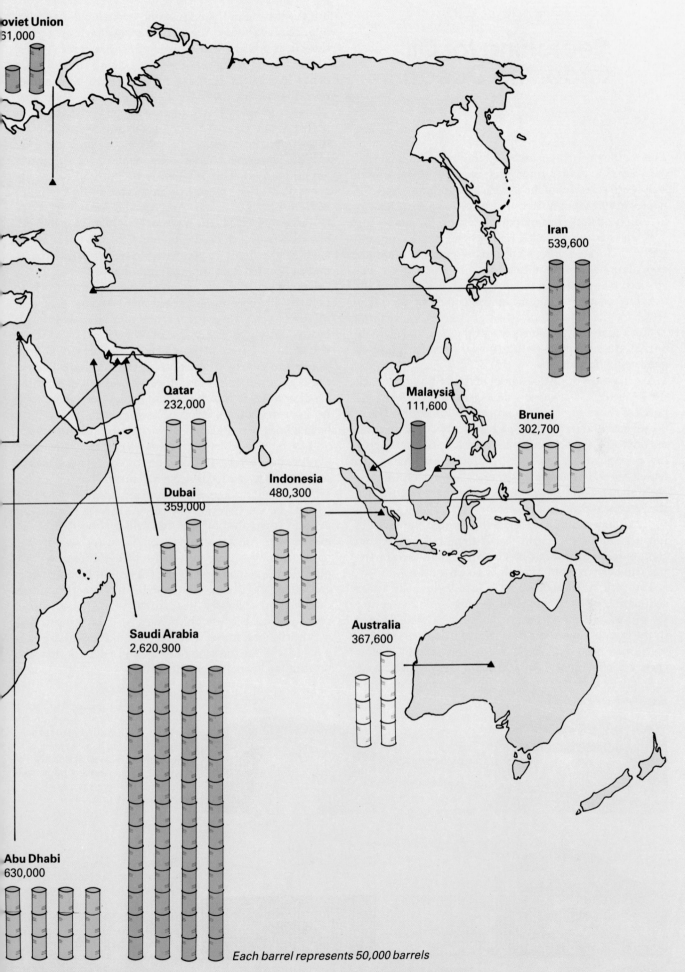

Soviet Union
61,000

Iran
539,600

Qatar
232,000

Malaysia
111,600

Brunei
302,700

Dubai
359,000

Indonesia
480,300

Saudi Arabia
2,620,900

Australia
367,600

Abu Dhabi
630,000

Each barrel represents 50,000 barrels

221

Searching for Oil under the Ocean

There are several reasons why the intensive effort to discover the world's remaining reserves of petroleum will be directed to offshore rather than land areas. To appreciate these it is first necessary to look at the way in which offshore oil and gas fields are formed.

It was beneath the sea that all the Earth's oil and gas were formed. Many millions of years ago animal and plant matter that had sunk to the seabed was covered by sediments which covered and crushed the organic matter to depths of thousands of feet. The spontaneous heat generated during this crushing process and the presence of catalytic minerals converted this animal and plant matter into liquid or gaseous hydrocarbons.

An oil reservoir is not a pool of liquid lying beneath the seabed. It is trapped under great pressure in the pores of rocks such as sandstone and limestone. Often it escapes, and natural seepages of oil both at sea and on land are still common today. But frequently the layer of porous rock – the sedimentary basin – in which the oil or gas is trapped has a layer of nonporous rock above it and salt beneath it. In these conditions, the hydrocarbons cannot escape. This creates potentially commercial oilfields. Drill through the layer of impermeable rock known by drillers as caprock and the pressurized liquids, or gas, will move toward the hole and try to make their way to the surface.

The task of the geologist is first to identify these sedimentary basins and then to judge whether or not they contain hydrocarbons. All oilfields were formed in the tropics, and their presence in other parts of the world has been explained by the comparatively new theory of continental drift. This theory also is helping in the plotting of the course and eventual destination of drifting oilfields. According to the continental drift theory, there are three types of sedimentary basins. Those on the margins of plates moving toward each other are called convergent. Those on continental margins moving away from each other are divergent. Those that were fairly distant from the margins of plates at the time of their formation are called plate interior basins.

Plate interior basins have yielded the greatest amounts of oil. This is because they are in relatively accessible areas, mostly on land, including the giant Middle East fields and those on the American continent. Divergent basins have so far yielded far less oil, but discoveries have been made in divergent areas such as the fields in the North Sea, off West Africa, and in Mexico both on land and offshore.

Geological detectives have been hard at work to identify little or nonexplored sedimentary basins of all three types. There are few unexplored plate interior basins, although vast tracts of Siberia and its offshore waters have been little probed. Convergent margin basins are nearly all offshore in the northern Pacific, the equatorial Far East, off New Zealand, and in the Mediterranean. Many of these basins have proved to be poor petroleum prospects, although there have been some good finds in California and the Far East.

It is on the divergent basin types that the main exploration interest is now centered, but the task is not easy. The major divergent margin basins are all offshore and in some of the harshest areas of the world ocean: in the Arctic; off the Antarctic continent; around Australia; off the Atlantic seaboard of North and South America; north of Norway; and in the monsoon-swept Indian Ocean. The technology required to develop these resources will be a far cry from the early days of offshore drilling, but the young offshore oil industry has grown fast and advanced greatly. The biggest challenges still lie ahead. The way such challenges are being met in the North Sea is a prototype for the many technological feats that will be standard practice in the future.

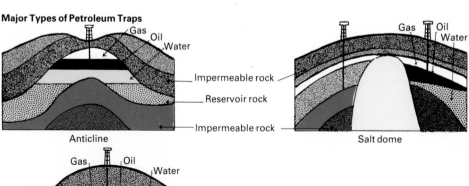

Major Types of Petroleum Traps

Anticline

Salt dome

Fault

Unconformity

Left: the four geological locations, called traps, in which oil occurs in nature. In each case, there must be a layer of impermeable rock above the petroleum to prevent its escape through the open pores of the rock layer that holds it.

Right: although any map of North Sea oil exploration goes out of date as it is made, this one gives an indication of the frequency of discoveries. The oilfields shown are enough to provide Britain with three quarters of its oil needs.

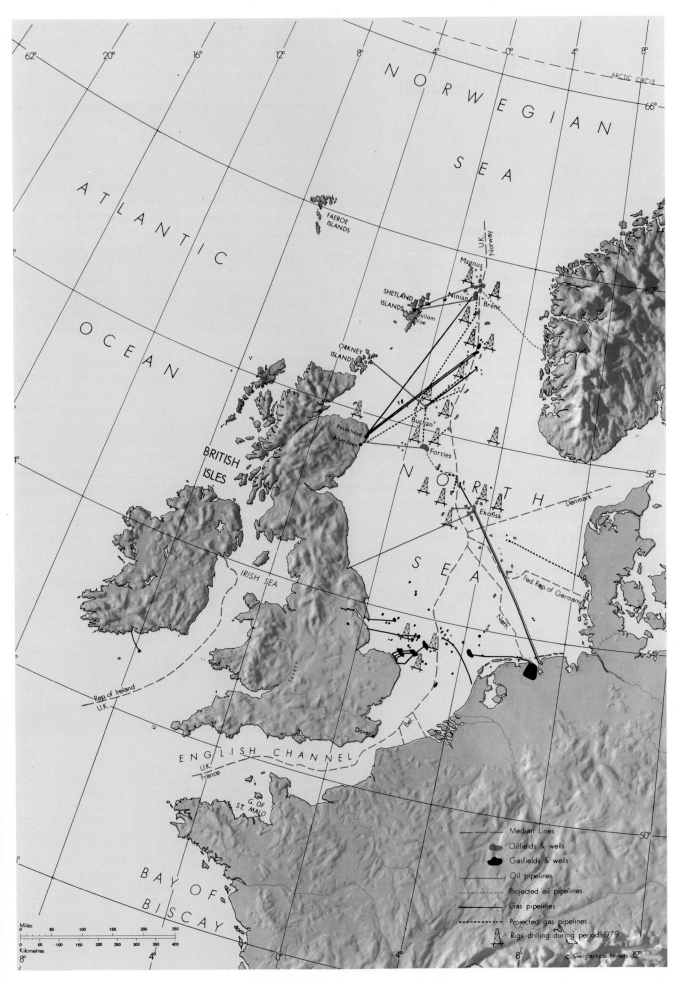

NORWEGIAN

SEA

ARCTIC CIRCLE

66°

ATLANTIC

OCEAN

FAEROE
ISLANDS

U.K.
Norway

62°

Magnus

SHETLAND
ISLANDS   Ninian   Brent

Sullom
Voe

ORKNEY
ISLANDS

BRITISH
ISLES

Peterhead
Aberdeen   Buchan

Forties

N   O   R   T   H

58°

Ekofisk

Denmark

IRISH SEA

SEA

Fed. Rep. of Germany

Neth.

Rep. of Ireland
U.K.

Severn

Dover   Bel.

ENGLISH   CHANNEL

U.K.
France

54°

G. OF
ST. MALO

50°

BAY   OF

Miles
0   50   100   150   200   250

Kilometres
0   50   100   150   200   250   300   350   400

BISCAY

—————— Median Lines

⬤ Oilfields & wells

⬤ Gasfields & wells

—————— Oil pipelines

············· Projected oil pipelines

—————— Gas pipelines

- - - - - Projected gas pipelines

⛏ Rigs drilling during period 1979

© Geographical Projects

First Steps in the Offshore Oil Quest

Oil exploration starts at a desk. A geologist looks at maps and formulates ideas about the location of sedimentary basins and the likelihood of oil being found within them. For land exploration the geologist often has some good clues: there are easily observable surface features; the geology of a region will often have been mapped; there will be the results of drilling for minerals, coal, and water; and sometimes there is evidence from past oilwell drilling.

The scientist assembles the facts provided by these clues under the headings of rock types, geological structures, geochemistry (which reveals the amount of organic carbon in a rock), and fossil content (which tells how old the rocks are).

In offshore work geologists have few of these clues available, and this is particularly true in previously unexplored regions. Their search is narrowed down to continental shelf regions where they concentrate on areas close to deltas because there are bound to be thick layers of sedimentary rock there. They can also assume that the geology of nearby land areas extends out under the ocean, but this assumption has to be made with care since it is often not the case – the northern North Sea is a good example.

Having assembled as many facts as possible, made a few guesses, and played some hunches, exploration over a broad area can begin. In little-explored areas a reconnaissance study to define a sedimentary basin is made by an aircraft that tows a magnetometer. This instrument detects and makes very precise measurements of tiny changes in the Earth's magnetic field. Different rock types have different intensities of magnetization and can thus be identified. Ship-borne magnetometers, with the probe towed beneath the surface, are rarely used on their own in oilfield exploration but have proved valuable as metal detectors. They are being used to detect buried pipelines and wrecks, for example.

An instrument called a gravimeter measures minute changes in the gravitational pull of the Earth. These changes are related to variations in the densities of the near surface rocks and can be used to interpret rock type and structure. This instrument is widely used in land exploration but rarely on its own for oil exploration at sea.

Having obtained a generalized geological picture of an area, the most likely parts of that area are subjected to a closer look. This is done by a seismic prospecting ship, which may also be equipped to obtain magnetic and gravity data.

A typical seismic ship operating in waters such as the North Sea is up to 250 feet long. It is most likely to have been purpose built, but can be converted from a freight ship or large fishing vessel. All seismic ships have one easily recognizable feature: tucked into a sheltered area at the stern is a big (10- to 12-foot diameter) reel, onto which can be wound a thick plastic

Above: a view from the stern of a seismic vessel showing the streamer in the sea. This streamer is a tube containing an array of pressure detectors. It gathers information for use in determining the existence of new oil fields.

Left: a seismic prospecting ship. This view shows the open hatch through which the streamer is paid out, although the streamer itself is not in place. Such a ship surveys primarily for oil, but may also collect magnetic and gravity data.

tube about 3 inches in diameter and up to two miles long. This is the seismic streamer, which consists of a polyurethene or PVC tube filled with cable oil, surrounding an assembly of pressure detectors and their connecting wires, supported by steel stress members. During a survey, it is streamed out from the stern of the ship about 30 feet below the surface where the pressure detectors, or hydrophones, relay their information back to the ship. As the ship moves forward at about four knots, an underwater energy source generates acoustic pulses at a depth of approximately 20 feet at 10 to 15 second intervals.

The energy used often to be supplied by explosives, but today the most commonly used source is a powerful pulse of released compressed air. The energy travels through the water and penetrates subsurface rocks to depths of more than three miles. The energy is reflected by subsurface strata, picked up by the hydrophones, and recorded in digital form on magnetic tape by highly sophisticated electronic equipment on board the ship.

On the completion of a seismic survey, the magnetic tapes containing the survey data are sent to a computer center. There the data is subjected to various computer procedures in order to convert it into a form in which it may be of use to the interpretative geophysicist.

One of the great breakthroughs in seismic exploration has been the development of a computer based technique known as common depth point (cdp) stacking. Information pertaining to the same depth point is obtained from a number of shotpoints by moving the ship only a small distance – typically a one-group interval – between shots. The computer gathers together all traces with a common depth point into a common depth family (cdf). Analysis of the relationship between the slightly different geometry of each trace due to the forward movement of the ship, and the variations in arrival time of energy reflected from a particular depth point that this produces, permits the average velocity of the energy wave between the surface and that depth point to be derived. All traces within a common depth family are then added (stacked) together with due allowance for the effects of the varying geometry. This has the effect of strengthening the recording of primary reflections of geological structures and of attenuating multiple and unwanted reflections.

Marine seismic propecting is that extreme rarity: an offshore task that is far cheaper, up to eight times so, and far more efficient than its land-based equivalent. This is mainly because a ship operates continuously on a 24 hour a day basis, weather permitting, and does not suffer from some of the time-consuming delays of land surveys such as cutting survey trails, drilling shot holes, and moving geophones and the cable.

As exploration moves into even more remote and hostile areas, the seismic industry will grow in sophistication. Computer power in the seismic industry is growing at three times the rate of that in ordinary commercial practice; seismic streamers and the related processing and recording equipment are being designed to handle 500 channels or more. Presentation of data attributes in full color and in three dimensions will make the results of the interpretative geophysicist even more precise.

Left: a closer picture of the seismic streamer, showing how it is prepared for operation. The open hatch from which it is paid out is seen at the left.

Below: the computer room on board the seismic vessel *Anne Bravo*. Data is processed by this system while the exploration for oil is going on.

Enter the Rigs

No matter how sophisticated the techniques and equipment that become available to the geologist and geophysicist, there will always only be one sure way to discover the presence of subsea oil – and that is to drill for it.

The first underwater drillings for oil were in the swamps of Louisiana, off California, and in Lake Maracaibo, Venezuela. But they were only an extension of land-based work, with drilling rigs built on wooden piles and connected to the shore by boardwalks. This technique is still used on a massive scale in the Caspian Sea, where piers curve out to sea with drilling rigs built from them on piles.

The first vessels to drill with no connection to the shore were inland barges, but they were confined to very protected waters. Fixed platforms were built, with their piles driven into the seabed 45 feet beneath the ocean surface. This technique was later refined to one that is still used today: most of the machinery, including engines and pumps as well as crew accommodation, are installed on a floating barge – a tender – moored to the platform. This means that the platform itself can be lighter and cheaper. After World War II, when offshore drilling began in earnest, there was a plentiful supply of old barges and ships ripe for conversion to tenders. The first commercially productive well to be drilled out of sight of land was developed in this way. It was in the Gulf of Mexico in 1947, and the platform built for that well is still in use today.

The disadvantage of using a fixed platform for exploration drilling is its very immobility. If no oil or gas is found, the entire structure has either to be dismantled or abandoned. This led the oil industry to develop truly mobile drilling units, and this development proceeded in three stages. At first, the surplus fleets of World War II were again tapped for this purpose. Ships were fitted with massive beams so that a drilling derrick could be slid out from the side of the ship. This obviously did not do much for the stability of the vessel so that the first drillships could work only in very calm weather.

At about the same time, drilling rigs began to be mounted on submersible structures which had big pontoons at their base. There were up to six columns supporting a deck that typically would be 180 by 150 feet. This rig had an obvious advantage over the drillship in that the layout of machinery, accommodation,

Left: the many drilling rigs of Lake Maracaibo in Venezuela. This was one of the first sites in the world where underwater drilling took place.

and so on could be planned; there was also much more room to fit everything in. This kind of drilling rig was extremely stable because when the submersible barge was towed out to sea and its pontoons were flooded, it sank to rest securely on the seabed with the working deck clear of the waves. When the drilling was finished, the pontoons were pumped out, the rig rose up in the water, and it was towed away. These were ideal drilling platforms, but they had one serious limitation, which was that they were operational in a water depth of only around 60 feet. A few of these units are still working in the shallow waters of the Gulf of Mexico.

The third development came out of the fixed platforms, and had to do with overcoming the immobility created by their need for legs to stand on. Why not, thought the engineers and naval architects, devise a unit that is mobile but that carries its legs around with it? The idea was pursued, and out of it evolved the self-elevating or jack-up rig. The first unit of this kind of rig to be built to drill for oil went into service in 1954. It was a barge with 12 legs that were carried above the sea surface in the hull when the unit was being towed to the drilling site. On arrival the legs were wound down through the hull until they touched the seabed in 40 feet of water. Air jacks then pushed the hull of the barge clear of the surface until it was free from the action of waves and formed an artificial island. Drilling then began.

It has to be remembered that these basic mobile drilling units were only developed in the 1950s. From

227

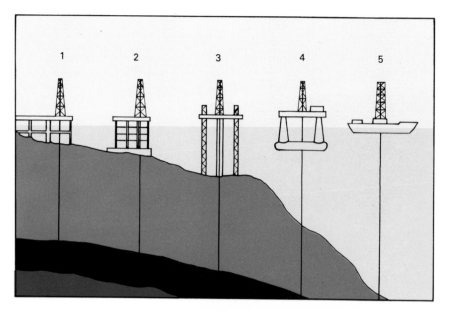

them evolved the designs of the exploratory rigs that are used today. It is always dangerous to make predictions about the course of technology and engineering, but it is extremely unlikely that any other fundamentally different type of exploratory drilling unit will be developed before the end of the petroleum era to supplant the three basic units now operating. These are the jack-up barge, the semisubmersible barge, and the drillship. That their basic designs have changed so little from those outlined above is a tribute to the pioneers of offshore oil exploration.

The first often perilously constructed mobile rigs worked close to land and in shallow waters. The results they obtained, in terms of the oil they discovered, soon made it plain that there were vast reserves of petroleum on the continental shelves of the world in areas that would take the oil industry right to continental margins and require drilling in depths of up to 5000 feet or more. Development was concentrated on increasing the depth capabilities of the three basic units for offshore exploratory drilling.

The jack-up drilling rig evolved on a logical basis. From the clumsy 12-legged early designs the number of legs on the units were reduced to five, then four, and finally three. The main development was in the design of safe, efficient jacking mechanisms – the means by which the legs are lowered from the floating hull down to the seabed and the hull is pushed clear of the waves. Electromechanical and hydraulic means were developed to achieve this, with the latter method now adopted as standard. The depth capability of this type of rig was soon extended to 300 feet, and there its disadvantages become apparent. To be able to operate in 300 feet of water, the jack-up rig needs legs that are nearly 400 feet long. The very nature of offshore exploration means that mobile rigs have to be moved to and from drilling areas often thousands of miles apart. A device that has three or four legs sticking up some 400 feet above its hull is an intrinsically unwieldy object to tow around the world's oceans. So legs that were capable of being disassembled for long ocean

voyages had to be devised, which was a lengthy and costly procedure. The shape of the hull of the barge also made the structure difficult to tow. Once the rig arrives on site there can be added difficulties. The seabed has to be carefully surveyed to insure that each leg of the rig achieves equal penetration, for example. Also the operation of jacking-in and jacking-out of the legs is fraught with dangers if a sudden storm blows up, and most accidents to jack-up rigs occur at this crucial time.

Various solutions to these problems have been and will continue to be proposed. Sails have been set on jack-up rigs to speed their progress across the oceans; ship-shaped hulls have been designed; entire jack-up rigs have been mounted on vast barges for towing across the oceans; and telescoping legs that will enable the rigs to work in depths of over 500 feet have been suggested.

The jack-up type of rig seems destined to work in maximum water depths of 350 feet. Here it will doubtlessly reign supreme as the workhorse of the offshore drilling industry. More than likely as the price of oil rises, it will increasingly be used to go back to shallow water areas which were previously regarded either as uneconomical or played out. New shallow water areas opening up for the first time will also prove highly suitable for jack-up operations. Perhaps the best evidence for the continuing popularity of this type of rig is provided by the fact that at the time of writing, there are no fewer than 53 jack-up rigs under construction around the world, against three drillships and four semisubmersibles. The jack-up rig clearly has a large part to play in the offshore oil exploration programs of the future.

It must be constantly stressed that development of today's mobile offshore drilling units moved in parallel. At the same time as the jack-up rig was being developed, it was noticed that the big submersible barges, restricted by the very nature of their construction to shallow waters, were little affected by the waves that rolled past them as they were being raised and lowered. Their big pontoons kept the units stable in the water.

This led to the development of the semisubmersible drilling barge, and it is safe to say that if it were not for this development, many of the world's major offshore oilfields would not have been discovered. Taking that statement one step further, it means that the whole economy of Britain and Norway has been transformed because these giant barges were capable of withstanding the rigors of the northern North Sea environment. There would be no North Sea oil if it were not for the semis.

The semisubmersible drilling rig consists of a platform mounted on between three and eight cylindrical legs attached to pontoons. The rig is towed into position in a light ballast condition, and the pontoons are then flooded so that the barge sits low in the water with the pontoons far beneath the worst effects of waves. This gives the whole structure a great degree of stability even in the worst storms. It is secured in position by up to 10 anchors, weighing up to 20 tons each, which are laid out by tugs. The semisubmersible can operate in water depths to 2000 feet. One of its disadvantages – moving its sheer 40,000-ton bulk – has been overcome by equipping it with a degree of self-propulsion so that it is capable of moving around the world with only one tug reinforcing its own power.

The semisubmersible's size enables it to carry enormous quantities of supplies of fuel and water, giving it a degree of independence from shore support.

The crew has ample living and recreation space, and there is room for the chef to practice the most exacting culinary arts. There is also room for meeting special preferences: on French-built and operated semis, some of the storage space is given over to tanks of the vin ordinaire without which no Gallic meal is complete; on some Scandinavian rigs, that same space has been adapted to include a sauna.

The semisubmersible seemed to many to be the perfect drilling unit, especially for the North Sea, and such vast numbers were commissioned in the early 1970s that the market was flooded with the giant vessels. The result was that many had to be adapted to act as floating production facilities, as floating hotels for construction crews, or as offshore firefighting units. In many ways in fact, this glut showed the very versatility of the semisubmersible barge and insured its role in the development of offshore oil and gas.

Above: the canteen of an oil rig in the North Sea. Crew members have full dining, living and recreation facilities for their non-working hours.

Opposite: a typical drill ship. Although drillships have the advantage of being more mobile than other exploratory rigs, such mobility makes them less stable in rough seas. This makes them unsuitable for use in the North Sea during the winter.

Left: a semisubmersible drilling rig. Semisubmersibles are constructed and positioned so that they have a great degree of stability during storms.

Against the superior stability and increasing mobility of the semisubmersible rig, the drillship seemed doomed to early extinction. The disadvantages of the drillship in having to have the drilling derrick skidded out from its side, had been overcome by designing it with a hole in its midships – the moonpool – through which drilling operations were conducted. But it still had the disadvantage of having a ship's shape which made it the victim of all the sea could throw at it. However, the drillship has found its own niche in the field of exploratory drilling which it will maintain for many years to come. The drillship is the perfect unit for pioneer exploratory work in remote areas in great water depths. In transit from one exploratory region to another, it functions as a conventional ship with its own propulsion and with no need for attendant tugs. It carries large quantities of supplies and can operate on remote sites independent of shore support for longer periods than any other type of drilling rig. Once on site, it is capable of drilling in deeper waters than any other rig; commercially operated vessels have drilled in water depths of over 5000 feet. The drillship is able to maintain station with no anchors through the use of dynamic positioning. Briefly, this technique involves the deployment of seabed transponders which are interrogated by the ship and its exact position relative to them computed at all times. If winds and waves conspire to push the ship off station, this is "sensed" by the computer which sends a message to propellers in the sides of the hull. These drive the ship forward, backward, or sideways a few feet to keep it in position. The fact that the ship is connected to the bottom by a pipe – the marine riser – that has only a certain degree of flexibility, shows just how sensitive the dynamic positioning system has to be.

What Happens on a Drilling Rig

Many people are disappointed when they visit an exploratory drilling rig. The helicopter trip is fun and the first sight of the rig, often the only object in miles of sea, is exciting. If the visit is to a semisubmersible rig, the sheer size of it cannot fail to impress.

It is on board that a feeling of let-down can set in. On a calm day the motion of the sea cannot be felt. There is the hum of machinery and the smell of diesel fuel. The crew members on duty move quietly about in overalls and hard hats and those off-duty are either asleep, eating yet another gargantuan steak, or watching the eighth rerun of an American football game on the video player.

While unexciting for the visitor, the quietness is as it should be. The massive structure in the middle of the ocean is there for just one purpose, and that is concentrated on the slowly turning section of pipe plumb in the center of the platform. The rig is "making hole" and all is well.

Visit the rig on other occasions and there will be a great deal more action, particularly during the first few days after arrival at the drilling site.

The exact location at which the rig should start to probe for oil will have been decided perhaps two years before by geologists. The location will have been very precisely fixed. Then the rig uses extremely accurate radio navigation or satellite fixes to position itself within a few feet of the chosen spot. If it is a giant semisubmersible, tugs scurry around sinking up to 10 anchors.

When the rig is held securely in place, "spudding-in" begins and this is another period of intense activity. A 2.5-foot diameter hole is drilled or driven to a depth of 200 feet or more. Steel casing is run into this hole and is cemented into place. A steel plate – the temporary guide base – is then lowered on wires to the seabed to mate with the top of the casing. The next step is to use the same set of guide wires to lower to the guide base a huge assembly of valves, cutters, and rams known as the BOP or blowout preventer stack. This can be up to 20 feet high. If the well "kicks," meaning that pressure builds up in the hole and begins to bubble dangerously back toward the surface, the valves are actuated from the control cabin on the rig to operate the rams. These close across the hole and prevent the gas or liquid from flying up to the rig floor. Sometimes the valves are not closed in time and the liquid or gas reaches the surface in a potentially perilous blowout.

With the first run of casing cemented in and the guide base and the BOP stack installed, drilling starts again with a smaller bit. More casing is run into the hole and cemented into place. The sequence is repeated using smaller and smaller diameter casing until the drill bit has arrived at its "target depth" –

Above: members of the drilling crew making up a length of drill pipe for landing the main guide base on the seabed. This is called "spudding in."

Left: a new drill bit being lowered. The bit is fixed to the end of the drillpipe and, when it arrives at its target depth, oil drilling starts.

the area in which it is hoped that hydrocarbons will occur.

The drill bit is fixed to the end of the drillpipe, small diameter tubing, each section of which is pulled upright under the derrick by a specialized winch called the draw works, and joined to the previous section of pipe as the hole becomes deeper. This drillstring is turned by a turntable (the rotary table) in the rig floor. As the drill advances, the cuttings have to be removed

and this is achieved by a drilling matter known as mud. "Mud" is a deceivingly simple term for a complex water or oil-based solution of chemicals which are carefully formulated for each stage of the well. The mud keeps the sides of the hole from caving in before the casing is installed. It also cools the drilling bit and displaces the cuttings, bringing them back to the surface via the marine riser.

When the drill bit becomes blunted on its downward

Right: engineers examining a core sample. This is one of the tests done to make sure of the presence of oil before full-scale operations begin.

Below: a blowout preventer (BOP) stack on a semisubmersible drilling rig. An idea of how big this piece of equipment is can be gained by comparing the size of the worker at its base.

path, another flurry of activity takes place on the rig as the entire drillstring – perhaps 10,000 feet or more – is lifted out of the hole, broken down, and then re-joined when the bit has been replaced.

The cuttings and any fluids brought up with the recirculating mud are carefully separated and, when the drill bit is approaching its target depth, these are closely studied by the petroleum geologist because they provide the first clues to the presence of hydro-carbons. Some of the biggest fields in the world have first been spotted by a very slight smear of oil on a tiny chip of rock.

Other tests, such as taking cores at the bottom of the hole or making radioactive and electric tests called "well logging," all help to confirm the presence or lack of hydrocarbons. One thing that never happens when oil is discovered today is the huge gusher. This would be a sign that the well was out of control and in extreme danger.

After all this effort and expense (it costs around $7 million to drill a well in the northern North Sea), what are the chances of discovering oil? Fairly slim. On a worldwide basis the average success rate is one in every 25 wells drilled. Even if oil is discovered on the first attempt in a new area, there is still much work to be done before the field can be judged to be commercial and development work can begin.

233

Developing an Offshore Field

Exploration drillers tend to become almost paranoid about anything that interferes with the steady routine of "making hole" because of the pressure to drill as quickly as possible and get on to another exploratory location. They seem to resent the well loggers who spend hours probing the borehole with electronic instruments; they regret having to stop drilling while divers inspect the BOP stack.

Drillers will therefore be among the first to welcome one of the latest areas of research in the industry, which is the development of changing the bit at the bottom of the hole without having to trip, or retrieve, the entire string. They will also appreciate the trend toward measurement-while-drilling (MWD) by which geologists will be able to obtain a constant record of what is happening at the face of the drill bit without having to delay the drilling operation.

The costs of exploration rise so sharply in remote, hostile areas that the money being invested in developing these timesaving techniques will bring swift and profitable returns. It will enable oil companies to assess more quickly the recoverable reserves of hydrocarbons in an exploratory well.

When a flow of oil is established from an exploratory well, it is carefully tested through a series of chokes that gives the drillers an idea of the size of the reservoir they have tapped. Before they can decide on the way in which that flow can be produced and brought ashore, they have to have far more information. So the exploration rig removes the BOP stack, caps the well with cement, and moves off to drill other wells in the vicinity to establish the extent of the reservoir. From this an assessment of the total reserves of the field can be made.

If the reserves are sufficient to justify production, the next set of decisions has to be made. The oil company has to decide upon the method that will be adopted for bringing the field into production, the way in which and exactly where the product will be brought ashore, and the way in which it will then be processed and distributed. Factors affecting decisions on any of these points include: the distance of the field from land; the water depth; the availability of shore-based processing facilities or transhipment terminals; and the political and taxation regime of the nation exploiting the oil resource.

Economists, geologists, and engineers have to juggle with the figures that are the result of such analyses, and then to decide whether or not their new-

found discovery is worth further investment. There is no rule-of-thumb formula for the economic viability of an offshore field. An offshore reservoir with X-million barrels of recoverable oil in the shallow, relatively calm waters of the Persian Gulf could bring millions of dollars of profits to the oil company that produces the field. In the northern North Sea, on the other hand, a field with the same amount of recoverable reserves might be totally uneconomical to produce because of its distance from land and the harsh environment.

It is not just a question of recoverable reserves, either. The shape of the undersea reservoir has to be examined to answer such questions as: How many production wells will have to be drilled to tap the full extent of the reservoir? How many platforms will need to be installed to drill that number of wells? What size should those platforms be? Should they be made of steel or concrete? How should the oil be brought ashore – by tankers from a loading facility installed on the field or by pipeline? How long will it take to build the platforms and pipeline and/or buoys, and how does this timetable relate to the potential profitability of the field? It must be borne in mind that valuable cash resources will be tied up for up to five years before a single drop of revenue-earning oil flows ashore, so a wrong answer can have stark consequences.

The questions go on and on, in fact, and even if the answers show that an oilfield is not economical to produce, the feasibility studies cannot necessarily be shelved. A sudden and sharp rise in oil prices by the OPEC nations could make a marginally profitable field a hot property. So the feasibility studies have to be brought out again, dusted off, and an inflation factor added.

All this happens very quickly after a discovery has been made. If these complicated preliminaries lead to a decision to develop a field, activity moves into high gear.

Above: an offshore oil rig under construction. The crew is working on one of the pile guides in preparation for piling in – that is, placing the support piles.

Left: crude oil and gas being burned off so that the well can be capped. Once this is done, other wells will be drilled in other parts of the field to determine whether the total reserves make further exploitation worth it.

Opposite: it's a long way down in this view from the drilling well tower. The helicopter base is seen at the bottom.

Platforms: Work and Construction

The most widely used structure for developing and producing offshore oilfields is the piled steel platform. It is built and installed in three sections: jacket, deck, and modules.

The platform jacket, which consists of tubular steel legs with cross bracings, is built on land on its side. When complete, it is winched down a slipway onto a barge and towed out to the oilfield. An alternative method is to build the jacket in a drydock and fit it with built-in flotation tanks that are removed as the jacket is installed. When launch day arrives the dry-dock is flooded, the jacket floats, and is then towed out to sea.

Whichever method is used, the towing and installation of steel jackets is an anxious business. Extremely accurate weather forecasts have to be obtained before the unwieldy mass of steel can be towed slowly to its required position, often hundreds of miles away from where it was built.

The jacket then has to be slid gently off the barge, and valves actuated by radio from a control vessel to enable buoyancy tanks to flood at a controlled rate

so that it assumes a vertical position and sinks gently to the seabed.

Immediately, giant barges move in and crews leap aboard the jacket, which is pinned to the seabed by the few piles that have been installed on its corners prior to float-out. The launch barge or flotation tanks are then towed away. The piling barges remain, working day and night to finish driving in the many piles. The most powerful piledriving hammers in the

Left: a jacket section, towed by tugs, on its way to a field in the North Sea. It is designed to have two cylindrical oil storage tanks attached to the legs. These act as buoyancy tanks and enable the structure to be towed unsupported to the place it is needed.

Below: the jacket breaks through the surface of the sea as it is being set upright. This is a crucial part of fixing a platform in place. It is possible to complete the operation in 12 hours if weather conditions are perfect.

Opposite: a jacket section just prior to its launch. The white flecks visible all over the structure are cathodic anodes, which provide protection against corrosion.

Right: driving piles from a derrick barge in the North Sea. This jacket will need 32 piles.

Below: one of the tanks that will be attached to the jacket legs, to be used both for storage and to give buoyancy.

world – whether on land or offshore – have been developed to work in the North Sea.

Once the jacket section has been piled-in, the deck is floated out, lifted, and slotted into place on the jacket. Then come the modules, of which there can be up to 20 per platform. These are simply containers housing generators, compressors, pumps, living accommodations, mess and recreation facilities for the crews, laboratories, offices, and so on. Each module is

towed out to the jacket on a barge and lifted by crane into its allotted position. The helicopter landing deck and the drilling derricks – usually two per platform – are winched aboard last.

The hook-up phase follows. The modules are interconnected for such services as power, water, heat, light, air conditioning, and ventilation.

With hook-up completed and essential supplies brought to the platform by ship and helicopter, the platform is ready to start the next stage of the oilfield development program – development drilling. The barges depart, the installation crews move to another platform, and the serious business of drilling starts.

Steel platforms can be enormous. Those in the deep waters of the northern North Sea can have jacket sections weighing up to 25,000 tons. Individual modules can weigh 2000 tons. The costs of building and installation are enormous too. The North Sea's Forties Field, for example, has four steel production platforms. Total capital cost of development of the field, including platforms and pipelines, was about $1500 million. But in terms of the total recoverable reserves of the field, the investment is well justified: that huge development figure represents a cost of 81 cents per barrel delivered to the shore. True, the oil has then to be refined, transported, and marketed; but with oil prices edging inexorably up, there would seem to be little doubt of a satisfactory return on investment.

The commissioning of a large production platform is a major challenge to technology and human ingenuity, especially in insuring that the timetable for profitable production is adhered to. In areas such as the northern North Sea, slippage in delivery time of a major component by one month – be it of jacket, deck, or key module – does not necessarily set the whole project back by the same period. It could put back the development of an oilfield by at least nine months.

237

In the North Sea there is a "weather window" for the installation of major items of equipment, which is the spring and summer period of late April to mid-September. Outside that window the weather is usually too bad, or at least too unpredictable, to risk towing and installing millions of dollars worth of equipment. So if a piece of equipment due to be delivered and installed in September runs late, it can well miss the weather window.

A major task for oil companies and their project management consultants is to keep close tabs on all suppliers. The complications can be easily imagined when it is known that a jacket section may be built in Scotland, the deck in France, and modules in Britain, the Netherlands, or Spain. In the end it is the sea that has the final word on whether or not a structure, or even a whole production complex, will be installed and working according to planned schedules.

The number of development and production platforms per oilfield varies according to the size and shape of the reservoir. Some fields will have just one structure. Others will have six. The object is to insure that every section of the reservoir is probed by drill bits, so the platforms are strategically located with each one capable of drilling up to 60 wells on a deviated pattern. Deviated drilling means that instead of drilling vertically, as in exploratory work, the wells are first drilled vertically but then fan out around the platform at angles of up to 30 degrees some 12,000 feet beneath the seabed. This taps a circular area with a diameter of up to two miles with the platform at the center. Development drilling of this type proceeds at a frenetic pace: the platform may well have two drilling derricks and double crews that drill round the clock.

Soon the oil starts to flow, from one or two wells at first, building to a peak a year or so later as all the wells are brought onstream. The drilling derricks are then removed and the platform becomes a production facility with up to 150 men insuring its safe, continuous operation. Processing equipment on the platform separates gas from oil, removes water and sand, and often treats the oil to remove wax and emulsifiers that would otherwise clog pipelines.

The North Sea has seen the pioneering of another form of development drilling and production facility. This is the concrete gravity platform, so called because it relies on its own weight to keep it in place on the seabed. The massive structure, which stands in up to 470 feet of water, can be over 30,000 tons and its base measurement 550 feet by 430 feet. The base is built in a drydock and then towed to a deep water anchorage just offshore. As more and more concrete is added to give the towers height, the whole platform gradually sinks to its designed ballast depth. It is then towed out to the oilfield and allowed to sink into place on the seabed.

The gravity platform has a number of advantages over its steel counterpart: the deck section and many of the modules may be installed close to land, so avoiding expensive and potentially perilous lifting opera-

tions offshore; it does not require piling-in; it can store up to a million barrels of oil within its columns, a particularly useful feature when oil is taken to shore by tankers and bad weather curtails loading operations; and the risers bringing oil to the processing facilities from the seafloor are protected within the concrete towers.

At one stage in the mid-1970s there was a great rush to build concrete platforms for the North Sea and there will probably eventually be about 15 of them installed. However, their universal use is restricted because of the need to have a very stable seabed and because suitable sites for their construction – close to the oilfield and with deep water close to the shore – are limited. Some doubts have also been expressed about the long-term effects on concrete of prolonged exposure to the harsh environment of the North Sea. For the time being the pendulum has swung back in favor of the steel jacket system, especially since the manufacturers of these structures are increasingly demonstrating their ability to install them in the deepest waters. In 1979 the tallest single-piece jacket yet was installed in the Gulf of Mexico from a barge. The water depth was 685 feet and the 12,800-ton jacket was 708 feet high. A single-piece jacket for 935 feet of water is scheduled for offshore Texas, and a structure installed in three pieces stands in 1020 feet of water off Louisiana.

Right: the first concrete
gravity platform ever used,
built in the North Sea, 1975.
Its own weight keeps it in
place on the sea floor.

Opposite: a large module
being lowered onto the
platform deck. There can be
as many as 20 modules on a
platform, used for housing,
offices, recreation. They
also contain equipment.

Steel and concrete platforms will more than likely
continue to be the main means by which oil is produced
offshore. There will be variations on the basic designs.
For example, steel gravity structures with storage
capacity have been built and others are planned, and
some fields will be developed by tension leg platforms –
mobile units with vertical cables anchoring them to
the seabed in a way in which the positive buoyancy of
the platform provides the tension to keep the anchoring cables taut. There is also a move to install more

and more of the production facilities beneath the
surface.

As a footnote, it is interesting to speculate on what
will happen in 30 years or so when the oil from major
offshore areas such as the North Sea has been depleted. What will happen to the platforms? Will they
be recovered and towed back to shore? Will they be
blown up and left to litter the seabed? Will they just
be abandoned to remain as gaunt reminders to future
generations of the petroleum era?

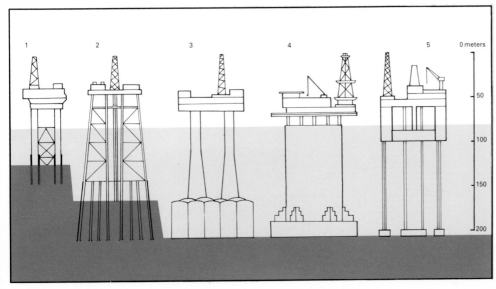

Left: examples of various
kinds of oil production platforms. **1** A straight-sided
steel tower. The most common and easiest to build, but
not the sturdiest. **2** A template structure. The broader
base is more stable but it is
costlier. **3** A ballasted reinforced concrete structure.
It can do double duty be-
cause its base can serve as
storage tanks. **4** An alternative design of **3**. Its one
large column stands up
better against the force of
waves. **5** A proposed tension
leg platform. The buoyancy
of the platform will provide
the tension to keep anchor
cables taut.

Semisubmersibles for Deepest Work

Sometimes an oil reservoir in deep water, far from land, does not contain enough hydrocarbons to justify building huge platforms and laying pipelines to shore. Until quite recently such reservoirs have had to be reluctantly abandoned. But rising oil prices and new technology have made the development of these marginal fields worth investigating. The techniques and equipment used have also proved invaluable in providing temporary production facilities on big fields so that the oil, and the capital from it, may start to flow during the four or five year period that elapses while permanent facilities are constructed. The North Sea has again been the proving ground for many of these developments.

Early production systems and marginal field developments vary from area to area and field to field depending on water depth and the size and shape of the reservoir to be tapped. Typically, however, instead of one or more fixed platforms and the drilling of deviated wells, a semisubmersible rig drills single vertical wells at planned points in the reservoir. On a fixed platform, the oil comes to the platform deck, passes through a block of valves known as the Christmas tree, then into a manifold to bring the separate flows together, and then to the production facilities described earlier.

In a typical early production system, each of the wells drilled vertically by the semisubmersible has what is called a subsea completion. This consists of a specially designed Christmas tree installed on the seabed over each well. Flowlines lead from each well to a manifold which sits on the seabed beneath the floating production facility – usually a semisubmersible rig – and the combined flow from each well passes from the manifold to that rig by a marine riser. After processing, the oil passes from the structure back down to the seabed by flowline and then to a mooring buoy at which it is loaded into tankers.

There are many variations on this basic system. For really deep water production, the oil can be made to flow from the manifold by pipeline to a production platform nearer the shore and in shallower water rather than to a semisubmersible/tanker loading combination situated on the field itself. Great cost savings are realized by this method because the platform is smaller and lighter.

Another variation is to use the semisubmersible rig both for drilling wells and as a production facility. On Buchan Field in the North Sea, a steel template has been installed on the seabed. This is basically a large flat frame with holes in it through which the semisubmersible rig drills four wells. The original exploration well, just a few yards from the template, is also being used for production. (On many fields the exploration well that discovered the field is abandoned because it does not fit into the pattern of wells that needs to be drilled to obtain maximum recovery from the reservoir.) Two satellite wells are drilled about a mile from the template and these are connected by flowline to the production facility. Treated oil then travels just over a mile to a buoy at which tankers load.

The Buchan Field development is costing its British developers about $275 million, a modest amount by today's offshore oilfield standards, but farther north

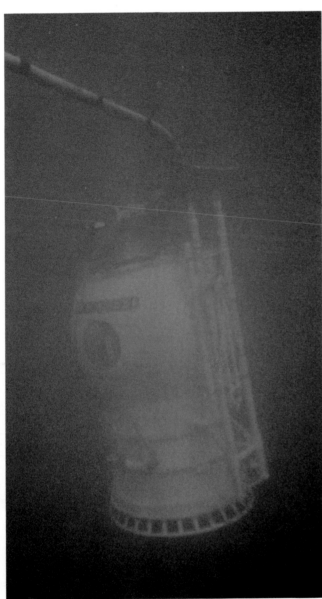

the same developer is spending a record $3000 million to develop its Magnus Field.

In Magnus, a central steel production platform will be installed in 613 feet of water and will drill 15 deviated wells in a pattern designed to tap as much of the reservoir as possible. Because the reservoir has a long and straggly shape, however, some seven satellite wells are being drilled by a semisubmersible rig beyond the area that can be covered by the production platform. These will be connected to the production platform, and from there a pipeline will take the oil from all the wells to connect with the Ninian Field pipeline farther south. This pipeline hits land at Sullom Voe in the Shetlands.

Another approach to seabed production is even more ambitious. The main proponent of this project is a Canadian developer. It starts out the same by drilling satellite wells at strategic points around the reservoir, but then the wellhead equipment installed on the seabed at each well is encapsulated in a "cellar." This is maintained at atmospheric pressure, like the central seabed manifold to which all the wells feed. A diving bell is lowered from a purpose-built support ship and "mates" with the wellhead cellars or the manifold. Engineers gain entrance into the wellhead through a lock in order to undertake routine maintenance work in an environment compatible to humans.

The main advantage of the system is that divers are not required to service the underwater equipment, and therefore conventional, well-proven land equipment can be installed in the cellars where it is not subject to the corrosive influences of seawater. A system of this type has been installed on the Garoupa Field off Brazil, and while it has had a fair share of problems, it will undoubtedly serve as the forerunner of many similar deepwater oil production facilities.

Above: the diving bell or service capsule that carries crew and equipment to one kind of seabed production center. It locks onto wellhead cellars that have a constant and acceptable environment.
Right: this diagram shows the interlocking system of drilling and production used in the Buchan Field in the North Sea. Oil from the subsea wells and satellite wells goes up to the same semisubmersible that does the drilling. After production is carried out, the oil is sent down the rig, across the seabed, and up to the loading buoy from which tankers pick it up for shore delivery.
Opposite: the one-atmosphere manifold center for deepwater oil production being towed out to the Garoupa Field in Brazil.

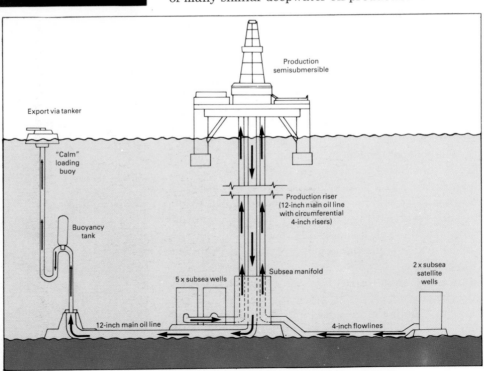

Export via tanker

"Calm" loading buoy

Buoyancy tank

Production semisubmersible

Production riser (12-inch main oil line with circumferential 4-inch risers)

Subsea manifold

2 x subsea satellite wells

5 x subsea wells

12-inch main oil line

4-inch flowlines

Transporting Oil from Sea to Shore

After discovering and getting oil out of the seabed, the next problem – and a major one – is getting the goods to market.

Oil and gas are hazardous substances and pose a threat to human life and the environment. Huge amounts have to be transported quickly yet safely. If pipelines transport one million barrels of oil a day, they are moving the equivalent of the volume of living space in 300 average houses, traveling at the rate of a houseful every five minutes.

The seabed pipeline is the safest way of transporting oil and gas and is the method most widely used. There are, for example, 18 pipelines taking North Sea oil and gas to shore processing plants in Britain, the Netherlands, and Germany. Some of them are installed in water depths of over 500 feet.

Before the pipe can be laid from the platform to the shore, the route has to be carefully surveyed to insure that it is not littered with boulders and wrecks, and that seabed conditions are suitable for burying the pipeline.

In shallow, relatively calm seas, flat bottom barges are used to lay pipe. But for deeper, more hostile waters such as the northern North Sea, semisubmersible barges have been developed. These work on the same principle as semisubmersible drilling rigs. They have two huge pontoons across which the work deck is mounted high above the water. These barges are the giants of the oil business – up to 550 feet long, over 190 feet wide, and capable of laying pipe in water depths of up to 1200 feet.

Steel pipe arrives from the mill at a coastal base to be prepared for its journey to the barge. It comes in 40-foot lengths which are shot-blasted, primed, and coated with coal tar glass fiber enamel. A concrete coating is then applied, but a short section at each end is left bare to enable the pipes to be welded together on the laybarge. The concrete gives the pipe the necessary weight to take it from the barge to the seabed, and also protects it against damage from trawl boards and anchors should it ever become unburied. Specially designed pipe hauling ships keep the laybarge constantly supplied with the coated pipe.

The pipelayer's dream is to have pipe snaking away from the barge to the seabed in a continuous line, 24 hours a day, from the moment the barge moves away from the production platform to when it reaches shore. Certainly the laybarge is designed to do this but, as ever, the weather usually prevents it.

The continuous process on board the laybarge is as follows: 40-foot sections of pipe are transported across the deck into the welding stations. There they are welded and an X-ray inspection made of the weld. The joint is then wrapped with metal and a layer of mastic applied around it.

A series of rollers grips the pipe and moves it toward the stern of the barge where it feeds down a ramp – known as a stinger – into the sea. The feeding continues on a carefully controlled curve to prevent cracking of the coating or kinks until the pipe rests gently on the seabed.

The laybarge deploys a number of anchors in a pattern and, as the pipe is laid, gradually winches itself along. Tugs assist in this operation. Checks are made to insure that the pipe reaches the seabed in good shape: ships with sonar and echo sounders plot its progress, and it is frequently inspected by submarines or divers.

When the barge approaches the shore, the water is usually too shallow to allow it to continue laying pipe right up to the beach. So a short section of pipeline – perhaps half a mile, but depending on how close the barge can approach – is laid from the shore and welded to the main line. In this case, the opposite process is

Below: a semisubmersible barge for laying pipe. Two immense pontoons support the work platform, from which pipe can be laid on the seabed in depths of up to 1200 feet. The laybarge is capable of operation on a 24-hour-a-day basis if the weather allows it.

used. Lengths of pipe are welded on the beach to form a continuous string which is winched out to sea by a small but powerful barge.

At the other end of the line, next to the production platform, the first section of pipe that has been laid has been fitted with a connecting device so that it can eventually be plumbed-in to the platform. This plumbing-in is achieved by mechanical couplings or by underwater welding. In shallow waters this is done by divers, but in deeper water it is done by hyperbaric

Left: testing equipment for "dry" welding underwater. Gas is pumped into the unit to drive out the water. This enables the diver, who still breathes from the surface support line, to weld in dry conditions.

Below: work in progress at a welding station on a lay-barge. Workers weld pipes of 40-feet long sections.

methods in which a chamber is fitted around the two ends to be welded, the water is pumped out, and the welding is carried out in the "dry."

When at last the connection between subsea oilfield and land-based processing plant has been made, there is still more work to be done before oil or gas can flow. A device known as a pig is pushed through the entire length of the line by compressed air or water to check that it has retained its oval shape and to clear it of any minor debris. The whole line is then flooded with seawater and pressure-tested to check for leaks.

Above: a diver inspecting pipelines under the sea. Pipelines have to be inspected frequently because of the many things that can go wrong in deep water. Inspections are also made by sonar equipment and submarines.

To give additional protection to the pipeline and its precious if dangerous cargo, it is then buried to depths of between one and three yards beneath the seabed. This process is usually undertaken by a bury barge, a flat-bottomed craft that tows a high-pressure sled along the seabed. The sled is fitted with jets through which water under pressure dissolves or cuts into the seabed, creating a trench into which the pipeline gently falls. Seabed currents can then usually be relied on to fill the trench with sediment and bury the pipe.

The sea is notoriously unreliable, as we know. Frequently pipelines become unburied and become vulnerable to having their concrete coating cracked by a trawl. The worst situation is one in which the seabed is scoured away from beneath yards and yards of pipeline so that a span is created and the pipeline flexes dangerously in currents and tides. The story is told of a diver who was put down in the southern North Sea to examine how severely a pipe had spanned. After some time on the bottom, the diver reported to the surface that he could see no sign of the pipe, which meant that it must still be buried. The surveyor on the ship disagreed and told him to keep looking. The diver grumbled but continued to plod diligently around the seabed. Suddenly the ship's crew heard a frightened squawk from him: the pipeline had spanned so badly that instead of looking down he should have been looking up. Out of the gloom the pipeline advanced toward him and passed over his head!

This points up why pipelines have to be inspected frequently, initially by ships with sonar and sub-bottom profiling equipment, and then by submarines and divers.

Start to talk about the costs of pipelaying and it quickly gets into the realm of incomprehensible figures. In the northern North Sea, for example, where weather can keep barges, aircraft, boats, and men sitting idly for days at a time, total costs can leap to £2 million for every mile of pipe laid. Some oilfields are so far from shore that a cost-per-mile of this order

Above: the Dutch-built buoy in the North Sea that is a collection point for tankers transporting oil to shore. The lower section is for storage, the middle one helps resist the waves, the top one is the mooring and loading unit.

would be completely uneconomical, particularly if the oilfield was small. In such cases the second main method of offshore oil transportation is used: tanker loading at the field.

Various types of tanker loading have been developed, but they share some of the following basic principles. A flowline is constructed along the seabed for a short distance from the production platform. From a seabed collection point, pipes or hoses lead up to a moored buoy. The tanker moors up to the buoy, hoses are connected from the buoy to the tanker, and the oil is pumped into the ship's tanks.

That sounds simple, but it is not. In the often severe

Above: the tanker-buoy transport system. Hoses are connected to the tanker from the buoy and the oil is pumped through them into the ship's holds. Tanker loading is the best delivery system for small, distant fields.

and rapidly changing weather conditions of the North Sea, the tanker must be moored by the bows to a flexible arm or swivel at the top of the buoy so that it is free to swing according to the dictates of wind and tide. But sometimes the weather is so wild that loading operations must stop. It is costly and time-consuming to stop the flow of oil from a subsea well, so various forms of storage have been devised. Sometimes these facilities are in the production platform, sometimes they are in the buoy itself. On Brent Field in the North Sea a huge Dutch-built buoy with the configuration of an angler's float is used. The lower section has storage and ballast tanks. The middle section is a cylinder

with reduced diameter at the waterline to present the least resistance to waves. The upper section is the unit to which the ship moors; this contains pumps, accommodation for a crew of 12, diving equipment, a helicopter landing pad, and cranes to handle the loading hoses.

The difficulties attending any offshore loading operation are readily apparent: hoses may kink and break, getting destroyed themselves and causing pollution; a prolonged spell of bad weather could mean that oil production has to shut down when storage facilities are exhausted; there has to be a convenient shore terminal at which the tanker may discharge. Oil producers prefer pipelines. But for the development of small fields many miles from shore, tanker loading will probably continue to provide a next-best solution to the problem of getting the product to consumers.

Caribbean and Gulf of Mexico

Anchovy   Hake   Grouper   Menhaden

Oilfields & wells
Gasfields & wells
Oil pipelines
Gas pipelines
Oil rigs drilling

Mullet

Offshore drilling for oil and gas

Shrimp

Snapper

Tuna

Venezuelan sardine

Support Services

In the late 1960s, when pipelines were being laid to bring ashore the gas discovered in the southern North Sea, a big pipelaying barge moored off the eastern coast of England near the port of Great Yarmouth. It was close to shore, sheltering from a spell of particularly bad weather, and early morning strollers must have marveled at the size of this great vessel.

On board there was little for the crew to do while they waited for the weather to lift, so the films shown in the recreation room made a pleasant break in the routine. Naturally there was understandable consternation when the bulb in the projector unexpectedly blew and there was no replacement on board. An urgent call was made to Houston, Texas and a supply of spare projector bulbs was sent express air freight to New York and from there to London. From London's airport a representative of the pipelaying company collected them, went to Great Yarmouth by car, and finally got them to the barge by boat. All that time and just a few hundred yards away from the barge's mooring, the local photographic dealer of Great Yarmouth had ample stocks of replacement bulbs for the Japanese-made projector.

That story may well be apocryphal but it points up the way in which the oil industry had to rely on head office to supply nearly all its needs when it first began operating on an international basis. Of course, the earliest operations were conducted in areas where shore facilities were often minimal – off Africa, in the Persian Gulf, and in the Far East. So the operators of rigs and barges took with them as much as possible in the way of supplies and established shore bases with small boats to transfer crews and to ferry out consumables like drilling fluids, water, and fuel oil. Major supplies and items required unexpectedly – like replacement projector bulbs – had to be shipped from wherever the home office happened to be.

Ferrying was all very well when operations were conducted close to shore and in shallow, relatively calm waters. But with the development of oil and gas resources in deep rough waters far from shore, such as the North Sea, it became necessary to have much more elaborate back-up support.

The ideal shore base to support offshore operations in such areas must meet a number of requirements. It must have a sheltered harbor with adequate depth of water at all states of the tide to enable supply ships to operate at any time of the day or night. It will also have what the oil industry would call "an established infrastructure." This means a nearby airport, good road and rail links, hospitals, and availability of housing, warehousing, and offices. Given these conditions, the oil companies and drilling contractors are able to move in their key personnel with their families. The vanguard of offshore operations is usually American, and many American children of the next generation will grow up with as much ex-

Opposite top: a hotel near the airport in Aberdeen. One of the changes that came with Aberdeen's fast growth was an increase in hotels.
Opposite above: mobile homes help meet the housing shortage in the fast-growing town. These are centrally heated and fully furnished.

Left: an aerial view of part of Aberdeen harbor in Scotland. Aberdeen has become a boom town since the North Sea oil industry has been centered there.

perience of overseas travel by the time they reach university age as most people get in a lifetime. Two- or three-year tours of duty are normal, so that an American child could live in Canada, London, Aberdeen, Singapore, and Athens during his or her childhood. A worldwide network of American schools has been established, with very similar curriculums, so that children are able to take up their schooling easily each time the family transfers to a new area.

The arrival of the oil industry in an area can bring far-reaching economic and social changes – some good and some bad. Aberdeen, on Scotland's northeast coast, is called the "offshore capital of Europe" and is a good example of the ways in which these changes work. Before the oil industry arrived in the Granite City in the early 1970s, it depended on the traditional industries of fishing, whisky distilling, shipbuilding and shipping, and on the farms of its hinterland. Unemployment, as in the rest of Scotland, was high and there was increasing movement of people to the south and overseas. The city, while not declining, was static.

Then came the oil industry. The northern North Sea exploration effort, based largely on Aberdeen, was incredibly successful. The average worldwide success rate in terms of the ratio of wells drilled to commercial

discoveries is around 25 to one. In the exciting early days of the northern North Sea, that ratio was a much higher eight to one. As a result, Aberdeen attracted oil companies like a magnet.

Soon brightly painted ships in the harbor made a pleasant change from the sight of the rusting remains of what had once been a proud trawler fleet; low-rise industrial complexes sprang up on the outskirts of the city; roads were improved; more hotels were built; and the spartan airport facilities were improved, even if the chances of taking off or landing on time decreased in direct proportion to the additional number of planes made available to cope with increased demand for flights.

Those were physical changes, but there were other, more complicated pressures. Property prices soared to become the highest in all of Britain. (There was even a brisk trade in castles.) Young couples not earning the high salaries paid by the oil industry found it impossible to buy a house. Traditional industries suffered initially as people left to join the oil industry. And there was always the nagging doubt, as Aberdeen enjoyed its new-found prosperity, about what would happen when the boom ended. Would Aberdeen be left a ghost town?

Fortunately, many of the industries that arrived to provide temporary support for offshore operations liked what they found and have stayed, establishing manufacturing facilities to serve not only the North Sea but the Middle East, Africa, and other areas.

Visitors from all over the world now go to Aberdeen to study the effects offshore oil has had on the city, and to learn how it has come to terms with the new industry.

What does the visitor see? The impressions start at the airport where American southern drawls and Scottish brogues mingle and are frequently drowned by the roar of a 30-seat helicopter taking a crew out to a rig or platform. Once in Aberdeen proper, visitors make their way through the granite-lined streets to the harbor which has been drastically improved to enable supply ships to sail in and out on any state of the tide. It is worth the visitor's while to spend some time looking at these sturdy ships; they are the lifeline of the offshore industry.

Supply ships constantly ferry essential supplies from shore to rigs, platforms, and barges hundreds of miles out in the North Sea. They collect drilling mud and cement from brightly painted silos on the quayside; fresh water and diesel fuel is pumped into their tanks; replacement equipment and spare parts are loaded in containers onto the low flat deck behind the wheelhouse structure.

As the ship approaches the rig or platform, the skipper has to use all his skills to turn the craft to back toward the structure, using the additional power provided by auxiliary propellers mounted in the side of the ship to aid maneuverability. The captain works closely, via radio, with the crane driver on the rig who picks up his mooring lines and passes them back down

Left: a supply ship moored alongside a rig in the North Sea on a calm day. Gas can be seen being flared off.

Opposite: cargo being unloaded from a supply ship on a rough day. The job of unloading has many hazards on an oil rig.

Below: a view up the legs of a rig, showing how fish gather to scavenge or to escape fishing boats.

again so that the ship is securely fastened to the structure. Then the crane driver swings the big hook down again and again to pick up the deck-mounted cargoes. This can be a dangerous task for the ship's crew as they struggle, often at night, on a sea-washed deck to attach the wildly swinging and heavy crane hook to the deck containers. This activity will usually be watched by the skipper from his vantage point aft of the bridge, although from time to time he may well pay some attention to the boat's fishing lines. Fish are attracted to offshore structures. The scavenging species can get rich pickings from galley scraps thrown from the rig and are safe from trawls and purse seines because fishing boats are not allowed within half a mile of an offshore installation. So supply boat crews catch fish to eat on their homeward journey, and frequently supplement their incomes by selling the results of a good night's fishing in port.

The larger supply boats double as anchor handling tugs for semisubmersible rigs. They tow the rig onto location and then run out the huge chains and anchors.

250

Each rig can have up to 10 anchors and each anchor will weigh up to 15 tons. When drilling is complete, the ship returns and uses all its power to break the anchors out of the seabed.

Supply ships are capable of operating in the worst weather that the notoriously bad northern North Sea can throw at them. But sometimes the weather is too bad for even these tough vessels to moor to the rig or production platform once arrived there. If the spell of such bad weather is protracted, essential supplies soon run out. On an exploration rig the "downtime" – those periods when the drill bit is not making hole – costs thousands a day. On a production platform the problem is even more serious because the flow of oil may have to be stopped. If a platform on a big North Sea oilfield is producing up to 300,000 barrels of oil a day and that oil sells at around $25 a barrel, then each day that the platform is shut down because the supply ship cannot get alongside costs the oil company $7,500,000. There is a faint irony in the fact that these colossal losses may be incurred because the supply

ship cannot deliver – of all things – fuel oil.

When platforms are being installed and pipelines being laid, our harbor visitor will see scenes of intense activity. Besides supply boats there will be pipe carriers. These look like big supply boats except that they have a high freeboard on the after deck to accommodate hundreds of joints of pipe. They pick these up along the country's east coast from yards where the bare steel pipe is coated with mastic and concrete. The ships then run it out to the pipelaying barges that are gradually making their way to shore.

The visitor will see still bigger ships, their aft decks crammed with equipment for the underwater phase of construction work as well as the subsequent inspection and maintenance tasks. Often they are fitted with helicopter landing pads (helidecks) and from the large gantry at the stern can handle diving bells and small manned or unmanned submarines.

Even Aberdeen Harbor is not able to accommodate the true giants of the offshore industry – the pipe-laying and crane barges. To see these the visitor to

Right: a supply base in
Aberdeen. The pipes that
will be used for transporting
oil from field to shore are
seen in the foreground.

northeast Scotland must drive 30 miles north along
the coast to the vast harbor at Peterhead, constructed
in the last century by the prisoners from the local jail.
There the visitor might see one of the new generation
of offshore crane barges. Even from pictures it is not
possible to gain an impression of the sheer size of
these vessels. One, the *Balder*, carries the largest
revolving crane in the world.

The underwater floats that give *Balder* its stability
measure 360 feet long by 79 feet wide by 36 feet high.
The columns connecting these floats to the deck are
each 61 feet wide. The deck, which is nearly 140 feet
above sea level, is bigger than a soccer field.

Balder and other vessels of this type are so stable
that they are able to work in the North Sea all the
year round on a variety of construction tasks, the

Opposite: a supergiant offshore crane
barge, the *Hermod*, seen lifting modules
onto a platform. Like its sister ship,
the *Balder*, it carries one of the world's
largest revolving cranes.

Left: a support vessel of the kind
used in the North Sea, moored alongside a platform. It is equipped to serve
divers and firefighters and also provides
a helicopter deck on top.

chief of which is lifting deck sections and modules onto jacket sections.

What the visitor will not see at any of the supply bases that dot the east coast of England and Scotland is a single drop of what all this massive support effort is ultimately all about – oil. In Britain the precious product comes ashore at four points. At one of these – Sullom Voe in the Shetlands – is located the massive complex responsible for processing half of Britain's crude oil requirement – over 1.5 million barrels a day. Costing over $1600 million, it takes oil from all the northernmost fields in the North Sea, an area known as the East Shetland Basin.

There currently. is an ambitious project to run a network of pipelines around all the major producing oilfields that also have quantities of natural gas. This gas bubbles to the surface with the oil and is separated on the production platform. It is then used to drive machinery, returned to the wells to maintain the pressure of the reservoir, or simply flared off. It has been estimated that the equivalent of 15 percent of Britain's gas supplies were being burned off in this way at one stage. With the rocketing price of energy, it cannot any longer be allowed to waste valuable fuel by simply burning it, and the lessons learned in how to deal with this problem will stand the oil industry in good stead as it moves to even more remote and hostile regions of the world.

The offshore oil industry is in the paradoxical position of being able to see the end of its life clearly while knowing that the next few years will require it to summon up every technological and human resource to wring out the last few drops of this precious liquid. Such efforts will not be wasted. More than any other marine activity, the offshore oil industry has taught the world about many aspects of the sea, including geology, oceanography, marine engineering, naval architecture, and chemical processing.

Chapter 10

Diving
to the Depths

The human body is not built to perform efficiently under water, but as yet there is no machine that can substitute entirely for its unique gifts. This is the paradox that has to be faced by anyone engaged in diving. The work that divers do – from discovering sunken cities in the Mediterranean to repairing pipelines in the North Sea or taking underwater photographs in the Pacific – requires special nerve, skills, and training. It also requires special equipment that modern technology is more and more able to supply. Diving is often dangerous work. The ills and accidents that can afflict divers can sometimes cause disablement for life – if, in fact, the result is not outright death.

Divers are exposed to nitrogen narcosis, oxygen poisoning, and the dreaded disorder commonly known as the bends. In spite of all this – and with the help of new equipment – the depths barriers to diving are being steadily broken.

Opposite: alone in the depths of the sea, a diver faces great physical and psychological hazards. The sea fights human intrusion – but people go on diving ever deeper.

Frail Body against Powerful Sea

The human body is ill-equipped to go underwater. Just duck your head beneath the surface of the bathwater and you soon appreciate that fact. You cannot breathe, your eyes sting, and the water gets into your ears. Stay in the water for any length of time and you not only get very cold but also your skin takes on the puckered look of a prune. This happens just in a bath. Get to even swimming pool depth and all sorts of unpleasant additional problems to underwater adventures soon make themselves felt, some immediately, some in the long term after prolonged exposure to the underwater environment.

Problem one is pressure. The human body is by nature equipped for a dry land environment. At sea level the body is subjected to a pressure of one atmosphere, which is 14.7 pounds, on every square inch. Pressure increases with depth in seawater by one atmosphere for every 33 feet. At 33 feet, pressure is double that of the air above water. At 1000 feet there is a pressure on the total body surface corresponding to a weight of six tons, or over 450 pounds per square inch. Pressures of this magnitude are sufficient to crush such a relatively frail mechanism as the human body. They do not because divers insure that they are taking in a breathable mixture at the same pressure as the water surrounding them. Because the solids and fluids in the body are virtually incompressible, only the air spaces need to be neutralized against the crushing pressures of the water outside.

Left: diving in the shallows requires little more than a device such as a snorkel for a supply of air, a face mask, and fins.

Opposite: a scuba diver in action. Scuba diving gives great freedom of action with easy-to-carry equipment for the supply of air.

Even a nondiver can test this fact. You can stay underwater for a considerable time by breathing through a tube that extends above the surface. But you are able to do this only in a very shallow depth, perhaps a couple of feet. In a greater depth, the pressure of water on your rib cage is too great to let you expand your lungs sufficiently.

The next step, if you are still keen to explore underwater, is to take a deep breath and swim downward, perhaps aided by flippers on your feet. As you go deeper, the volume of air in your lungs is reduced by the water pressure. Get to depths of over 100 feet and the water pressure really gets to work, squeezing blood into the lung air sacs and quickly killing you. For all practical intents and purposes, therefore, "swim" diving is limited to a depth of 30 feet at most. However, Aristotle wrote of the ancient trade of divers who, unprotected and clasping rocks to enable them to sink swiftly through the water, dove to harvest sponges at depths of 100 feet. In more modern times there are also records of male and female divers sinking to great depths to recover substances as diverse as coral and pearl-bearing oysters. They are exceptions, learning special skills and tricks of the trade developed from generation to generation.

For lesser mortals, swimming at depths greater than 30 feet requires a supply of a breathable mixture at the same pressure as the surrounding water. Diving apparatus has been developed to supply gas at the right pressure for the depth to which the diver swims.

The most common type of breathing apparatus has given its name to a sport. It is known as scuba diving, scuba being short for self-contained underwater breathing apparatus. (Professional oilfield divers who use altogether different equipment refer somewhat contemptuously to sports divers as "scoobie-doos.") Scuba divers carry on their backs one or two cylinders containing air pressurized to as much as 200 pounds per square inch. A hose from the cylinders leads to a mouthpiece that is often built in to the face mask. The diver inhales through this mouthpiece, which incor-

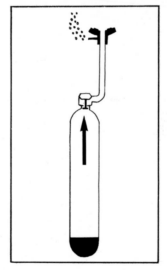

Above: a Greek vase dating from around 500 BC shows a diver about to enter the sea. In their search for sponges, the ancient Greeks are known to have reached depths of from 75 to 100 feet.

Left: an aqualung or scuba. Technically called an open-circuit demand set, it works by allowing an air flow only when the diver inhales. Gas waste is exhaled into the water in bursts of bubbles.

porates a demand valve. This valve gets it name because it provides divers with the right amount of gas at the right pressure at the right time as soon as they inhale. The valve has a diaphragm of synthetic rubber, which is exposed to the water pressure on the outside and which has an air-filled chamber on the inside. The diver inhales the air in this chamber through the breathing tube, and the water pressure then pushes the diaphragm down to fill the partial vacuum created. In its downward movement the diaphragm pushes a piston away from a seal, so permitting air from the diver's cylinder to flow into the air chamber and build up enough pressure to push the diaphragm back up

257

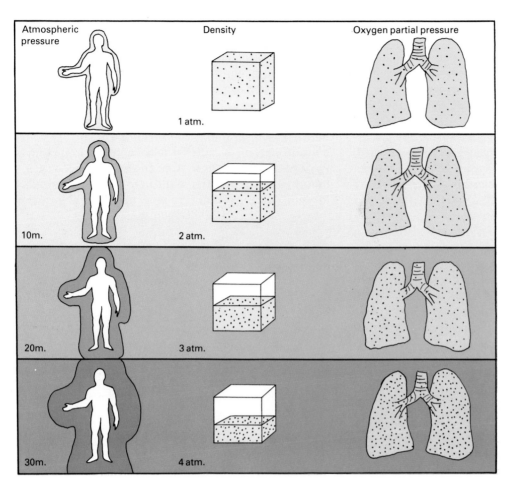

Atmospheric pressure	Density	Oxygen partial pressure
	1 atm.	
10m.	2 atm.	
20m.	3 atm.	
30m.	4 atm.	

Left: a diagrammatic illustration of how increasing pressures in a dive down to 30 meters (98 feet) affects the body. The first column shows how much the surrounding pressure on the body increases as the depth increases. The second column shows how much the gas breathed is compressed within the lungs, filling them with four times the amount of air as at the surface. The third column shows how oxygen (black dots) and nitrogen (red dots) exert an independent pressure on the lungs, called "partial" pressure. Partial pressure of each is nearly doubled with every doubling of depth and each presents a separate physiological hazard to the body.

again. The diver exhales the air into the sea through a non-return valve, sending up those little clouds of bubbles so characteristic of skin divers.

In warm waters scuba divers can wear nothing more than a swimsuit, flippers, mask, and cylinders. In colder waters warmer clothing is required, and this usually takes the form of a "wet" suit made of foam neoprene and fitting closely. It allows just a film of water to penetrate between suit and body, so increasing the insulating properties of the suit.

For prolonged work underwater, in opposition to sport, more substantial protective clothing is required.

Most people are familiar with the deepsea diving outfit consisting of a windowed copper helmet (or hard hat) and rigid shoulder structure connected to a baggy suit and lead-soled boots. Air is fed from the surface to fill the whole helmet so that the diver does not have to use a mouthpiece and can have the positive luxury of being able to talk to the surface by a telephone built into the helmet. Although still a great favorite with many divers who have to work in cold and dirty water – perhaps in a harbor clearing lines that have fouled ships' propellers – this standard diving dress has been largely superseded by constant volume equipment that consists of an inflated suit of synthetic rubber topped by a plastic helmet.

Pressure on the body is not the only hindrance to deep diving. Even the air we breathe, and which we take for granted on the surface, becomes a dangerous mixture once it is inhaled underwater via compressed

air bottles or from a supply pumped from the surface. This has to do with partial pressures of the gases that make up ordinary air. At sea level, at one atmosphere, air consists of 21 percent oxygen and 78 percent nitrogen. Dive to 33 feet where the pressure has increased by one atmosphere and the partial pressure of oxygen in the air breathed becomes 42 percent of one atmosphere.

Strangely enough, this life-supporting gas can become poisonous, and there are safe lower and upper limits of partial pressure of oxygen. If its partial pressure drops below 16 percent of one atmosphere, unconsciousness comes suddenly. If the oxygen content is above two atmospheres of partial pressure, violent convulsions afflict the body and death is swift. For diving, then, the partial pressure of oxygen must be kept lower than about 50 percent of one atmosphere. In practical terms this means that no one can go more than 200 feet deep when diving on ordinary compressed air.

The other, and major, constituent of air is nitrogen. Harmless when breathed on the surface, it becomes deadly underwater. Nitrogen is an inert gas which, at high pressures, dissolves in the blood and begins to interfere with the function of the nerves. This creates the dangerous condition known as nitrogen narcosis, sometimes called "rapture of the deeps." In its mildest stage it is a little like the first step of overindulgence in alcohol: a pleasant warm feeling comes over the diver and cares seem to fade. But soon come hallucina-

Above: a scuba diver wearing the protective clothing known as a "wet" suit. Water enters the suit at its edges, such as neck and cuffs, and is retained against the wearer's skin as a fine layer, or film, by the close fit. Body heat warms this film of water in providing insulation.

Below: a diver preparing to go in the water wearing a helmet or hard hat. The helmet, which has a front window, is attached to rigid shoulders on a baggy suit. Lead-soled boots complete the outfit. Not as widely used as it once was, this dress still seems to typify diving to most people.

tions, lethargy, and loss of control of movement. Underwater that can quickly spell death. Nitrogen is also a heavy gas, and as its pressure increases it places a strain on the lungs, making breathing difficult.

Oxygen poisoning and nitrogen narcosis both begin to have their effects at depths of around 200 feet, which rules out the use of ordinary air below that depth. Any breathing mixture must, of course, contain some oxygen, but the amount of the gas in any mixture has to be carefully regulated and reduced as divers go deeper. Around 1000 feet they must have a gas mixture containing only two percent of oxygen. The other 98 percent of the breathing mixture is made up of helium, a light inert gas with no narcotic properties. At first this gas seemed to be the perfect substitute for nitrogen, but it appears that every attempt to adapt the human body to an environment as alien as underwater produces as many problems as it solves.

The problem of breathing difficulties at depth was solved because helium is less dense than nitrogen. But the very lightness of the gas created its own problem. Being even more soluble than nitrogen in blood and tissues, it is absorbed very quickly. Because body tissues are not of uniform consistency, the helium permeates some faster than others. This has an unpleasant effect on a diver if the pressure of the breathing mixture is increased too quickly. When helium was first used in diving mixtures, divers were lowered straight to working depths of around 300 feet. They became dizzy, trembled, had pains in their joints, and fell over. The body tissues, absorbing helium at different rates, were simply out of balance. It was soon learned that the only way to overcome this is to increase the pressure very slowly, holding it steady at various levels for several hours while the body adjusts.

It can take up to 24 hours to get the diver pressurized and ready for work.

Helium is expensive, divers are paid a lot of money, and emergencies on the seabed often require quick action. So efforts continue to find ways of reducing the period of preparing divers for their deepsea work. One line of research has been to re-introduce a very small amount of nitrogen to the breathing mixture. Nitrogen, as we have seen, has a narcotic effect that can counteract the tremors induced by too rapid absorption of helium. In some laboratory experiments, divers in pressure chambers have been compressed to the equivalent of 1000 feet in depth in just over half an hour instead of the usual 24. There is opposition to using nitrogen, however. Opponents say that all the nitrogen does is suppress symptoms, not remove them, by inducing a mild narcosis that, mild or not, could be extremely dangerous.

Having withstood the descent to deep waters with a minimum of trembling, dizziness, nausea, or simply drowning, the human frame is beset by even more problems when it has to be brought back to the surface and is exposed to atmospheric pressure again.

Under pressure the inert gas, whether helium or nitrogen, dissolves in the tissues and blood. This is the process of compression. As the body depressurizes – a process called decompression – gas comes out of solution. Extreme care has to be taken to insure that this decompression does not take place too quickly. A too-rapid ascent to the surface makes the gas form bubbles in the blood and tissues and causes the serious illness known as decompression sickness, or more graphically, the bends. The bubbles in the blood can not only cause excruciating pain in the joints and then paralysis, but can also result in crippling or even death.

The only way to avoid decompression sickness is to reduce pressure slowly so that the gas bubbles do not form or, if they do, that they do not grow to a dangerous size. On really deep dives this decompression pro-

Left: these divers are undergoing decompression during the ascent from shallow depths. They hold onto a rope, pausing for intervals in accordance with a decompression schedule printed on a small board attached to the wrist. They are fully decompressed when they reach surface.

Opposite: divers in a decompression chamber. Decompression is a slow and delicate process, and divers must spend long hours of inactivity while returning to normal. This part of a diver's life as much as any demands a certain kind of special resistance and adaptation.

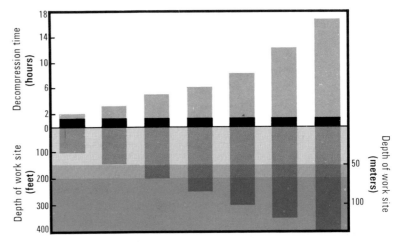

Right: this diagram shows the decompression time necessary during ascent from various depths after being down for a one-hour work period. The transition from air to oxy-helium breathing mixtures is shown by a change in tone between the depths of 150 and 200 feet.

cess is very slow. After spending an hour at a depth of 600 feet, a diver will need to decompress for nearly 40 hours. At deeper depths, the decompression time stretches to days.

Extensive worldwide research has been carried out by navies, universities, and commercial diving companies to draw up tables showing the amount of time needed to be spent in decompression in relation to the depth reached and the time spent at the depth. Many were the mice, monkeys, and goats that perished in the late 1950s and the early 1960s in the cause of learning how to bring divers back to atmospheric pressure faster but in safety.

One of the highly successful techniques to emerge from all this research was that of saturation diving. This is based on the fact that after a number of hours at a particular depth, the body tissues can absorb no more gas. They become saturated. When that point is reached the divers can remain pressurized for as long as necessary. They require no longer in decompression after 12 days than they would after 12 hours. Obviously, divers cannot spend 12 days at a time in the water – they would die of thirst or in a number of other unpleasant ways. What happens is that they are kept on the surface under pressure for a particular length of time – perhaps up to six weeks, including decompression time – and make occasional excursions to the depth for which they have been pressurized. Divers refer to this uncomfortable but highly lucrative part of their work as being "in sat" (in saturation). Saturation requires the mobilization of great amounts of sophisticated equipment.

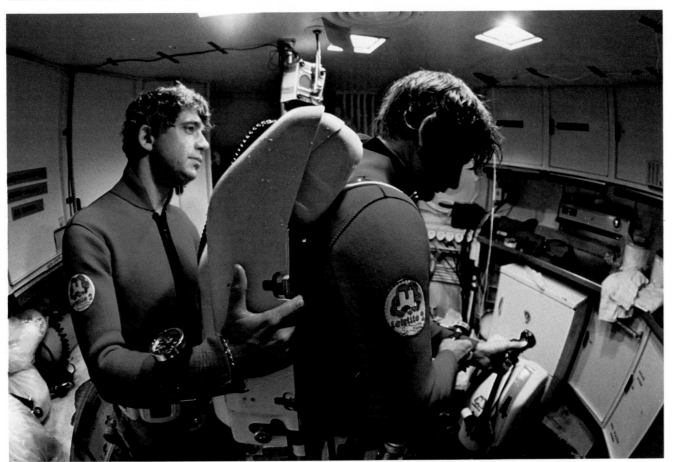

Going Deep and Staying Down

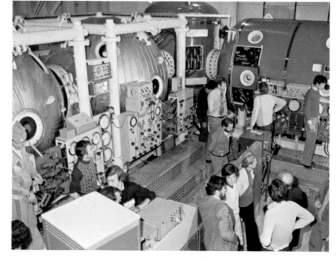

Why has there been a headlong rush to get deeper and deeper into the oceans, with all the dangers and discomforts that this entails? For it has certainly been a rush. At the beginning of this century diving had not changed much for hundreds of years. Then came the invention in World War I of self-contained breathing apparatus for escaping from submarines. World War II saw the introduction of compressed air cylinders that freed the diver both from the helmet and from dependence on the surface for breathing supplies. The main thrust of research came from navies, chiefly the British and American, to whom divers made a useful contribution in the waging of war. Private researchers, interested variously in physiology, in breaking depth records, and in publicity, also experimented with breathing mixtures, decompression times, and various types of equipment.

Until the early 1960s very few people had broken the 500-foot depth barrier. Aside from going to great depths "just because they were there," there was little point in it.

The oil industry changed all that. Suddenly the commercial impetus was provided to push divers to unprecedented depths to undertake essential work in exploration drilling, pipelaying, and underwater construction. Time and time again throughout the 1960s and 1970s the records for depth and time spent at depth were smashed. These records were being made mostly by French and American commercial diving companies.

Many of the problems associated with great depths are physiological, and therefore diving is experimentally simulated in dry chambers on land. Human subjects are pressurized under close medical supervision to the equivalent of 2000 feet in depth.

In commercial diving it is necessary not only to be able to stay alive at great depths, but also to undertake often strenuous work. So the next step in experimentation with techniques and equipment is to simulate ocean depths in huge waterfilled tanks. In

Below: how the SDC (submersible diving chamber)/DDC (deck decompression chamber) system works. From the locked position (far left) the SDC is unlocked and lowered to just below the surface for a seaworthiness test. The third sketch shows the divers at work while their co-worker inside the bell feeds them breathing gas by means of their umbilical hoses. The last sketch shows the SDC again locked into place and the crew inside the DDC. There they can either decompress or remain in saturation.

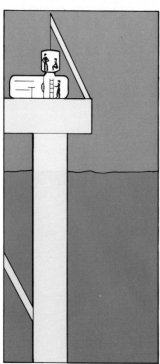
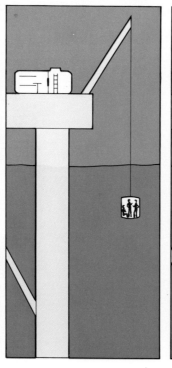
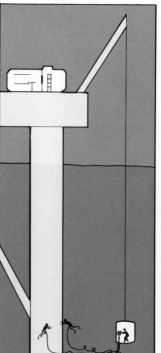
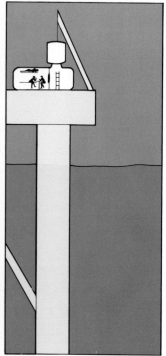

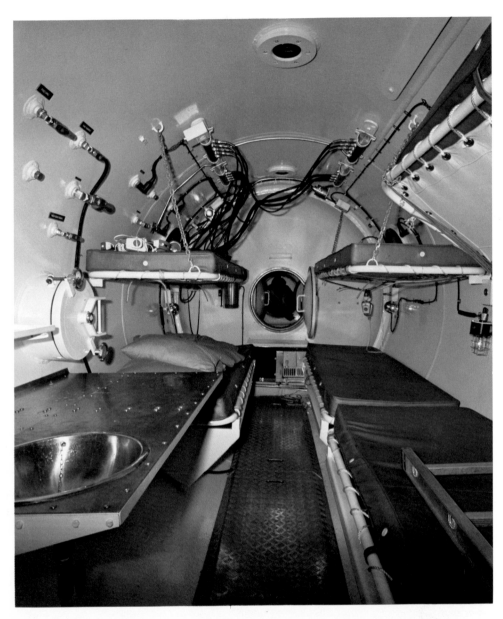

Right: an interior view of a typical DDC on board an offshore drilling rig. It is made as comfortable as possible for a confined space.

the United States Navy's Ocean Simulation Facility in Florida, for example, there are six chambers in which divers can carry out a whole range of work activities in realistic sea conditions. The biggest of these tanks is 50 feet long and nearly 15 feet wide.

For a diver in saturation on an oil rig, conditions are not nearly as spacious or comfortable as in one of the experimental simulation chambers. On an exploration rig, production platform, construction barge, or diving support ship, home for a month or more at a time is a cylindrical chamber known as a deck decompression chamber (DDC). It is usually about 25 feet long and 7 feet in diameter, lying on its side. It will have four or six bunks, toilets, showers, a supply of books, stereo headphones, and two-way communications facilities with the outside world. Once the divers have settled in, compression begins. The gas mixture in the chamber is gradually changed from normal air to one which steadily admits more helium while reducing the partial pressure of oxygen. This continues until the divers are compressed to a "depth" corresponding to that at which they are going to have to

work. Once they have gone into saturation, they cannot leave the chamber for the rig outside. Nor can anyone enter except in the direst emergency, and then through an air lock. If the divers were merely to step outside, it would spell almost certain, painful death by the bends. Food, magazines, books, and other supplies are sent into the chamber through small air locks.

When the time to work comes, the divers in the DDC hear the clunk of the submersible diving chamber (SDC) mating to a lock above their heads. This will be pressurized exactly like the DDC. When the divers are fully dressed, they climb up into the SDC through an entry lock. The divers sit on benches in the SDC, which is a spherical bell with observation ports. When the bottom hatch has been securely closed, they feel a lurch as it moves into the air, pulled up on cables by a swinging hydraulic A-frame. This moves the SDC out over the side of the rig and lowers it perhaps 100 feet toward the sea surface. The powerful external lights are switched on and the divers hold on tight as the waves crash into the bell for a second or two as it slides beneath the surface. As the bell approaches the work

263

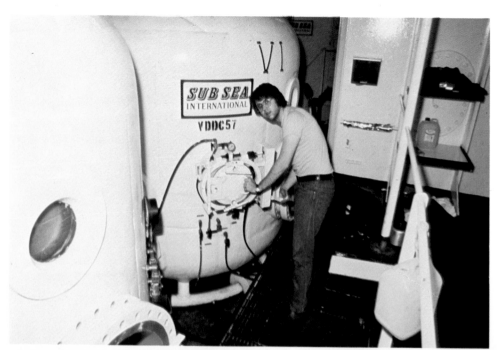

Left: a hatch, called a medical lock, gives communication between those outside and inside a DDC. Food, books and magazines, and anything else is passed through the hatch of the air lock.

Below: the control room of an SDC/DDC unit. Divers are monitored constantly during a dive in a two-way communication system.

site, the divers are in constant contact with the surface via a telephone link, giving maneuvering instructions. Some bells even have small electric motors and propellers to give them a certain amount of independent sideways movement.

Instructions from the SDC are monitored on the surface in a control cabin. The divers' voices have to go through an unscrambling device because one of the strange properties of helium is the effect it has on the vocal chords. After breathing it for a while the voice takes on a squawking tone that is difficult to decipher without the aid of a mechanical decoder. The control cabin also has gauges that enable the operator to control the exact mixture of gases going into the DDC. On a short dive the SDC uses gas from bottles stored on the outside of the bell, but for long periods underwater the gas is pumped down a hose from the surface.

When the SDC is in position, the bottom hatch is opened and the divers move out into the water. However many team members there are, one always stays behind in the chamber to keep the others in sight and pay out communications cables and cords, called "umbilicals," which keep the divers supplied with gas.

The divers working outside the SDC are cossetted against cold. They have to be, because one of helium's other drawbacks is that it is an extremely good conductor of heat. When divers are saturated with the gas, the heat flows away from their bodies – and this flow increases with depth. The first thing divers put on for heat retention is long cotton underwear. Often on top of that they wear an electrically heated wool suit. Then comes a nylon suit with a hood, and finally a nylon/neoprene diving suit. A helmet covering the entire head goes on top; this has a microphone, earphone, and a mouthpiece with a demand valve into which gas is fed from the SDC.

A great deal of research has been conducted into the problem of keeping helium-breathing divers warm underwater, and is still going on. A cold diver is an inefficient diver – and in the depths of the sea any inefficiency can spell death. One approach, already mentioned, is a set of underclothes in which heating filaments are embedded and which get electric power through a cord from the SDC. The circuit is low voltage so that there is no risk of electrocution, but it is difficult to regulate the heat over the body. Divers therefore often find that parts of them freeze while others scorch. Another disadvantage is that the delicate filaments are prone to break under the rough treatment that divers' clothes get.

Another approach is to give the diver the equivalent of a sophisticated hot water bottle, with warm water constantly flowing over the skin. The idea is said to have stemmed from a trick developed in the early days of North Sea diving when divers frequently worked clad only in standard wet suits. One of the divers explained that they drank a lot of hot tea before going down. About 20 minutes later, when they were beginning to get cold, they stood on their heads and urinated over themselves. It was "lovely and warming," they

Above: an SDC about to be locked onto a DDC on an oil rig. It takes careful operation of the winches to lower the SDC gently and accurately onto the locking hatch, through which the divers will transfer to the DDC. They will seldom stay in the DDC less than a week.

said. The idea was adapted by pumping hot water between the skin and a rubber suit worn next to it. The water enters at the waist and exits at the gloves and boots. Warming as it is, this can be uncomfortable on long dives, and there are also great difficulties in getting the temperature exactly right. Because the water has to flow through a long hose from the surface, it is hard to adjust the upper and lower temperature limits to the fine extent that can prevent it either from being tepid or scalding.

On completing the work, the divers return to the SDC, the hatch is closed, and the bell is lifted to the surface. The divers, still kept at the full pressure of their working depth, again hold tight as the SDC is hauled through the waves – known as the "splash zone" – and up to the deck of the rig. There the winch operator works carefully to lower it gently onto the hatch of the DDC. The divers transfer through the hatch, strip off their wet clothes, store their diving equipment in the transfer area, and move into their living quarters, which are maintained at pressure. There they stay, whiling away the hours with books and music until they are called upon to dive once more. When their stint has finished, they go into decompression. Their progress is carefully monitored over a week or more as the pressure is slowly reduced and the constituents of the breathing gas are changed. Sometimes there are two DDCs on a rig or barge so that a second team of saturated divers is ready to go into the water while the first team is undergoing decompression.

Many offshore operations do not require the divers to be in saturation. In this case they do not have to undergo compression in the DDC but are free to use the usual facilities of the rig or barge until their services are called upon. Then they get into their diving gear and enter the chamber which is at atmospheric pressure. Sometimes the bell is not pressurized at all – the divers simply stay in the bell and make observations through the ports before being hauled to the surface again. On other occasions only a short dive is needed. In this case the divers enter the bell at atmospheric pressure and undergo compression on the way down, perhaps taking an hour before the hatch is opened and the divers swim out. When the dive is finished, the bell is brought to the surface and the team transfers to the DDC which has already been pressur-

ized to the equivalent of their working depth. They undergo decompression lasting only a few hours, and are then able to leave the chamber.

Sometimes the divers do not even use the SDC. This happens when there is work to do very close to the surface, such as inspecting a rig for possible damage just below the waterline. Many divers dislike this aspect of their work and often with some justification: they have to don back-packs of compressed air cylinders and get lowered by the rig crane in a cradle toward the sea. Seen from the top of the rig, the divers look particularly frail and unprotected against the side of the vast structure and the forbidding sea beneath them. At the sea surface they have to get through choppy seas, taking great care not to be swept away in swirling tidal currents.

The divers who actually work in the water from an oil rig or any other offshore installation are only the tip of an organizational iceberg. Heading the team is a nondiving supervisor who decides whether condi-

Above: an SDC about to enter the water. At the point of entry, the divers inside have to hold on tightly because of the force exerted against the bell.

Right: the A-frame in the background is used to lower and raise the SDC into the water by swinging it over the side.

266

Left: another method of lowering the bell without an A-frame. It is done through a tunnel amidships, called a moon-pool. Air is pumped through the bottom of the moon-pool, which breaks up the force of the sea.

tions are right to dive, takes instructions from the personnel on the rig on the exact nature of the work to be done, and relays orders to the divers. The team leader is responsible for insuring that the equipment is maintained, and must also keep a watchful eye on the general physical and mental well-being of the team, particularly when they are in sat for weeks at a time. Confinement in stuffy cramped quarters leads to boredom and friction, which can swiftly build up over the seemingly most trivial incidents.

Oil drillers, eager to the point of obsession to "make hole," often keep divers standing by just in case anything happens to interrupt the routine of the rig and which their services can help to put right. This is especially true of exploration rigs. As a result, it is not unknown for a diving team to spend the whole of a two-week tour of duty on a rig without ever having to go into the water. The supervisor has to make sure they are kept busy, usually by assigning essential maintenance work on equipment and letting them help out the rest of the rig crew with such general chores as painting, welding, and cleaning.

Diving, then, is not a glamourous trade. For a few hours of work underwater, the diver spends days in spartan surroundings doing nothing. Yet there is something fascinating in it for many who go deep into the seas to work.

Right: a diver at work, attached to the bell by the hose known as an umbilical. It is truly a lifeline because it keeps divers supplied with breathing gas. One person always remains in the bell to operate the gas supply and keep a watchful eye on how things are going.

267

What Divers Do

Divers can seldom expect a lifetime career. If they are lucky, their span of work will last 20 years; usually it will not be more than 10. Perfect good health is more important in diving than in most other jobs: a simple chest cold, an ear infection, or a sprained ankle can reduce a diver's earning power to zero overnight in the middle of the diving season. Neither can divers expect constant work because employment is cyclical. The only divers who work year round are those of the top echelon whose services are so in demand worldwide that they go from area to area. The smallest downturn in deepsea exploration or construction can see the diver unemployed for weeks at a time.

A diver's work is often dangerous, and always there is the nagging worry about the accident that could cause permanent disablement – or even death. Even if a diver manages to have an accident-free career, there is another potential danger at the end of the line: prolonged exposure to pressure over the years, especially if decompression procedures have been lax, can bring on the potentially crippling disease of bone necrosis.

So why do divers go deep under the waters? Part of the answer is that many find it worthwhile in spite of the hazards. Another part of it is that the financial reward can be great, especially in the offshore oil industry.

Related to deepsea diving as a profession is diving as a sport. Scuba diving may be mainly for fun, but it often happens that sports divers fill a useful role as well. For example, they may straighten wires that have tangled around the propellers of pleasure or commercial craft; search for and recover objects that have fallen into the water; or locate ancient wrecks and artifacts. This latter activity has grown into something of an industry for some sections of the sports diving community, and highly organized expeditions are made to search for treasure long buried by the waves. In many nations the recovery and sale of antique objects are banned by law. In others, archaeological diving activities may only be carried out under the supervision of qualified archaeologists in order to protect the site – be it wreck or sunken city – from plunderers.

Underwater archaeological work has grown at a

Top: scuba divers at play in a Malta harbor. This is diving as sport.

Left: scuba divers at work. They are recovering drums from the seabed.

great rate in recent years with the realization that much evidence about the ways in which previous civilizations lived, traded, and fought often lies beneath the water. Archaeological expeditions mostly concentrate on the great trading routes between colonies, which are littered with wrecks, and on coastal sites of antiquity, particularly in the Mediterranean, where the rising sea level, earthquakes, or volcanic eruptions have submerged settlements. Often there is only a trace of a ship's wooden hull or other wooden objects left, but exciting clues to the ways of the past can be found by the recovery of metal and ceramic objects as widely varied as wine flasks, cannons, belt buckles, and the fastenings used to join the planks of a hull.

Major underwater expeditions today are launched by universities and archaeological societies, and they draw on a wide range of subsea technological skills and equipment to conduct surveys, excavations, and recoveries that are every bit as scientifically organized as archaeological digs on land. A major under-

Above: two scuba divers doing a job for science by collecting water samples. The sample containers can be seen on the back of the diver in the foreground. They are using underwater scooters to get around faster and easier.

Right: a marine biologist on a study dive. Many marine scientists find diving the most practical way to learn about the sea and sea life at first hand.

water "dig" may be initiated as the result of a seabed survey. Side scan sonar identifies small objects standing proud of the seabed – perhaps a projecting spar that indicates the presence of a buried wreck, or a suspiciously straight line that might mean a long-lost harbor. Sub-bottom profilers send beams of sound into the seabed sediment to pick out more details, and a magnetometer towed above a likely area tells the surveyors of buried metal objects. Anything revealed by these surveys will be carefully marked on a chart, its position being fixed by sextant, theodolite, or short-range, highly accurate radio positioning. Because it is much easier to see seabed features – particularly in the generally clear waters of the Mediterranean – from the air, the search is also conducted in the sky. Photographs from fixed-wing aircraft and helicopters show a great deal, but the very best photographs are taken from tethered balloons on windless days when the sea is unruffled. Often the shape of an entire underwater city is revealed by photographs taken in this way.

When the archaeological site has been discovered and charted, the divers move in. Often a wire or plastic grid is laid over the site so that the exact point from which anything is removed may be carefully logged. The divers work carefully, picking through soil and stones to uncover objects that are then sent up to the surface. One ingenious transport method is to use strong canvas or nylon bags to give buoyancy to an object. We have seen that compressed gas increases in volume as it progresses upward through the water. When a bag is tied to an object, divers give it a squirt of air from their aqualung and it starts slowly to float

Above: a marine biologist taking pictures underwater. The invention of the aqualung by Jacques Cousteau in the early 1950s revolutionized scientific study of the sea. For the first time, it enabled marine specialists to examine at close range the animals and the phenomena that they had had to speculate about before. Scuba diving has become such a common method of marine study that almost every student has to expect to do it. The technology of underwater photography has also kept pace so that the results are often beautiful as well as useful.

to the surface, like a parachute in reverse, billowing out as the air inside expands. A few large bags linked together and floated in this way can lift objects weighing several tons.

Many archaeologists have learned to dive in order to be able to direct underwater excavations themselves. Students and scientists in other marine disciplines have also found it advantageous to dive deep in the environment they have chosen to study. Fisheries scientists armed with cameras and underwater writing paper look at the configurations taken up by trawl nets as they are towed along the seabed and at the ways in which fish try to escape from them. Geophysicists correlate the interpretative results of surveys with sonar and profilers by actually sitting on the seabed and sampling the soil with their hands. Biologists obtain remarkable close-up photographs of plankton, fish, and mammals. Many police forces have full-time diving teams available to search rivers, canals, and gravel pits for discarded weapons, stolen property, and the often more grisly results of murder.

All these are people who have taken up diving to complement their existing careers. For those who have made diving a full-time career, the work is often more routine and monotonous. Hard-hat and constant-volume-suit divers may spend weeks at a time constructing an underwater wall as part of a harbor development project, working in thick silt through which even the most powerful lights will penetrate only a few inches. They may be called in merely to inspect the foundations of a whole series of bridges in a cold fast-flowing river. They may have to clear blocked culverts of rotting vegetation, general litter, or untreated sewage.

The two jobs that the professional shallow-water diver is called upon to do most frequently are cutting and welding. Heavy cables have to be cut and burned away when they have entangled the propellers of big ships; piling has laboriously to be cut away on cofferdams that have been installed on a temporary basis to

Right: a diver on an archaeological assignment holds an ancient amphora recovered from a site off Turkey.

Below: a scientist working from an underwater habitat, the *Tektite*, developed in the United States. Habitats began with experiments by Jacques Cousteau, Ed Link, and Captain George Bond. They have power, fresh water, and emergency communications from shore facilities by means of umbilical cables – much like the umbilical hoses that support divers from SDC bells. *Tektite II* was put into use in 1970 for underwater scientific research.

support a harbor construction project; the underwater sections of decrepit piers have to be leveled off so that they are not a hazard to shipping; and parts of wrecks that project dangerously toward the sea surface have to be removed.

For work of this nature the diver uses oxy-hydrogen or oxy-arc cutting equipment in much the same way as it is used on the surface. Another particularly favored method is that provided by the thermic lance, a rod or cable that, fed by high-pressure oxygen, burns at an extremely high temperature of over 10,000° F. It is able to cut through steel, rock, and reinforced concrete. (On land the thermic lance is a favorite with safe crackers.)

Shallow-water divers will also have a good working knowledge of explosives and their underwater applications. They use explosives in marine civil engineering work when shipping channels need to be blasted through rocks or when unwanted piling needs to be removed. When wrecks are a hazard to shipping and cannot be removed by conventional salvage methods, the diver is called in to place explosive charges to level the ship. On a 1000-ton steel vessel, the diver makes his way to the fore part and places up to 200 pounds of explosives. Another 500 pounds goes admidships, and 300 pounds in the stern. If it is a steam-powered ship, a third large charge may be needed in the boiler. On bigger vessels, strings of explosives are laid along the whole length of the deck and hull so that the ship cracks open and sinks to the seabed.

Not all divers' work is destructive, of course. They frequently use underwater welding equipment to patch up ships temporarily when there is not a readily available drydock, and they also use underwater bolt-firing guns, powerful enough to penetrate steel over an inch thick, to place large patches over a hole or in general underwater construction work.

All the diver's skills and equipment are brought into play in marine salvage work, when the contents of a ship, parts of a ship, or the entire vessel has to be brought to the surface.

If a salvage company is interested in recovering cargo – perhaps precious metals or the safe of a passenger liner – teams of divers will judge the best way in which to reach the hold. If the ship is in a more or less upright position, the hatches can often be opened fairly easily with the aid of some judicious cutting or a shrewdly placed explosive charge. The divers can simply walk into the ship and attach ropes to the cargo, which is then lifted out by a surface crane. If the ship is on its side, however, it is sometimes simpler to cut a

Above: a diver using underwater welding equipment. There is frequent call on divers to repair ships temporarily by welding when a drydock is unavailable.

Below: this diver is using a high pressure water jetting hose to clear away marine growth around a rig. This must be done before a proper inspection can be made.

hole in the bottom and gain access that way. Often the cargo of a ship is not worth recovering but if the vessel is large enough – perhaps of oil tanker size – its big manganese bronze propellers have enormous value even if they are only used for scrap. Again, cutting or blasting is used to free the object, which is winched to the surface.

Salvaging of complete vessels is an art and many of the techniques used are closely guarded secrets of the few companies that specialize in the work. The broad principles are known, however. Small vessels such as dredgers and fishing boats that have sunk near the coast in tidal waters are brought ashore gradually. Divers loop ropes through the propeller aperture and under the ship near the bow, and at low water these are hauled taut on a salvage vessel or pontoon on the surface. As the tide rises, the vessel is lifted clear of the seabed; then it is pulled closer to the shore until, caught by the falling tide, it grounds again. The procedure is repeated until there is enough hull exposed at low water for temporary repairs to be made. The ship is then towed to a drydock.

In deeper waters or with big vessels, the divers first make an inspection. Then they attempt to make the ship as watertight as possible by plugging holes with wood, by bolting on plates, or by filling cracks with concrete. Vast hawsers are then looped under the ship. To achieve this the divers use another valuable underwater tool – the jetting hose. This blasts high pressure water into the silt, sand, and mud into which the ship has settled, displacing it so that the diver digs a tunnel right under the ship through which a "messenger" wire – a thin light rope – can be passed. The

messenger is attached to the heavy hawser, which is pulled through the tunnel by powerful winches on the surface. Pontoons, known as camels, are then sunk in suitable positions and the divers make the hawsers fast to these. Some holes in the ship are left unplugged in order for hoses to pump air into the hull to push out the water. The camels are also pumped full of air, and this is usually enough to give the vessel positive buoyancy to lift it clear of the seabed. Another salvage technique involves pumping the ship's hull full of tiny bubbles of polystyrene whose buoyancy lifts the ship clear.

The diving elite is the section of the profession that breathes helium and is associated with the oil indus-

Below: close-up of an underwater magnetic particle inspection. The piece to be tested for defects is magnetized, which shows up cracks by flux leakage. A magnetic powder is then applied to reveal the extent of any faults.

try. During an offshore oilfield career, divers will probably see all facets of offshore operations. They may well start an oilfield career on a semisubmersible drilling rig or a drillship. One of the hazards of this job is boredom, because most of the work is on emergencies and much of the time is spent in waiting for something to happen. Drilling contractors have become so experienced in working in deep hostile waters that emergencies are becoming rarer and rarer, but a

diving team is always on hand in case of need. What kind of emergencies arise? It could be that a BOP stack does not land perfectly on a guide base and a guide wire gets tangled, requiring a diver to cut it loose. A surface-actuated hydraulic circuit controlling connectors can fail, requiring a manual override to be operated on the sea floor. A BOP ram may develop a leak and have to be changed undersea. Problems can also be experienced in connecting the marine

Above: a piece of salvaged equipment that has been brought up by means of buoyancy bags. Made of strong nylon or canvas, the bags are tied to an object and given a squirt of gas. They then float upward and, because of their buoyancy, can carry very heavy objects.

Opposite: this diver is inspecting a rig leg for cracks by means of ultrasonic equipment.

Left: a diving team and the buoyancy bags they used for salvage recovery after surfacing.

274

riser to the BOP stack. In an extreme case, if the rig should get blown off its location before there has been a chance to recover the BOP stack, the divers will more than earn their money in cutting away the tangle of wires and damaged metal that may well be left behind on the seabed.

Boredom will probably never be a concern when the diver works on a pipelaying or pipe-jetting barge. We have already seen that the pipelaying barge is a continuous-production factory, with work going on day and night when the weather is right. The work of a diving team in this connection is varied, ranging from very shallow water tasks concerned with checking the profile of the stinger and that of the pipeline as it snakes into the water, to deepwater tasks requiring a bell. Typically these would be securing the first-laid section of pipeline to an anchor on the seabed next to the platform to which it will eventually be connected, and inspecting the pipeline on the seabed. Later the diver may move to the bury barge to undertake a range of inspection tasks involved with jetting the pipeline into a seabed trench.

If an oilfield has one or more platforms for production, divers are an essential part of the development activities. They will spend long periods in saturation while they tie-in the pipeline to the vertical pipe that

form actually begins to produce oil. Platforms have to be inspected regularly, and the diver will often have to stay in saturation to perform the skilled tasks required. Divers take still and movie pictures of areas where stress might be producing cracks – usually where legs and bracings all meet at a giant welded section called

Above: a diver going into the water from a pipelaying barge to do an important job in getting oil to shore.

takes the part-processed oil from the platform deck to the seabed line. If the platform is feeding a single-buoy mooring at which tankers load, they will have to check that the buoy's anchors are securely embedded in the seabed. They will have to be on hand to see that the buoy is installed safely and then connect up the flowlines from the platform to the seabed manifold beneath the bell, as well as securing the underbuoy hoses by which the oil flows into the buoy itself.

The diver's work is not finished even when the plat-

a node. Often they will use ultrasonic or X-ray equipment to check for cracks too small for the eye to detect. They will also check the sacrificial anodes that protect the platform from corrosion and advise when these require repair.

The work described is just a part of the many tasks that divers perform. But is the frail human frame, even equipped with power tools and other aids, always the right instrument to use? Are there alternatives – and are those alternatives better?

275

The High Risks
of Deepsea Diving

It has been said – although with what statistical justification is unclear – that more divers come to untimely ends in fast cars and through midnight leaps from bedroom windows than ever perish underwater. That somewhat cynical reflection is probably based on the fact that over a year the number of deaths in the diving profession does not appear, at first glance, to be very high compared to the losses in other high-risk industries such as fishing and mining. But when those

than any other work. One possible explanation for this is that when a diver dies, there is a great deal of publicity. Diving still has enough of an aura of swashbuckling romance about it to fire the often morbid imagination of a public immune to the kinds of accidents that happen every day in the construction industry, for example. Government departments, oil companies, and diving contractors are under a certain amount of pressure to maintain safety standards and this has led them to be particularly vigilant in the framing and implementation of diving legislation and working codes.

In spite of all efforts, divers are badly injured or killed. Why? More and more the industry is having to accept that diving is inherently dangerous and that the only law which would effectively stop all underwater accidents is one banning diving altogether.

Diving is dangerous mainly because of the un-

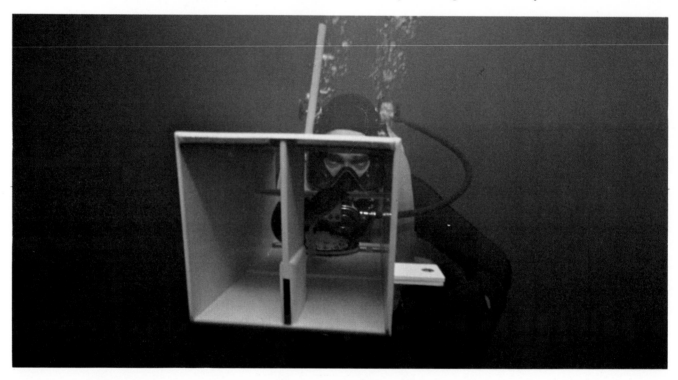

figures are computed in percentage terms – the number of deaths per hundred employees – diving frequently emerges as the most dangerous job of all.

Government departments of all nations with major diving industries, together with the companies that employ divers, have worked together for many years to improve safety standards in the industry. There is a mass of legislation on the design and operation of every piece of equipment that the diver has to use, from underwater wrenches to decompression chambers. There are any number of codes of practice for the diver to follow, ranging from detailed instructions on how to set underwater explosive charges in the safest way to advice on the type and amount of food that should be eaten before diving. In most countries divers are required to undergo stringent medical check-ups at regular intervals.

There is probably more safety legislation on diving

Above: a diver using a navigational aid that is fitted with a compass and distance counter. No matter how well equipped a diver is or how much care the support ship takes, the unexpected and dangerous can always happen. For the frail human body, an accident often means death.

expected event that no legislation or code of practice could possibly foresee. In a repetitive factory routine it is easy to produce a code that helps machine operators handle emergencies. Underwater, it is only possible to tell divers how to be prepared in case of an emergency. Even though the job they are doing may be perfectly routine – such as tightening nuts, welding, or chipping away at marine growth – it is not possible to legislate for the sudden emergence of a harmless but ghastly looking eel that can cause a diver to drop his welding torch in momentary fright. The consequences of what would be a simple slip on land can be fatal when it happens underwater.

So the risks are high. But is it necessary for these risks to be taken every day? Is the diver's work so essential that the price, in terms of lives, must be paid to complete simple underwater engineering tasks?

Even if diving could be made considerably safer, does the cost of putting the diver underwater always

ine can reproduce: they can search and find small objects as no machine can, and they can perform delicate manipulative tasks that no machine is capable of.

There is always likely to be room for the deepsea diver, even though many of the most dangerous and tedious tasks may one day be performed by machines.

Left: two divers working together at least dispel some of the terrible loneliness of being in great depths. It does not necessarily mean greater safety.

Below: it is probably the dark as much as anything else that can fill a diver with anxiety about danger.

pay off in terms of efficiency? Many authorities argue that the diver cannot be cost-effective. A vast proportion of a diver's mental and physical energy is expended merely in staying alive, they say. Only a small proportion can be devoted to the actual job in hand. Divers worry about the cold, the bends, the equipment on which their life depends, the degree of trust they can put in the vast support crew. In terms of capital equipment, breathing gases, and back-up personnel, it costs a fortune to get a diver to the seabed.

All these factors have led designers and engineers to look at ways of replacing open-sea divers by machines or by encapsulating them against the worst rigors of the underwater environment. It must be stressed that this is not a "man-versus-machine" argument. For all the underwater tasks that can be taken over by machines, some still demand the diver's unique gifts. People have hand and eye coordination that no mach-

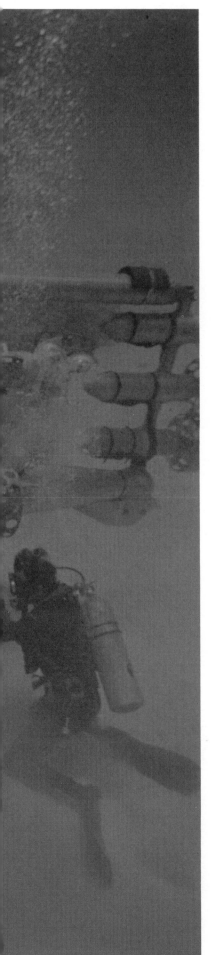

Chapter 11

Underwater Equipment

Inherent dangers and inefficiencies in having people work virtually unprotected at great depths in the sea have led the diving industry to look at other ways of putting workers below the surface. One way is simply to provide a suit of armor with flexible joints. Another is to make underwater vessels. These include submarines that can stay underwater for months at a time, bathyscaphes that can sit on the ocean floor with miles of water above them, and small submersibles that enable anyone – not just professional divers – to observe and work at great depths. Finally, there is the completely different approach that seeks to keep people out of the water altogether and replace them with remote-control vehicles ranging in type from "swimming eyeballs" of football size to big machines capable of undertaking heavy work.

Opposite: more and more the trend is to replace humans with machines for working in the deepest parts of the dark cold ocean. Their capabilities are amazing.

Protective Suits for Deep Divers

In 1968 in a small town about 30 miles from London, two underwater equipment designers and engineers started examining the commercial possibilities of developing a range of manipulators for performing simple subsea tasks. The two men, Mike Borrow and Mike Humphrey, were co-directors of a company and had been involved in the underwater business for some time. They had seen the intensive development both of deep diving, spurred by the offshore oil industry, and of manned miniature submarines – submersibles – which were being used for a variety of underwater research and, increasingly, for commercial projects.

Both men were aware of the problems faced by the deep diving industry as it strove to develop ways of working at deeper and deeper depths. Saturation diving was becoming commonplace in oilfield work, but the strain that this put on divers, equipment, and

Above: a diving suit invented in the mid-18th century by a German named Kleingert. It allowed movement only by the arms and legs and required the wearer to carry a very large compressed air machine.

Left: a French painting of the 13th century depicting Alexander the Great at the bottom of the sea in a glass barrel. The story of Alexander's dive to the ocean floor in the 4th century BC to observe marine life is probably only legend, but there are many pictures of this subject.

Below: a diving suit developed by Edmund Halley about 1690 and used by John Lethbridge for regular undersea work about 15 years later. Lethbridge was face down in an air-filled barrel for half an hour at a time as he dove for sunken treasure. His porthole was only four inches round.

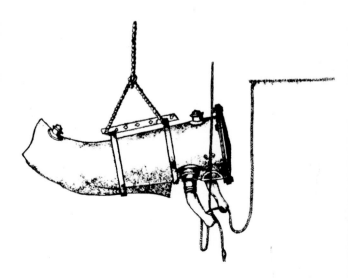

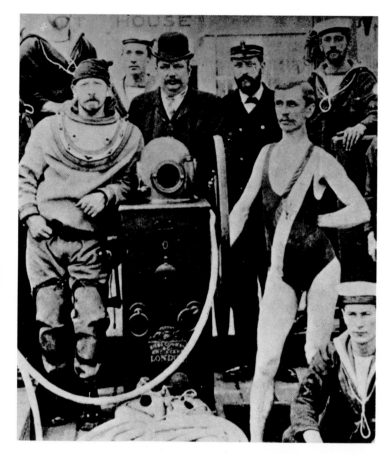

Above: an early diving suit found on the seabed in the south of France. Little is known about its invention or use.

Left: a photograph taken in 1891 shows a diver wearing a suit designed for work at a depth of 30 feet. There was no communication with anyone on the surface from these suits which, however, allowed some freedom of action.

the finances of oil company clients was beginning to be appreciated. It was argued that many simple tasks could be performed just as well, if not better, without having virtually unprotected divers performing at less than top efficiency and at great cost hundreds of feet beneath the surface.

In the course of their research on manipulators, Borrow and Humphrey turned to one of the standard works on diving, *Deep Diving and Submarine Operations* by R H Davis, published in 1962. The book had a number of accounts of a nearly forgotten diving method: the atmospheric diving suit or ADS.

In essence, the ADS or one-atmosphere suit encapsulates the diver in a pressure-proof metal shell with joints at shoulders, elbows, wrists, hips, knees, and ankles. These joints enable the operator to walk and perform simple manipulative tasks. Divers breathe air at atmospheric pressure from oxygen cylinders, and because they are not subject to the pressure of seawater, they do not incur the time penalty of compression or decompression. They are safe from the bends, from nitrogen narcosis, from oxygen poisoning, from drowning (unless the suit springs a catastrophic leak), and from cold. Of the ADS designs described in Davis' book, Humphrey was particularly interested in one that had last been used in 1937, designed by Joseph Peress. He thought that the Peress suit seemed to incorporate the best engineering ideas.

Later Borrow and Humphrey had the opportunity of meeting Joseph Peress, then 72 years old and not particularly interested in their enthusiastic attempts

to revive what he considered to be a long-dead concept. Why had the idea died? And how far had Peress' invention taken deepsea diving?

Peress had come to England via Paris in 1912 and, although he worked as an aircraft designer, he was also interested in underwater work. He designed and made the first workable ADS in 1921, and along other lines, he patented a design for an underwater jet thruster in 1928.

Peress was by no means the first to develop atmospheric diving suits. Their history could be said to have started about 200 years before. In 1715 an Englishman by the name of John Lethbridge had made regular dives to 60 feet of depth inside an elongated wooden barrel made for him by a local cooper. The barrel was lowered horizontally with Lethbridge in a prone position, looking out through a small port on the underside. His bare arms projected into the water, clamped firmly just above the elbow by leather seals. Lethbridge made expeditions 60 feet down for periods of up to half an hour.

The first designs for metal suits – none of which, it seems, were ever developed – appeared in the United States in the 19th century. It was not until the early years of this century, however, that the first complete metal one-atmosphere diving suits began to appear.

The main spur to their development was the salvage of valuable cargo from ships that lay in deep waters beyond the limits of air divers. All the earliest suits had one great disadvantage: the deeper the diver descended, the more the effects of water pressure were felt on the suit's joints, and many stiffened to the point of complete immobility at depths of 200 feet or more. Nevertheless, some suits were used on salvage work, probably the most famous being the one developed by the British company of Neufeldt and Kuhnke. It was used in the recovery of tons of gold and silver from the *Egypt*, which had sunk in over 400 feet of water in the Bay of Biscay after a collision in 1922, although the diver in it could only observe and direct rather than work actively. Through the telephone in the suit, the operator told a crane driver on the surface how to positive a grab in order to lift cargo, place explosives, or rip open plates. The operation was an overall success to the extent that three-quarters of the precious cargo was recovered. But it began to sound the death knell for the one-atmosphere suit.

In the meantime, Joseph Peress had improved on his ADS of 1921 and had developed another that, on trials, enabled successful dives to 500 feet. It was used to locate and work on the wreck of the *Lusitania* off the south coast of Ireland in 1936. Despite this, interest in the use of such suits was fading. Their application was too limited, confining the diver to a role as observer only. By then there was a far easier

and cheaper way of lowering one or more people to great depths: they were simply lowered in a closed diving bell at atmospheric pressure. The bell overcame the need for the exacting and expensive engineering involved in building a suit.

In the late 1930s the British Royal Navy evaluated the Peress ADS and commented that it "was found to come up to the claims of the inventor, but no use could be made of it as the Navy has no requirements for deep diving."

So that was effectively the end of the era of the ADS. If the Navy had required a deep diving capability at that time, most likely there would not have been a 40-year technology gap, because Joseph Peress had gone a long way to solving the main problem of that generation of suits: the joint stiffening problem. The Peress joint consisted of an annular piston moving in an annular semihemispherical cylinder to achieve angular movement, and it worked on the principle of using the effect of the outside water pressure to help achieve movement, rather than fighting against it.

By the time Mike Borrow and Mike Humphrey met Joseph Peress, he was past having an interest in diving suits. But they managed to revive such interest and

Above: the diving dress developed by the English company of Neufeldt and Kuhnke, shown in trials by the British Navy in 1947. The suit provided telephone communication with the surface.

Left: Joseph Peress photographed in 1931 with his atmospheric diving suit. Peress is given credit for the modern development of high-technology suits.

Above: an underwater chamber manufactured in the early 1950s for the British Navy and used for salvage work. It was for observation only.

he became a consultant to the company they established to develop a new generation of ADS. They knew that the development program would be greatly speeded if they could find the original Peress suit. This had last been seen by him in the late 1930s when it had been seized in lieu of a bad debt from a company that had put some money into developing it for sponge gathering in the Mediterranean deeps. There was every chance that it had been cut up and used as scrap metal, but the Borrow/Humphrey/Peress team went on an optimistic search.

And they found it! Even though it had lain forgotten in the corner of a workshop in Glasgow since 1937, it was intact when discovered, having survived World War II and much else besides.

Borrow and Humphrey were not the only team interested in reviving the idea of the ADS. Papers had already been published in the United States suggesting that re-development would be a good idea, and there had even been some tentative designs put forward. So the development of the new suit had to be conducted in conditions of tight security, and this applied to the shipping from Glasgow to London of the original Peress suit. Imagine the chagrin of the design

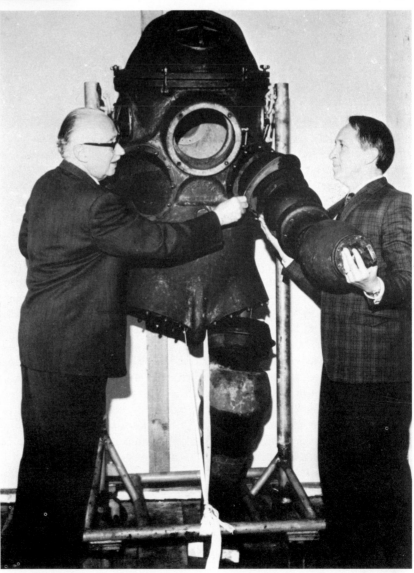

Right: the original Peress diving suit photographed in 1968 with the inventor (left) and Mike Borrow, developer of a new generation of suits for deep diving. Borrow tracked down the lost suit in Scotland.

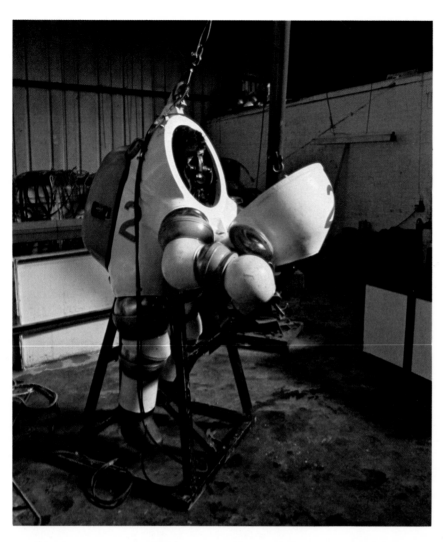

Left: the Jim suit, named after the chief diver who worked with Peress on salvaging the wreck of the *Lusitania* in the 1930s. The hinged dome is open, revealing the diver's mask.

Opposite: a diver in Jim during the final series of tank tests in 1971. The one-atmosphere suit gives the wearer a large degree of free movement and provides normal surface pressure.

Opposite below: a Jim-suited diver being lowered into the water from a rig off the coast of Brazil. Jim does away with the need for decompression and expensive mixed gases. The time saved is remarkable. Where a saturation diving team would take 15 days for a 1000-foot dive – 10 of them for decompression – Jim can do it in one day in some cases.

team when the suit arrived on a flatbed truck in a packing case that proclaimed on its side in foot-high letters, "Lusitania Diving Dress."

The suit had survived the years and its journey extremely well, showing only minimal corrosion. There was even some oxygen in one of the life support cylinders.

The suit was tested in a 10-foot-deep tank at Borrow's factory and confirmed all the hope that had been placed in it. The developers were sure that modern materials and machining techniques would enable them to make new versions incorporating a slightly different body shape and joints that would be able to work to extreme depths without stiffening.

It was decided to build two prototypes, and by November 1971 the first ADS was ready for testing. It was called Jim 2 after Jim Jarret, who had been Peress' chief diver and mechanic during the *Lusitania* dives in the 1930s. A series of trials took place in tanks and at sea, culminating in a "dive" in a hydrostatic pressure tank to 1000 feet.

For a diver used to working from a bell, the test dives with Jim were a novel experience. Here is what the test diver underwent. The operator climbs into the suit through the hinged dome at the head. There are no special preparations, such as undergoing compression. The diver wears ordinary clothes –

perhaps jeans and a light sweater. Inside the suit there is for his use an oro-nasal mask which supplies oxygen via valves from high-pressure cylinders fitted outside the suit in a backpack. The slight chill that is felt when first entering the suit soon goes as the diver is lowered to the seabed and the temperature reaches a comfortable level.

When the dome is fastened down, the operator has good visibility through four ports and, as the lifting crane takes the strain of its 1000-pound load, he can make checks on the temperature and oxygen capacity from indicators mounted close to the head. There are headphones inside the mask and a microphone around the throat. The lift cable has a telephone link to guarantee completely clear communications at all times, although a back-up through-water communications link is also installed.

Once on the seabed the operator can move about with ease because weights on the outside of the suit have been adjusted before the dive began to give it just the required amount of buoyancy. Observers of Jim in a tank are usually surprised at the way the diver is able to lean forward until the "hands" of the suit touch the bottom, lean backward until the backpack touches, climb a ladder, or recover footing after a fall. Once some interested observers watching such actions at Borrow's test facility commented that the

operator was obviously a man of considerable strength. When the suit was brought to the surface and opened up, Mike Borrow's slim 18-year-old daughter emerged – to their surprise.

After working on the bottom for some time, the operator is bound to become fatigued, but in much the same way as anyone performing physical work for over four hours would, no matter what the environment. There is ample oxygen available for work periods of such length: the life support system operates for 24 hours if required and there are two important back-up safety systems. If the cable from the surface should become snagged, it can be disconnected from inside the suit. The operator is then able to jettison the outside ballast weights, also from within the suit. The suit then rises slowly to the surface at about 1.5 feet per second.

The trials of the first two new Jim suits were successful and commercial work soon began. One job, done in 1974, illustrates all the advantages of the ADS. A United States tanker had lost its 18.5-ton anchor and 2500 feet of chain off Tenerife in the Canary Islands. It was worth hundreds of thousands of dollars, but more important than its cash value was that it would have taken over a year to get a replacement and the ship wanted to get under way again quickly. What was required was for the anchor and chain to be located, and for a line to be attached to them so that they could be hauled to the surface. The problem was that the anchor was in 350 feet of water. To get a diver to perform that simple task ordinarily would have required the mobilization of a ship complete with full deep diving spread, including bell, handling frame, winch, and deck decompression chamber. Although it would have been possible to get the equipment delivered by air, heavy as it was, there were no suitable ships on which it could have been mounted for many hundreds of miles.

The advantages of using Jim suits for a job of this nature were apparent from the start. Two suits were packed, in pieces, on a scheduled flight from London, together with a winch, at 10 am in the morning. By 4:30 pm the same day they were being assembled on board the support vessel – an easily available 70-foot long converted yacht. The operator, whose only previous experience had been in a 20-foot deep tank, took an exploratory dive to the seabed in 175 feet of water and, despite the silty bottom and strong currents, soon became accustomed to the conditions. He was brought up to the surface and dived again. On this dive he located the anchor and chain and secured a messenger line to it. A heavy lifting chain

Below: the most recent of the Borrow-Humphrey-Peress diving suits is the Sam (on the left), shown next to Jim. It has a very light body made of aluminum or glass reinforced plastic and its joints are movable.

was later secured and fed to a launch that stayed on the site until the tanker arrived to haul the anchor aboard. The support ship and Jim moved to another location off Las Palmas where the same task was repeated to recover a Norwegian tanker's anchor.

With a conventional diving spread, the divers would not have been able to make repeated dives to and from the anchors so swiftly unless a full team of six or so had been used, possibly having to operate within the very expensive process of saturation diving. Jim operators are able to dive to a site, make an inspection, come back to the surface to pick up equipment, inspect drawings, or discuss the full scope of the work with engineers, and then climb into the suit again and go back to work. They do not require massive back-up in terms of equipment and people, and when the job is finished, they do not have to sit for hours – sometimes days – decompressing. Another great advantage is that they are always warm.

The offshore oil industry had been responsible for pushing conventional diving to physiological and technical limits, and was now ready for the development of other ways of performing deep water work. So it was obvious that the industry would soon take a great interest in Jim suits for simple seabed tasks, especially when fitted with manipulators at the end of each arm. Various types of manipulators have been developed, and these can be changed quickly to meet the needs of a particular job.

Below: a diver testing the Sam suit in a tank, showing how the manipulators can be used. There are various kinds of manipulators that can be changed relatively easily by the operator to meet the needs of a particular job.

Above, left to right: the original Jim suit, 1930. Design number 13, completed in 1969. The prototype of design number 15 dating from 1978. The proposed design of the latest version made in 1980.

Records soon began to be broken by Jim throughout the world in terms of depth (over 1200 feet so far); cold (well below freezing point); and endurance (six hours by one operator). The company established by Borrow, Humphrey, and Peress went on to refine and develop the suit further. The latest version, called Sam, has a body made of aluminum or glass reinforced plastic. Sams are lighter than Jim and each joint is made up of a series of elements rather than a massive single unit.

In 1979 there were 12 Jims and Sams in operation throughout the world, with others under construction. Both the early and late contributions made by Joseph Peress to deepsea diving have a special place in its history.

The Jim and Sam suits are designed to work either on the seabed or on a firm surface such as an underwater catwalk or specially constructed staging. They are not designed to move from one point to another in midwater, and this has led to the development of other kinds of atmospheric diving suits.

The first of these is Wasp, developed by a British company in 1977. In this machine, Jim's walking, climbing, and bending capabilities have been traded off against the ability to work in midwater. Wasp has a cylindrical shape with a curve at the head that

287

terminates in a hemispherical window to give the operator a wide field of vision. This window hinges back to allow the operator access to the suit, which in air weighs 1100 pounds – the same as the first Jim suits – and is just under 7 feet long. Wasp is lowered into the water on a cable that also provides power for the suit's movements and incorporates a telephone link to the surface. The operator inserts his arms into the articulated joint structures, and from these is able to operate a range of interchangeable manipulators and power tools that are fitted at the "hands."

Although operators do not have to use their legs for walking, as in Jim, they need have no fear that their muscles will atrophy for lack of movement. They have to operate switches with their feet to give the movement in midwater that is the big selling point of Wasp. This is done by means of a thruster, powered from the surface and mounted on each side of the suit's body. By working the foot switches the operator can rotate the suit through 360 degrees in the vertical plane as well as being able to move forward or backward. The suit usually sits in midwater inclined at about 30 degrees forward from the vertical, but another foot switch triggers pumps that will swing it forward to a near-horizontal position and then back again. Even in the darkest water divers know which way the suit is facing and at what angle; dials showing the exact attitude and trim are fitted so that they can check them with a quick glance. Other dials give them a constant and reassuring check on their life support systems. Wasp operators do not wear an oxygen mask but breathe air that is automatically circulated by a fan throughout the whole suit, the waste carbon dioxide being absorbed in a soda-lime cannister. This gives them air for eight hours underwater. Should the system fail, there is an emergency breathing system incorporating a face mask, which allows them to breathe for anything up to 54 hours. The divers are also able to jettison the umbilical if it becomes snagged and to use battery power to provide

Above: a diver inside a Wasp suit being winched back up to the surface. The Wasp is designed for working off the bottom in midwater. Using thrusters controlled by the foot, the diver can maneuver in any direction.

Right: a diver in Wasp is shown inspecting platform anodes in the Gulf of Mexico. This suit enables divers to work comfortably and with a high degree of movement in depths up to 2000 feet.

them with thrust to extricate themselves from a dangerous position and, by discarding weights, rise to the surface.

Even more sophisticated developments have followed hard on the heels of Wasp. Another British company has produced Spider (Self-Propelled Inspection Diver) which not only offers high midwater maneuverability through no less than six thrusters, but also has hydraulically powered manipulator claws on each hand that can rotate through 360 degrees in either direction. The unit, dating from 1978, also has built-in suction pads by which it can attach itself to structures while the operator carries out a range of inspection and maintenance tasks, even to the extent of wielding a high pressure water jetting gun. It has a built-in TV camera linked to the surface but with a monitor inside the suit so that the operator can be instructed on what position to take to get the best pictures. There is enough space inside the suit to enable the operator to wield a hand-held camera while doing everything else required.

Some experts have remarked that, although it is highly desirable to have so many features built into a single one-atmosphere suit, there is a danger of over-taxing the operator. Despite the undoubted advantages that the one-atmosphere suit has to offer, it takes a person of a certain mental attitude to have the confidence to descend to the ocean depths in a closely fitting steel suit. There is a great feeling of isolation, more than that which a deep diver in a bell, with a companion close at hand, has to endure. Jim, Sam, Wasp, and Spider operators must at all times try to overcome the fear of becoming trapped, alone and at depth, despite all the back-up safety systems built into the suit. Great depths have been reached and great advances have been made in deep water diving, but many experts today admit that there might well be a psychological depth beneath which it will be difficult to persuade anyone to dive. Vehicles with crews is one answer to this problem.

Left: the one-atmosphere vehicle called Spider, from Self-Propelled Inspection Diver. Spider has powerful lights, a camera video system, and its support system can be easily mounted on a small vessel, rig, or platform.

Submarines and Bathyscaphes

With diving bells now being fitted with propellers and atmospheric diving suits having thrusters, the distinctions between "bell," "suit," and "vehicle" are becoming increasingly blurred. There are, however, three distinct types of free-standing vehicles – that is, those not supported by cable from the surface. These are the submarine, the bathyscaphe, and the submersible, and the jobs they do dictate the differences in their design and construction.

The submarine is mainly an instrument of war, the design of which has culminated in the nuclear-powered vehicles that now roam the world's oceans at submerged speeds of up to 45 miles per hour, each carrying a crew of 150 men and having the ability to stay submerged for up to three months at a time. Each has a dozen or more intercontinental ballistic missiles on board, representing a destructive power capable of wiping out entire cities thousands of miles away.

For all the frightening fire power of these war machines, the construction of their hulls and the way in which they control their descent and ascent have changed little since the first powered vehicles began to make their appearance in the late 19th century. The submarine has a near-cylindrical hull strengthened by steel ribs. On either side of this pressure hull are large buoyancy tanks in which valves are opened so that they flood when the submarine dives. The submarine dips downward, its angle of descent controlled by small adjustable "wings" on the hull. When it reaches its desired depth, the valves are closed. Minor adjustments to trim are made by pumping water or air into small tanks inside the pressure hull. When the submarine ascends, compressors pump air into the buoyancy tanks, forcing out the water.

The crew members of a submarine breathe oxygen produced by the combustion of chemicals and the air is kept clean by mechanical and chemical filters.

The greatest depth to which a submarine can dive is about 3000 feet. To go deeper than that requires a different hull shape and a different approach to achieving control of the rate of ascent and descent. The best shape for resisting extreme pressure is a sphere, and this has been the hull shape adopted for the few craft of this type that have been built and that have penetrated the deepest parts of the world ocean.

The first craft of this type, built in the United States in 1930, was called *Bathysphere*. Although it was lowered by cable, it is included in this section because it was the natural ancestor of so many later, free-swimming craft. It was designed by Captain J H Butler and financed and constructed by Otis Barton, an engineer who made the first dive in the craft with biologist William Beebe.

Bathysphere was 4 feet 9 inches in diameter, cast out of 1.25-inch thick steel, and with a total weight of 5400 pounds. It had three ports made of 3-inch thick fused quartz, and its 14-inch diameter opening was sealed by the simple expedient of a 400 pound steel plate bolted on with 10 nuts. The sphere, with oxygen tanks inside for life support, was simply dangled at the end of a single cable that also carried a telephone link and electric power for lighting.

Beebe's sole reason for submitting himself to encapsulation in this steel ball was to observe marine life at depths greater than anyone else ever had. This he achieved on a series of dives in the 1930s, culminating in a world record off Bermuda on August 15, 1934. The two men in *Bathysphere* reached a depth of 3028 feet – over half a mile down. The American public took Beebe to its heart, and on one of his dives a national radio network made a live broadcast of his telephoned observations.

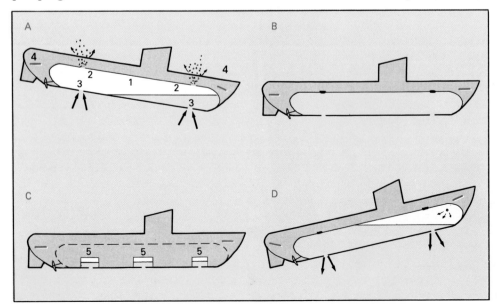

Left: how a submarine dives and ascends. (Blue indicates water.) In **A**, it is seen that buoyancy tanks (1) are filled with air and the valves (2) are shut. To dive, valves are opened and water floods the tanks through openings in the bottom (3). Winglike planes (4) control the angle of the dive. In **B**, the valves are closed so that flooding stops. In **C**, the vessel is seen in balance, maintained by pumping water or air into tanks (5). In **D**, the ascent is shown. A compressor forces air into the buoyancy tanks to replace water. The planes control the angle of ascent.

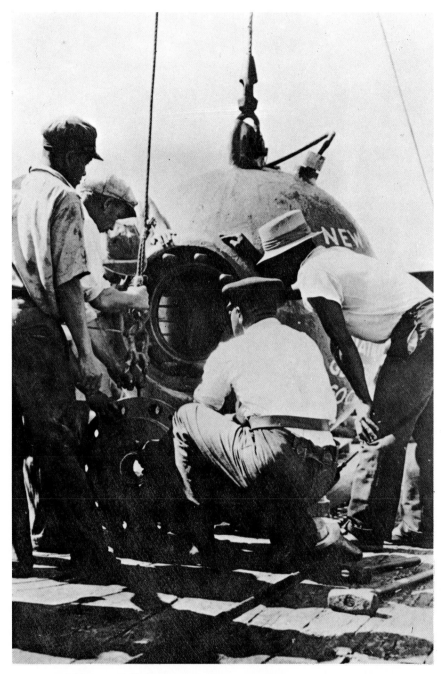

Above: the bathysphere, one of the first deepsea vessels to carry people, breaks the surface as it is hauled up. It had just completed a record-breaking dive of 2,510 feet on August 11, 1934.

Left: crew members of the stand-by barge *Ready* opening the hatch to release scientist William Beebe and engineer Otis Barton from the bathysphere.

The Beebe/Barton record stood for 15 years until Barton himself, in an improved sphere called *Benthoscope*, dived to 4488 feet off Santa Cruz, California.

Then came the bathyscaphe *Trieste*, which not only surpassed the achievements of *Bathysphere* and *Benthoscope* but also established a depth record that will never be exceeded unless someone finds a deeper point on the ocean floor.

At the time Beebe and Barton were making their series of dives, the Swiss scientist Auguste Piccard had been breaking records in an entirely different direction by ascending in a balloon to 10 miles above the Earth's surface. He then turned his attention to the deep ocean, determined to observe at first hand the creatures of the depth.

Piccard's first bathyscaphe – a name he coined from the Greek words for "deep" and "boat" – was *FNRS-2*. From this came the ideas for *Trieste*, named

after the Italian city that provided Piccard with much of the equipment for the craft. After a series of trials, *Trieste* reached a depth of over 10,000 feet in the Mediterranean in 1953. But the cost of operating the craft frustrated Piccard's attempts to reach even deeper dives and his record was short-lived. In 1954 two French Navy officers, Georges Houot and Pierre Willm, surpassed his record by diving to 13,365 feet. They also announced that they were going to build another craft and dive to the bottom of the Challenger Deep, nearly 36,000 feet beneath the surface of the Pacific.

Fortunately for Piccard, the United States Navy began to take an interest in bathyscaphes at about that time, and after a series of trials in which the value of the craft for making a whole range of physical and biological studies were proved, *Trieste* was sold to the United States Navy in 1958. Preparations were

Above: the *Trieste*, the bathyscaphe that made history by touching bottom on the world's deepest trench in the Pacific. It was crewed by Jacques Piccard, son of the inventor of the bathyscaphe, and Donald Walsh, a lieutenant in the United States Navy. The *Trieste* had been owned by the American Navy for two years at the time of the record-making dive in 1960, when it went down to 35,800 feet.

Below: an interior view of the cabin of the *Trieste*. The cabin is heavier than water and can withstand sea pressure. It is closely linked to a float, which is lighter than water and, when filled with gas, gives the vehicle its lifting power. Piccard's use of the lighter-than-water float was based on the principle of the lighter-than-air balloon and it revolutionized the design of diving vessels.

swiftly made for it to preempt the French attempt.

Seen floating on the surface, *Trieste* looks completely different from *Bathysphere*, which was a simple sphere. *Trieste* looks more like a submarine, over 50 feet long with a deck, rails, and a submarine-like conning tower. Fundamentally, however, it was the same as *Bathysphere*, equipped to carry only two people in a single sphere 6 feet in diameter – only 1 foot 3 inches bigger than the Butler/Barton vehicle. This sphere, tiny in relation to the overall size of the craft, formed what looks like a blob on the bottom of *Trieste*. All the rest of the craft was built around this 28,665-pound steel capsule with the sole aim of getting it down into the ocean depths and back again. This was the principle difference between *Trieste* and *Bathysphere*, for *Trieste* was designed to go down and back on its own, independent of surface support except for towing to and from the dive site. The comparatively large structure around the sphere, consisting of a series of compartments filled with a total of 70 tons of gasoline, was merely to provide buoyancy. Air would not have been suitable to

provide buoyancy, as in submarines, because of its compressibility. Gasoline is less compressible and is far lighter than water. However, tanks at either end of the craft contained air to give it buoyancy on the surface. When these were flooded, *Trieste* began slowly to sink, weighed down by no less than nine tons of iron shot pellets. The rate of descent was controlled from within the sphere by a system of valves that released gasoline from the floats. When the two occupants wished to leave the ocean floor, they released this shot by means of electromagnets and the craft began its ascent. *Trieste* had oxygen cylinders to provide air and silver-zinc batteries to give electric power for observation lights and instruments.

The preparations for *Trieste's* big dive included being fitted with a specially strengthened observation sphere. The vehicle then undertook a series of deeper and deeper excursions into the Pacific off the island of Guam. The aim was to get to the bottom of the Marianas Trench to the Challenger Deep, the deepest known point in the world ocean, 200 miles from Guam.

On January 23, 1960 Jacques Piccard, son of Auguste, and United States Lieutenant Don Walsh climbed down the 13-foot ladder inside the access

Below: the *Trieste* in the Atlantic waters off the coast of Boston, Massachusetts. Before the United States Navy bought the bathyscaphe in 1958, Jacques Piccard had made a record dive to 12,110 feet in the Mediterranean west of Naples. After the purchase, Piccard teamed up with American scientists, engineers, and naval officers to achieve new record descents, culminating in 35,800 feet in 1960.

tower to *Trieste's* sphere. The hatch was closed, the instruments were checked, the access tower was flooded, and the dive was on. Five hours later *Trieste* touched down on the bottom of the world: 35,800-feet – or nearly seven miles – beneath the surface at the foot of the Marianas Trench. *Trieste's* powerful searchlight probed through darkness that had never before been broken and picked out a fish swimming idly away. It was about a foot long and resembled a sole or flounder. If any other justification had been required for the great dive, the proof of the presence in the deepest part of the ocean of relatively complex organisms as represented by this vertebrate provided it.

After its record dive, *Trieste* went on to make a series of dives for the United States Navy, including one to look for the striken submarine *Thresher* in 1963. *Trieste I* was not preserved and some of its parts were incorporated into *Trieste II*. But apart from making observation dives to the bottom of the deep ocean, there was little scientific scope and certainly very few commercial applications for such vast vehicles requiring so much logistical support. What was required were lighter, simpler vehicles with midwater maneuvering capabilities and the ability to take samples and perform other simple underwater tasks. They were not long in coming.

The innovations and work of Auguste Piccard on bathyscaphes provided the inspiration and a mass of design engineering detail for the development of this new class of undersea craft.

Small and Large Submersibles

The safety officer on the dive made by Piccard's bathyscaphe off Dakar in 1948 was a French naval officer named Jacques-Yves Cousteau. He saw many of Piccard's ideas being put into practice at first hand and soon began to incorporate them into his own plans for an underwater vehicle.

Cousteau was not interested in exploring the deepest parts of the ocean but rather in extending the range of scuba divers so that they could observe and film at greater depths for longer periods of time, in safety and with a greater degree of comfort. This required a lightweight, highly maneuverable, self-propelled vehicle unencumbered by a cable to the surface but without the vast gasoline floats of the bathyscaphe.

The result of his design took shape in 1959 as the *Soucoupe Plongeante* or *Diving Saucer*, a vehicle that was able to take two people to depths of over 1000 feet at one atmosphere.

Cousteau's design was the first submersible, and it was an outstanding success. Today there are over 100 little vehicles of this type operating all over the world for scientists, navies, and, more and more, for the offshore oil and gas industry. They all incorporate the ideas formulated by Piccard and Cousteau.

The first essential of all submersibles is a pressure hull. In nearly every submersible this is spherical because a sphere gives the best strength-to-weight ratio. The preferred material has almost always been steel. Other materials have, however, been used to provide the essential combination of lightness and strength. The United States submersible *Alvin* has a pressure hull of titanium. Another United States craft, the *Aluminaut*, not only has an aluminum pressure hull, but is also cylindrical instead of spherical.

Some small submersibles have hulls of acrylic material that does not react favorably to great pressures but is nevertheless highly suitable for shallow water work. Many of the latest generation of crafts have their hulls terminating in hemispheres of acrylic material, which offers the advantage of almost complete all-round visibility for the crew.

Glass and ceramics have also been considered as hull materials. A glass sphere has the remarkable property of actually gaining in strength the more it is subjected to pressure and, like acrylic, its full transparency is an underwater boon. However, it is difficult to produce a large glass sphere that has no flaws of the type that could cause a catastrophic collapse at great depths. One glass-hulled vehicle has been built so far, and when a method of producing a flawless sphere is perfected, glass will very likely become a firm rival to steel as a hull material.

Behind the pressure hull is the battery pack, which drives the thrusters and the other vehicle components. Nearly all submersibles have lead acid batteries similar to those used in a car. To provide power for a submersible, banks of them have to be linked together,

Left: Jacques Cousteau looking out of his minisubmersible, the *Sea Flea*. One of the world's most famous ocean explorers, Cousteau spearheaded the development of light self-propelled underwater vehicles.

Left: *Alvin*, a submersible designed in the United States. It was this vehicle that located a lost H-bomb off the coast of Spain in 1966.

Below: *Aluminaut*, another United States submersible. It is different from others by virtue of its cylindrical shape and lightweight pressure hull. It can carry four to six people to a depth of 15,000 feet.

which means that they constitute the largest single load the craft has to carry. Even on a small, two-person vehicle, battery weight is likely to be 2.5 tons out of a total weight in air of perhaps seven tons. Although lead acid batteries are cheap, rugged, and easily discarded in an emergency to give the vehicle buoyancy, they are at the same time the principal limitation to the present range of submersibles. This is because their power output in terms of weight is low and they have to be recharged after each dive by the submersible's surface support ship. Speed has to be sacrificed in order to keep down the weight of batteries, since to double speed requires the power system to produce a staggering eight times more power.

Because of this power restriction, very few submersibles have a cruise speed of much more than 1.5 knots for an eight-hour period underwater. If the machine has to battle against a current – even one as slow as one knot – endurance is cut dramatically.

Research is therefore concentrating on developing other power sources, of which the current favorite is a fuel cell similar to those used to help the Apollo spacecraft land on the moon. The fuel cell is an electrochemical device that consumes hydrogen and

oxygen to produce electricity, with water as a by-product. Experiments in the United States on the big submersible *Deep Quest* – which weighs 50 tons, is 40 feet long, and has a crew of four – have yielded remarkable results. Its fuel cell, weighing only 275 pounds and occupying less than six cubic feet of space, produces 30kW of power. It has produced seven times as much electricity as that yielded by batteries before they need recharging, and has been tested to a depth of 5000 feet. Operators of submersibles through-

out the world are watching the *Deep Quest* experiments with keen interest in the hope that the fuel cell could be the breakthrough that overcomes the biggest limitation of these crafts.

After hull design and power supplies, an obvious feature of any submersible is life support. The system used is simple and safe. Cylinders feed oxygen into the cabin at a controlled rate and the air is circulated by fans. Air exhaled by the occupants is "scrubbed" of carbon dioxide by being passed by the fan through a tray containing a chemical absorbent. The pilot keeps a close watch on dials that show any possible build-up of pressure or carbon dioxide.

The main reason for operating a submersible is to allow people to see things at first hand underwater in relative comfort, compared to diving, and at atmos-

pheric pressure. It means that scientists, engineers, and business executives can descend to the depths of the ocean just for a look, provided that they have no great objection to being confined in a very small space. The value of simply looking cannot be overestimated. Biologists can study creatures in a habitat only slightly disturbed by the presence of their craft, rather than having to make scientific guesses about the environment and habits of a fish from the often-mangled remains of a solitary specimen that has been dragged up by a trawl. Geologists can look at complete formations underwater rather than having to make deductions about the nature of seabed features from sonar traces and grab samples. Engineers can inspect pipelines, cables, the underwater sections of oil production platforms, or subsea wellheads to

make direct comparisons with the original blueprints of them. Even just looking sometimes requires additional aids, however. For this the submersible can carry floodlights on a frame mounted at the forward end of the vehicle, some of them with pan and tilt units so that they can be trained on objects of interest. Such aids are usually used sparingly because lights consume a great deal of power.

Observers can arm themselves with notebook, pen, and portable tape recorder to record what they see, but often they will need to have a more permanent record to be able to show to experts on the surface. So submersibles are fitted with a range of still, movie, and TV cameras, some of which have light intensifiers to probe through the darkness to pick out details that the eye cannot see.

If the submersible is working over a wide area of seabed rather than at a specific point on a structure or pipeline, the crew needs to know where they are. Early underwater crafts were fitted with a compass and trailed a line with a float so that a surface ship could follow the float and pass navigational instructions to the pilot via an acoustic telephone link. Today's submersibles have adopted more sophisticated methods. They usually have an echo sounder that tells the pilot how far the craft is from the seabed and the surface. There is also a forward-looking sonar that warns of approaching large objects, and a side scan sonar that produces a trace of seabed features on either side of the craft. If the craft is surveying within an area up to 10 square miles, its surface support ship often lays a network of transponders that are slightly different from the type fitted in the submersible. The support ship can determine both its own position and

Above: a commercial research submersible called the *Johnson-Sea-Link*, which can manipulate small pieces of equipment with great precision. Like *Nemo*, it also has an acrylic pressure hull.

Right: an operator at the control panel of one of the United States underwater launch and recovery platforms. These platforms service scientific submersibles.

that of the submersible, exactly and relatively, by computation of the transponders' answers to interrogations.

Early submersible pilots and observers soon became aware of the frustration of being able to see but not to touch. This brought forth a range of equipment that can be bolted onto submersibles to enable them to perform simple tasks underwater. Manipulators can pick up samples or break off pieces of rock and drop them into a basket mounted on the front of the craft. Other tools do a variety of jobs.

However, submersibles are not equipped to undertake heavy work. This is because they are so light in water that they tend to drift when exerting force. Although some highly sophisticated tools have been developed with which the vehicle can probe, grab, and manipulate, none yet approaches the nimbleness of human fingers. A compromise has been devised in the shape of the diver lockout craft which combines the features of the one-atmosphere submersible with those of the submersible diving chamber. The diver

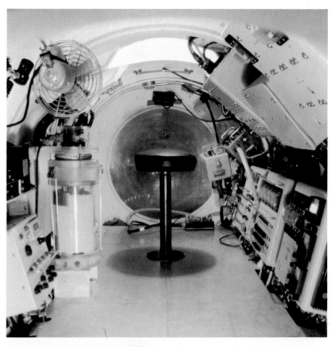

Above: an interior view of a diver lockout submersible looking forward from the dive chamber to the pilot's compartment. This lockout craft can carry a crew of two and two divers. It can dive to 1000 feet in depth.

Left: a full view of a diver lockout submersible on the deck of a support ship. This vehicle carries three crew members and two divers. It weighs 15 tons.

lockout submersible has two accommodation compartments. The forward one houses the pilot and observers and is equipped in exactly the same way as the conventional submersible. The aft compartment is in effect a submersible diving chamber capable of being pressurized to a depth equivalent to that at which divers need to work. Divers are transported to their work site at ambient pressure. They leave and reenter the submersible through a bottom hatch. When they reach the surface, still at ambient pressure, they are transferred to a deck decompression chamber on the support ship.

A big advantage of the diver lockout submersible is that it can deliver divers directly to widely spaced work sites, while a ship-and-chamber combination would involve bringing the divers up to the surface each time they wanted to move a quarter of a mile or so. The submersible can also provide improved lighting for the job and additional power for hand-operated tools.

Lockout submersibles cost in the hundreds of thousands to build and, because they are large themselves, need a big support ship from which to operate.

The question of surface support is a problem in the operation of all manned submersibles. A ship has to be purpose built or specially adapted to handle the craft over the stern by means of an hydraulic A-frame. A wire from the A-frame to the top of the submersible lifts the underwater craft from the deck, swings it out over the stern, and lowers it into the water. Divers then release the wire. When the submersible surfaces, it has to maneuver itself close to the stern of the ship and divers have to enter the water again to attach the lifting wire. It is easy to see that this is a risky operation when there are strong tidal currents or when the sea is choppy, so submersible operations are very dependent on good weather.

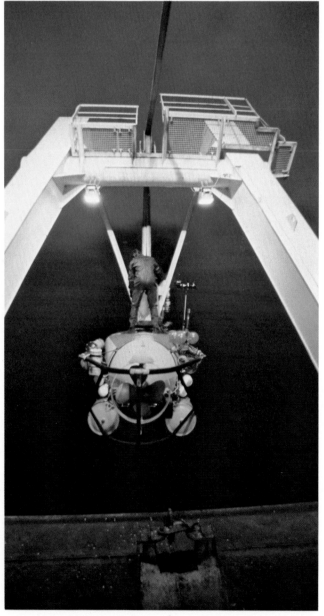

Replacing Divers with Machines

People can swim out of bells in rubber suits, walk across the seabed in armor, or be crammed into a little submarine. Any method by which they enter the underwater world is expensive and carries its full share of risks. So it is that a parallel line of development seeks to keep people out of the water altogether by creating a range of unmanned vehicles.

At their simplest, unmanned submarines are what the offshore oil business calls "swimming eyeballs." The most successful of these is the *RCV 225*. (RCV stands for Remote Control Vehicle). Thirty of these have been made in the United States for worldwide operation, usually in connection with offshore oil work. The *RCV 225* is about the size of a medicine ball and is fitted with a TV camera, lights, and thrusters to give it great maneuverability. It is lowered from a ship or rig in a cage and, when close to the area in which it has to work, it swims out of the cage and is controlled from the surface by a power and signal-carrying cable. TV pictures are constantly relayed back to the surface where they are taped for future study. Meanwhile, the live monitor on the surface console allows the vehicle to be moved in close to inspect structures and pipelines.

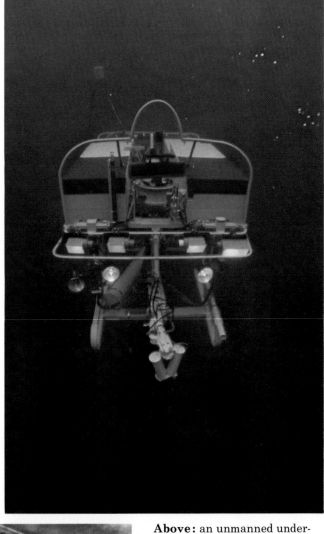

Above: an unmanned underwater vehicle known as Curv, from Cable-controlled Underwater Recovery Vehicle. It was designed for the United States Navy to salvage torpedoes during tests of the weapons.

Opposite: the *Smartie*, another remote control subsea vehicle. It does surveying work in the North Sea oil fields.

Left: the *RCV 225* on the deck of a ship. This is one of the simplest of the unmanned submersibles, known as a "swimming eyeball." It does inspections with a TV camera.

The *RCV 225* has also become the diver's friend. At first treated with suspicion, it is now welcomed for the light it provides when a diver is engaged on delicate work and for the general assurance it gives as a connection with the surface in case of trouble. Divers have been heard to refer to the *RCV 225* as their "watchdog" and, on completion of their work, to tell the surface operator to call the vehicle back into its "kennel."

Other unmanned underwater vehicles – and there are over 30 types – vary in size and capabilities. Many are similar to the *RCV 225* but have additional visual equipment such as still cameras, mounted singly or in pairs, and 8mm or 16mm movie cameras, often on pan and tilt heads. The quality of TV pictures obtained underwater is not generally very high, but vehicles of this type enable the operator to use the TV camera to pick out areas of particular interest and move the vehicle in close for taking high-quality color shots.

At the top of the unmanned vehicle tree come the big units that not only take films and send TV pictures

back to the surface but also perform simple underwater tasks. For example, the surface operator can guide the vehicle to the leg of a steel structure to which it attaches itself by means of a suction pad. Powerful water pumps then jet away marine growth and abrasive wheels clean the steel, which is photographed and subjected to ultrasonic and other tests for flaws and cracks. Vehicles of this size can also be fitted with sonar and sub-bottom profiling equipment to make inspections of miles of pipelines. Unlike divers and manned submersibles, they can work day and night and have a constant source of power from the surface.

However, unmanned vehicles have the big disadvantage of being dangled at the end of a cable – and cables are notorious for breaking, for being subjected to drag in currents, and for becoming tangled around underwater obstructions. So it is projected that the next generation of unmanned crafts will have no cable and will draw heavily on the microelectronics revolution by having its own computer. The operator will slip in the program for the job required, and the vehicle will swim off to take pictures and perform the types of tasks outlined above. It will return to base when its work is completed or upon a signal from the surface – the electronic equivalent of a shepherd controlling a flock of sheep by whistling to a dog.

In the fairly recent past, too much time has been devoted to arguments among the various proponents of free diving, atmospheric diving, or manned and unmanned submersibles as to which is the most suitable for underwater work – and, in fact, which would be mostly likely to replace the others. Today, as work in the oceans is conducted at greater and greater depths and as the tempo of activity increases, there is a growing realization that there will be a continuing need for every type of system. Mastering the skills of performing underwater work – by whatever means – will be an increasingly challenging and rewarding field of endeavor.

Chapter 12

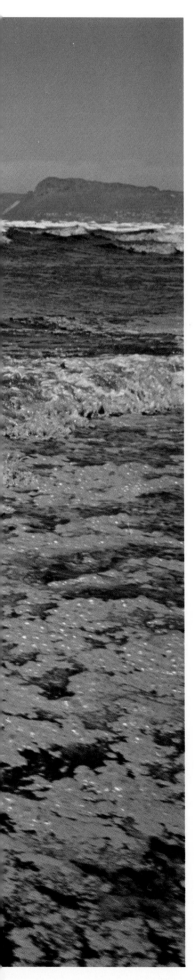

Problems
and Prospects

Who owns the huge resources of the deep oceans?
This is a prickly question that must have an answer if
the seas are to help supply the needs of the world in
the future – and already a great deal of the territorial
waters are a matter of dispute. Why should we be
concerned about pollution of the seas? In fact, there is
little future in the oceans – or in the very survival of
this planet – if we willfully turn the seas into a
dumping ground for all that we find inconvenient,
distasteful, or too dangerous to dispose of in any other
way. How can we get the best out of the world's ocean
in the future? Based on today's technology, it is
possible to look 50 years ahead to a world that is
making the most of the ocean's great potential.

Opposite: deep and vast as the
world's ocean is, it is in danger of
being polluted to the point of
killing all life in it. This is a
problem mankind must solve soon.

Who Owns the
World's Ocean?

The big day has arrived. You have spent many years and many millions in research and development of a manganese nodule mining operation at a site 12,000 feet beneath the Pacific Ocean, some 500 miles from shore. You have carefully assessed the extent of the resource, developed harvesting techniques, established processing plants, and run pilot trials. Now you are all ready to start mining, and from the stern of your ship your nodule collector heads toward the ocean floor for the first time. Very soon now you will start recouping your investment. You are feeling quite satisfied.

Then over the horizon appears another ship which you watch with curiosity. As it comes closer you look at it through binoculars, and as it heaves to you note with consternation that it is another nodule mining ship under a foreign flag. On its deck are people observing you with the same alarm you are feeling.

What do you do? You consider the options open to you. You could ignore the vessel and let it operate along with you, but that would ruin the economics of

"your" resource. In international law it "belongs" neither to you nor to the ship that is operating near you. Put another way, it "belongs" to both of you and to anyone else who has the technology and money to come along and exploit it. Politicians, law experts, and diplomats have been debating the question of ownership of deep ocean resources "beyond the limits of national jurisdiction" for about 15 years, and have made very little practical progress. Conferences on the Law of the Sea are solemnly convened in capital after capital all around the world, year after year. Each time, they carefully debate the question of deep ocean resource ownership. Each time, each clique of nations postulates its own self-interested solution to the problem, all of which all the delegates have heard the year before. Agreement on any solution is, inevitably, not reached, and the delegates pack their bags and arrange to meet the following year in yet another capital.

The stately diplomatic dance goes on, and it would be almost entertaining were it not for the fact that the means to exploit deep ocean resources, in the shape of manganese nodules, now exists. Can the debate continue to be purely academic when there are ships tied up at the dockside awaiting a conference decision before they sail? In the case of manganese nodules, that situation is nearly upon us and before long somebody will put the debating abilities of international lawyers to a severe test simply by going out and starting to mine.

Right: a grab going over the side of a ship during mining operations for manganese nodules. One of the troubling questions is: how do people share the resources of the oceans if it is agreed that they belong to all?

your operation which is geared to exploiting most of the nodules in that area. You could report the rival operation to your government. That would not work either, since no government has jurisdiction over deep ocean resources. Could you complain to the United Nations? Hardly. UN members are still arguing about who owns what in the oceans. Blast the vessel out of the water? The consequences of such an act are too risky, likely as they are to invite retaliation in the very near future.

In fact there is nothing at all you can do to protect

It all used to be much more simple. Before the advent of fishing fleets able to roam the world, and before the need arose to exploit minerals on and beneath the seabed, there was really no need for international agreement on who owned which particular stretch of marine territory. There was a generally accepted convention that each nation had a three mile stretch of water along its coasts within which no other powers had rights other than the innocent passage of their ships. The main purpose of this rule was to provide military security, so the three-mile limit had logic

behind it: it was the maximum distance over which it could be reasonably expected to hit a ship with a carefully aimed cannonball from a shore battery.

This arrangement worked fairly well for hundreds of years. Gradually, however, the patterns of world distribution of food, minerals, and other essentials began to change as the world population grew at an alarming rate. Nations sought to supplant their often meager land-based resources by greatly increased exploitation of the oceans. At first they concentrated on the areas close to their shores, but then they began to go farther and farther afield as they developed the technology to do so.

It was the exploitation of living resources that started the great ocean grab. A nation with huge stocks of fish close to its shores would watch in alarm as more and more foreign vessels came to scoop away what was often one of its prime natural resources. The country took moves to protect that resource. In seeking to protect fish stocks, the politicians would consult their fisheries scientists to insure that any extension of fisheries limits encompassed the whole area in which a species or group of species bred and lived. Seeing that its fishery resources extended to the width of the Humboldt Current, which was about 200 miles from the coast, Peru in 1947 not unexpectedly imposed a 200-mile exclusive fishery zone for itself. In 1958 Iceland, with "nursery" grounds for cod close to the shores, imposed a 12-mile limit – and policed

Above: a painting of the Dutch fleet that destroyed unprepared English warships at anchor in the Medway river in 1667. The two countries were at war over the control of sea trade, and Britain suffered an ignominious defeat because of the neglect of its navy. Wars over trade and sea power have been frequent throughout history.

Below: fishing vessels at sea. Rivalry in the world's fisheries is a source of grave problems today. Nations often dispute other nations' protective fishing limits.

this area rigorously. This and other unilateral extensions of fisheries limits on Iceland's part led to some dangerously bitter confrontations between Icelandic gunboats and the trawlers of other nations, notably Great Britain. The trawlers were given naval backing in their determination not to accept the new limits, though compromises were eventually reached.

By 1958 the politicians, lawyers, and diplomats had begun to come up with something governing the question of the law of the sea. Four Conventions were adopted by the United Nations. They related to the high seas, to fishing and conservation, to the territorial sea, and to the continental shelf.

The convention on the high seas laid down some general freedoms, such as the right to fish the high seas, fly over them, and sail on them. There was a stumbling block, however, in the question of what is meant by "high seas." Presumably it means the waters beyond the limits of national jurisdiction. But where does national jurisdiction end? That is where it starts to get complicated because national jurisdiction is tied in with the accompanying convention on the territorial sea. This convention establishes that the sovereignty of a state extends to "a belt of sea adjacent to its coast" and "to the air space over the territorial sea as well as to its bed and subsoil." It entirely begs the crucial question of what the exact breadth of the territorial sea should be.

Claims began to be made for territorial seas ranging from three miles to 200 miles, and everything was made still more complicated by some nations claiming exclusive zones for certain undertakings such as fishing and the enforcement of pollution regulations.

The 1958 convention on the continental shelf seemed to be far simpler in concept and implications. Basically it said that a nation had the exclusive right

to explore and exploit the resources lying on or under the seabed (not the waters above it) adjacent to its coast to a depth of 656 feet (200 meters) or "beyond that limit, to where the depth of the superjacent waters admits of the exploitation of the natural resources...."

In other words, a country could exploit resources to 656 feet or to whatever area its continental shelf extended.

What happens if the continental shelf is shared by

Left: the green parts of this map show the area of the continental shelves around the countries of Europe to a depth of 656 feet, the level which present international law gives nations exclusive rights over marine resources. Gray area shows European sea space extended to 13,200 feet.

Opposite: the North Atlantic divided up between nations on the equidistant point rule. This provides that the shelf boundaries of coastal states are to be "equidistant from the nearest points of the baselines from which the breadth of the territorial sea of each is measured."

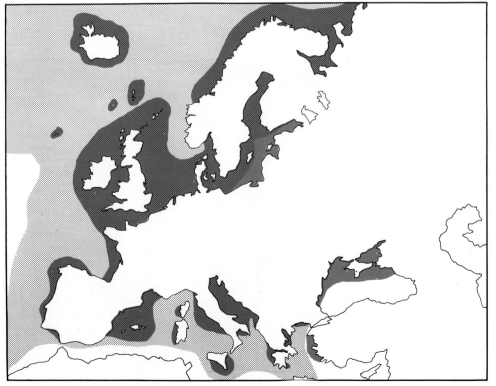

Left: fishing boats in the port of Karachi, Pakistan. One of the countries that claim a 12-mile territorial sea limit, Pakistan has also claimed a 50-mile exclusive fishing zone – that is, a zone where the Pakistan government has sole rights, not to fish, but to regulate fishing.

Below: the Grand Isle sulfur mine, 7 miles out in the Gulf of Mexico. Sulfur, like oil and gas is one of the undersea resources that make the ownership of the seabed an increasing matter of world concern.

a number of nations, as in the North Sea? The convention settled that by laying down firm rules for the drawing of median lines so that each nation got its fair share according to the length and shape of its coastline.

This all looked good in theory in 1958 when most people were mainly concerned with fish. But the offshore oil and gas industry was getting into its stride, and its rapid development both geographically and technologically has blasted great loopholes in that 1958 convention. Here are some obvious examples:

The convention left open the definition of the outer limits of the continental shelf, saying in effect that if a nation had the technology to exploit the seabed in 15,000 feet of water, then it had every right to do so. In 1958 it was inconceivable that such technology would exist, but now it does. In theory, then, and in keeping with the convention, the United States and Japan, having both developed deepsea mineral dredging capabilities, may now claim vast areas of the Pacific Ocean seabed and subseabed resources to themselves, based on a simple median line principle. This is clearly inequitable and unworkable. The future of the oceans simply cannot be decided solely on the basis of who has the most advanced technology.

The median line principle also has flaws. It was all

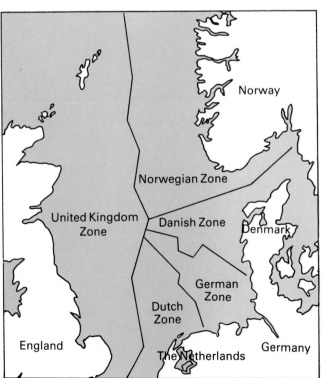

307

very well when the only resources of the continental shelf were sand, gravel, other relatively low-value minerals, lobsters, and crabs. But with the advent of deepwater drilling for oil, various nations started looking very seriously at their rights under that principle. Rockall is a clear case in point.

Rockall is an aptly named island of uninhabitable rock that projects above the stormy waters of the North Atlantic some 280 miles off the northwest coast of Britain. The question of its ownership had never been seriously discussed. Nobody particularly wanted it, but it was vaguely assumed that it was part of Great Britain. Then came the offshore oil and gas industry, and the geologists began to make encouraging noises about the likelihood of hydrocarbons in the area of the rock. Britain promptly claimed a 200-mile "exclusive economic zone" around the rock island. The Republic of Ireland, Denmark (in respect of the Faroes), and Iceland also promptly laid claim, and a glance at a map shows Rockall to be neatly spaced between them all. Although Britain had formally annexed Rockall some years before, it had done nothing about the territory. Now, however, it began to fuss about this lump of rock, devoting more time and money to Rockall than it did to many of the more deprived and highly populated areas on the mainland. Surveys were made and it was formally incorporated into the county of Inverness. In 1972 the Royal Navy installed a navigation beacon on Rockall. All this for an island whose land nobody is interested in.

Happily, the question of the ownership of the seabed around Rockall, which is where the real interest lies, will almost certainly be settled by negotiation, or at worst by a judgment of the International Court which will be accepted by all the parties concerned. Other disputes about oceanic resources, besides the Anglo-Icelandic cod wars, have reached the gunboat level. Disputes over oil and gas drilling rights have flared up between Israel and Egypt and between Greece and Turkey, even though existing political and territorial differences have done much to fuel these confrontations.

The original Law of the Sea conventions adopted by the UN were drawn up in a genuine attempt to impose a sense of order on the exploitation of the oceans, and the people who drafted and worked for the acceptance of those conventions cannot be blamed for having had so much of their work rapidly rendered ambiguous, if not irrelevant, by an explosion of technological capabilities. Now that it is known that the capabilities exist to exploit the very depths of the oceans, there is a chance to draw up a meaningful international convention for exploitation that will have a reasonable lifespan and that will be equitable to the have and have-not nations alike.

Inequity is the crux of the present series of slow-

Above: copper refining, Zambia. Many developing nations see ocean mining as a threat to their own mineral industries.

moving debates about the exploitation of deep-ocean resources. The developing nations, without the technology to exploit the deep oceans, have expressed fears that they will lose out completely on resources that should, they say, be "the common heritage of mankind." Among the developing nations making protests are those which at present supply the vast bulk of the minerals that are so avidly sought by the hopeful harvesters of manganese nodules. The harvesters are, of course, the developed nations that depend on the developing nations for supplies.

The droning arguments of the international conferences have produced at least one positive idea: that there should be an international authority to oversee the exploitation of deep sea resources – which in effect means manganese nodules. The idea unfortunately became submerged in a whole series of sub-arguments about the exact powers of the proposed authority. Would it just issue licenses? Would it actually become involved in mining? Would it distribute profits to developing nations and to those with no coastlines? If so, on what basis would the profits be split?

The bickering continues, but there are signs that the nations which have invested heavily in deepsea technology, notably the United States, are now looking for a return on their investment and are unlikely to tolerate the dithering much longer. They may well force the hands of the international lawyers, politicians, and diplomats by making unilateral claims to deep ocean resources and simply going ahead with exploitation.

The future of this planet depends a great deal on the way in which we use and exploit the oceans. It is to be hoped that the politicians – latecomers to knowledge about the potential of the oceans – are fully aware of that simple fact.

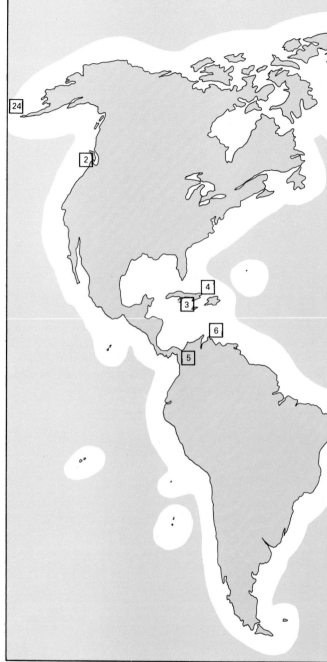

Above: this map illustrates the problems facing any international seabed authority of the future in controlling the use of ocean resources. It will have to administer all areas of the world sea not included in the 200-mile exclusive zones that will encircle almost every land mass, as seen by the outline on the map. It will also have to guarantee rights of passage through numerous straits that have no history of control in peacetime. The most important of these are shown on the map and the key opposite.

310

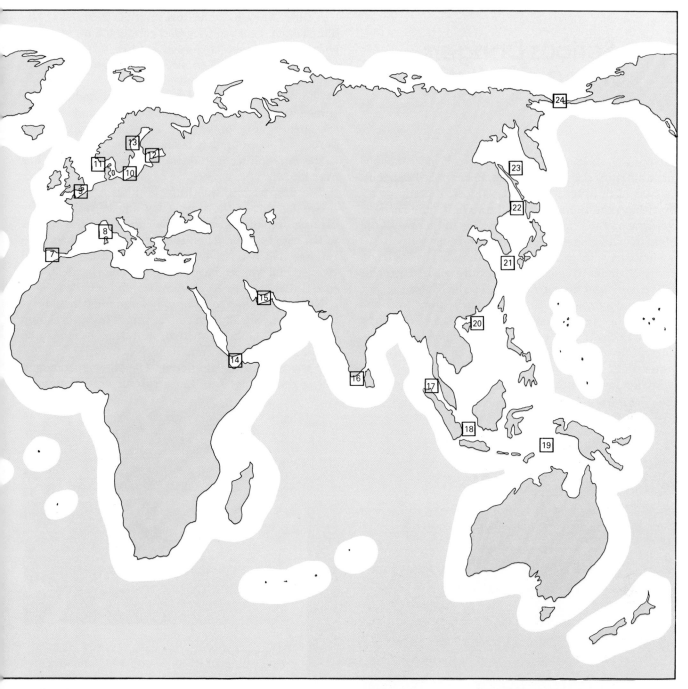

Straits	Sovereignty on either side	Straits	Sovereignty on either side
1 Robeson Channel	Canada/Denmark	14 Bab el Mandeb	France/Yemen
2 Strait of Juan de Fuca	United States/Canada	15 Strait of Hormoz	Iran/Muscat and Oman
3 Martinique Passage	France/United Kingdom	16 Palk Strait	Sri Lanka/India
4 Dominica Channel	France/United Kingdom	17 Strait of Malacca (North)	Indonesia/Malaysia
5 Aruba-Paraguana Passage	Netherlands/Venezuela	18 Strait of Malacca (South)	Indonesia/Malaysia
6 St. Lucia Channel	France/United Kingdom	19 Selat Ombai	Indonesia/(Portuguese) Timor
7 Strait of Gibraltar	Morocco/Spain	20 Lema Channel	China/United Kingdom
8 Strait of Bonificio	France/Italy	21 Western Chosen Strait	Japan/Korea
9 Strait of Dover	France/United Kingdom		
10 Bornholmsgat	Denmark/Sweden	22 Soya Kaikyo	Japan/U.S.S.R.
11 Öresund (the Sound)	Denmark/Sweden	23 Notsuke	Japan/U.S.S.R. Administration
12 Entrance to Gulf of Finland	U.S.S.R./Finland	24 Bering Strait	United States/U.S.S.R.
13 Entrance to Gulf of Bothnia	Finland/Sweden		

Serious Dangers of Heavy Pollution

On June 3, 1979 the drilling crew of the semisubmersible rig *Sedco 135* was going through the laborious process of pulling up over 11,500 feet of drillpipe so that they could change the blunted bit. The rig was working on the Ixtoc Field in the Bay of Campeche off Mexico, where there are enormous reserves of oil.

Suddenly something happened deep in the well. Oil and gas under tremendous pressure rushed up the well casing and hit the deck of the rig. There the gas came into contact with hot pump motors and enveloped the rig in a ball of flame. Most of the rig's derrick and winching equipment was ripped apart and hurled clear, to settle on the seabed in 165 feet of water.

Miraculously, none of the rig's 63 crew members was hurt. They had had just enough warning of the danger and, because they had conducted an abandon ship exercise just two days before, all the escape equipment was in top condition. The men left the rig in an orderly way, without panic.

The fire quickly took hold as the oil and gas continued to gush unchecked from the well, and soon the sea surface surrounding the rig was blazing just as fiercely. The fire in fact burned for over four months in the worst blowout that the offshore oil industry has ever known.

The first task in quelling the blowout was to get *Sedco 135* out of the way. Tugs had to sail into the blazing area of sea around the rig so that the anchor wires could be cut. Then the rig was towed away. So intense had been the heat that the huge hull had twisted and warped beyond repair. The rig was therefore towed out to very deep water and scuttled.

Next, a way had to be found of stopping the flow of oil and gas, as well as of preventing it from spreading all over the Gulf of Mexico, polluting beaches and ruining important shrimp fisheries.

It was a daunting challenge. The oil was streaming out of the damaged well at a rate of 30,000 barrels a

Above: Red Adair, known as a world expert in controlling oil rig blowouts, seen as he watches a video display of the wellhead during the tragic Ixtoc Field blowout in Mexico in 1979.

Left: a view of the flames of the Ixtoc blowout, taken from a barge. Although the accident ranks as the worst blowout in the history of offshore oil productions, none of the crew was lost.

day. To get some idea of this amount, it represents a volume of nearly 170,000 cubic feet. Half of this was burning and the rest was evaporating or spreading out over the sea in a four-mile wide slick. Once crude oil has been processed at the wellhead it does not ignite easily, but at Ixtoc it was spewing from the seabed mixed with gas.

One of the early arrivals at the scene of the blowout was trouble shooter Red Adair. He has become a legend in his lifetime for his courage and skill in putting out oil well fires and quelling blowouts when everyone else has failed. Adair wanted an inspection of the BOP stack on the seabed because, he said, if it were not too badly damaged, hydraulic lines could be attached to it and its valves and rams activated to close off the well.

An unmanned submersible was flown in from Houston, Texas and made TV inspections of the BOP stack, which Adair watched live on the surface. The oil and gas was escaping with such force that it created a vortex around the stack that stretched almost all the way to the surface. On one dive the submersible was caught in the vortex and was hurled over 150 feet up through the water to the surface.

The thick gush of oil around the BOP stack meant that visibility was poor. Adair could not get a clear enough picture to decide if the equipment was sufficiently undamaged for the hydraulic lines to be connected. Divers had to be called in to make a close visual inspection. This must have been terrifying work. The divers had to operate from the deck of a barge, one edge of which was actually inside the flaming sea. Temperatures on the deck reached 145°F. Divers and deck workers alike had to be kept under a constant cooling spray of water.

Once underwater the divers followed the cable of the unmanned submersible and then made their inspection with the aid of the craft's lights, but the vortex was a constant threat. One man paid for a momentary lapse of concentration with his life: he was sucked into the vortex and ripped to pieces on the BOP stack.

Finally, in late June, two hydraulic lines were fitted to the stack and pressure was applied to close down the valves and actuate the rams. It worked. The flow of oil stopped and the sea ceased to blaze, but it was only a temporary success. Within a few hours the oil began to burst out again, this time from around the base of the BOP stack where it emerged with such force that it scoured the seabed. Adair's team re-ignited the mixture on the sea surface so that at least some of it would not spread out into the Gulf. The small fleet of vessels fighting the blowout withdrew while new strategies were considered.

In the meantime the oil slick had grown to over 600

Below: clearing the beach of oil after the *Torrey Canyon* disaster. Oil tanker wrecks too often mean polluted beaches, like that on the coast of southwest England in 1967 when *Torrey Canyon* spilled 100,000 tons of oil.

miles long. Floating booms, designed to retain and soak up the oil, had been installed around the blazing area and at first successfully recovered some of the oil. But then the weather deteriorated and strong currents and 10 foot high waves reduced the efficiency of the booms considerably. Engineers at ports and harbors along the Mexican and Texan coastline prepared for the oil to come ashore and mounted protective booms across entrances.

Two jack-up drilling barges were brought in to try another approach. They were positioned some distance away on either side of the blowout, and began to drill angled holes, or relief wells, that would eventually link up with the rogue well. Heavy chemicals were to be pumped down the relief wells to bring the blowout under control. But it would take several weeks to drill the wells and the oil was still flowing unabated. Meantime, Adair and his team tried every trick they knew. Steel balls were dropped into the well to try to plug it and mercury was injected into it. All failed. Finally, in late October, a huge steel cone was dropped over the well so that the flow could be at least controlled until the relief wells had been drilled. The oil was also directed into tankers.

In November the flow from Ixtoc was finally stopped, but not before nearly 2.5 million barrels of oil had

Opposite: an oil slick drifts inshore at Kirby Cove State Park in California, the aftermath of a collision at sea. It takes a few months to remove the sticky oil even if it is tackled at once.

Right: the wreck of the *Amoco Cadiz* seen at low tide. This oil tanker grounded off northern France in 1978 and the beaches suffered severe pollution from its cargo of 222,000 tons of crude oil.

Below: a special floating boom is sometimes used to contain an oil spill and prevent beach pollution.

flowed into the Gulf of Mexico. The combined cost of the lost oil and the rescue operation was over $150 million. The rough weather that impeded operations paradoxically had one beneficial effect: it dispersed much of the oil so that it evaporated and was spread thinly. No short-term effects on the coastline or the ecology of the Gulf of Mexico have been observed.

The Ixtoc disaster is an example of marine pollution on a nearly catastrophic scale. Oil tanker disasters are another example, and the names *Torrey Canyon* and *Amoco Cadiz* are now synonymous with polluted beaches, dead seabirds, and ruined fisheries. *Torrey Canyon* ripped into rocks off the coast of Cornwall in southwest England in 1967, pouring 100,000 tons of crude oil into the sea. *Amoco Cadiz* grounded off Brittany in 1978, and much of its 222,000-ton cargo of oil ended up on French beaches.

Can anything be done to prevent disasters such as blowouts and tanker wrecks? The short answer is no. Accidents happen, and although oil companies and shipping lines adopt ever-more stringent safety procedures, there will always be the chance of mechanical failure or human error. As long as the world depends so heavily on oil for energy supplies, there is bound to be the occasional catastrophic spillage. Oddly enough, it is the rarity of disasters like Ixtoc

and *Torrey Canyon* that gets them into international headlines, especially because the pollution is immediately apparent. Environmentalists who call for the curtailment of drilling operations because of the likelihood of blowout perhaps do themselves and the rest of the community a disservice by diverting attention away from far more serious aspects of pollution. The amount of oil lost in blowouts and tanker accidents each year is literally a drop in the ocean. The *Amoco Cadiz* grounding was the biggest ever tanker disaster in terms of the amount of oil lost – nearly a quarter of a million tons. But each year something like 10 million tons of oil finds its way into the world ocean. It is washed into the sea by rivers after being deliberately dumped from coastal factories, or it is callously pumped overboard from tankers when they clean their tanks out. International laws are passed to stop these practices – but a look at many beaches shows that they continue.

This is insidious pollution, and it poses a far greater threat to the future of the oceans than dramatic accidents do. The sea is by and large an excellent dumping ground for domestic and industrial waste, which it disperses and consumes swiftly and efficiently. But this efficiency must not be taken for granted. It has its limit, and many scientists are now worried that the limit is being approached too quickly for comfort. Chemical waste poses the biggest threat. Industrial effluents are poured into the oceans, and artificial fertilizers and pesticides run off the land. Their effect is long-term, but this does not decrease the dangers: fish, particularly shellfish, have the capacity to absorb dangerous chemicals and, worse, to accumulate them so that they themselves eventually become poisonous. Equally, or perhaps more, threatening are nuclear fallout and the discharge of nuclear wastes. Nobody yet knows fully what effect these can have on the marine ecosystem.

There must be increasing vigilance to guard against drilling and shipping accidents. But insidious pollution presents the greatest danger of all.

315

An Ideal Future

As recently as the 1950s nobody had observed the greatest depths of the oceans. No satellite had orbited the Earth. The complex mathematical calculations now made on pocket-size machines would then have required a roomful of whirring electronic equipment. Such rapid and almost frightening changes make predictions dangerous.

So let us assume a straightforward advancement of technology when we look at the future of the oceans. Let us look 50 years into the future on the basis of an extrapolation rather than a prediction. We must first assume that the world has not been atomically devastated. We must assume that the world ocean has not been irreparably polluted. We must assume that disputes over ownership of oceanic resources have been settled between nations. It is only on such hopeful assumptions that any extrapolation can be made.

We will start at the shoreline of the 2030s. It is an age of greatly increased leisure and this is reflected in the development of large parts of the coast. There are marinas crammed with yachts, although they are probably powered by wind instead of gas, and many beaches are dotted with wind surfing boards.

Commercial development of underwater areas 50 years hence reflects a growing fascination with the world in which we live. Skin diving is still a major activity, but for those not fit enough or for those who merely prefer to take their pleasures in more comfort, there are many developments of the Japanese ideas of the 1960s for underwater parks. At the end of long piers are huge towers. People can enjoy the view and the sea air from the top, or else they can descend by elevator to the sea floor where they can have a meal while looking out through big windows at the thousands of fish of many species that swim lazily past. The passing parade of marine animals is attracted by powerful lights placed on the seabed and by food put out for them. Other, more ambitious projects feature entire underwater hotels, reached by submarine. Instead of waking to the sight of a Mediterranean sun reflecting off limpid blue waters, visitors will pull back the curtains to see a school of mullet regarding them with just as much curiosity as they are regarding the fish.

The school of mullet scatters as a submersible inches through them. Fuel cells are so cheap and reliable in the 2030s that a status symbol of the age is to have a personal underwater craft, perhaps even with sleeping and cooking space. A simple air lock means that people can transfer from the craft to underwater hotels and bars – perhaps even to the world's first underwater disco?

As we contemplate the underwater view, we are puzzled to see a scuba diver approaching. The puzzle-

Below: a view of the marina in Brighton, England, as it looks today. One of the big differences in marinas of the future could be that the yachts in them will be powered by wind, and wind surfing boards will dot the shoreline.

Above: a diver outside the habitat *Tektite* shows a big lobster to a colleague inside. In the future it might be possible for anyone interested, not just scientists, to watch and study marine life from underwater hotels.

Right: a diver in a one-person submersible used to move around underwater. Will such underwater transport be the private jets of the future?

ment is natural because we are 300 feet beneath the surface, which is outside the range of scuba diving. Our bewilderment turns to outright surprise when the diver gets closer. He is wearing a conventional wet suit and flippers and a simple mask to protect his eyes, but he has no backpack of air cylinders and no hoses. We find out that survival is assured because his lungs have been filled with an oxygen-saturated fluid, the oxygen content of which is passing through the lungs in the normal way. The diver is able to make a swift dive to 300 feet and straight back again without being concerned with decompression because the inert gas in any breathing mixture causes decompression problems, and there is no inert gas in this breathing mixture. This kind of skin diving is the logical conclusion of the experiments of the 1960s in which mice were attached to a machine that fed a warm, oxygen-saturated liquid into their lungs. They were pressurized to a depth equivalent to over 8000 feet and then decompressed very quickly – in fact to the equivalent of zooming to the surface at a speed of 2000 miles an hour – with seemingly no ill-effects. Now people are using the technique although it is not the most popular method. Few have actually taken kindly to the idea of having their lungs filled with liquid.

Some areas of the continental shelf are quite crowded with hotels, bars, submersible parks, and diving chambers. To get away from it all the rich have their own bathyscaphes in which they drop slowly to the bottom of the deep ocean. There the silence of thousands of years is broken by the clink of ice in cocktail glasses.

Fifty years on, the areas of calm water in the huge artificial lakes created by tidal power barrages are great favorites with anglers, yacht enthusiasts, and swimmers. Some parts are enclosed to protect the flatfish and shellfish that are being farmed there.

How has the space been found for all this recreation? In the 1970s there was intense pressure on the coastal zones of many nations from the often conflicting interests of industrial complexes, ports, and recreation.

The answer is to be found offshore, out of sight of land but still in the shallow waters of the continental

shelf. Huge artificial islands have been constructed, and much of the attendant ugliness of an industrialized society has been sited on these. The islands are simple if painstaking to build: a perimeter wall is built up by dumping huge rocks. Then waste material from coal tips or the fly ash from power stations is shipped out and dumped inside. Dredgers suck up sand and gravel from the seabed nearby and this is also used to help build up the land inside the restraining wall. The island has many uses. It has its own deep-water harbor at which huge tankers offload raw materials, including extremely precious oil. These are

Opposite: experiments with artificial gills for breathing underwater could lead to the day when people can go deep into the sea without any kind of backpack for air.

Right: an underwater park of the present allows people to observe marine life from a vessel deep in the ocean. The future promises bigger and more comfortable observation posts, including hotels.

Right below: an artist's impression of an industrial island, meant for possible development 45 miles off the Hook of Holland. The use of such islands would take some of the ugliness out of cities.

processed in factories and the finished products taken to ports on the mainland by smaller vessels. It is also a center for waste processing and disposal. Former oil tankers now carry domestic and chemical waste, scrap metal, glass, plastic, and such, from the shore. Much is recycled. Some is incinerated and the power generated by the giant furnaces helps fuel an air separation plant to produce nitrogen, oxygen, hydrogen, and other gases, as well as to make fresh water from the surrounding sea. There is also a nuclear power plant, sited many miles from shore. Power is piped from it to mainland cities by electricity cables buried beneath the seabed.

More power for the island is produced by huge aerogenerators strategically sited in the surrounding sea. Wave power devices bob up and down close to the island. They have the secondary benefit of damping down the worst effects of the storm waves that occasionally sweep toward the huge built-up island.

Other islands are used for airports, with high-speed trains taking passengers to the shore through tunnels beneath the seabed. Nearby, beneath the surface of the sea, are huge silos that store grain, fishmeal, oil, and other bulk materials. Submarine tankers, which have perhaps navigated under the North Pole icecap on their way from the Far East to Europe, moor up to these great containers and discharge their cargoes. Powerful pumps push the contents ashore by pipeline.

Go farther out to sea, passing a concrete oil production platform that has been preserved as a piece of industrial archaeology, and in the deep ocean manganese nodule mining is in full swing. Lumps of ore are being conveyed straight from the harvesting barge to tankers that bring them to processing plants on the artificial islands on the continental shelf. Off the coasts of tropical islands in deep water, ocean thermal energy conversion plants are at work, exploiting the differences in temperature between the surface and the deep water to produce electricity that is piped ashore by cables.

All over the world, ocean fishing boats are at work. The crisis brought about by the overexploitation of certain species 50 years before has come to an end. Quota systems have been introduced for all areas of the world ocean, and a huge number of species is caught. A worldwide, mutually agreed ban on fishing for fishmeal is in operation. Instead, any fish that is unpalatable is converted into a tasteless, odorless protein concentrate that can be added to any food or drink for direct human consumption.

The point about the concepts outlined in this glimpse of the future is that they are all technically feasible today. However, it will require more than technical and engineering skills for many of them to be put into operation. What will be required is a degree of international agreement, understanding, and goodwill that has not even been seen in the management of the land areas of this planet. The sea is huge and many of its mysteries wait to be unfolded to the benefit of all the people on Earth. There is still time for us to learn to use the power and resources of the oceans for lasting good. But the sea also holds the key to destruction if it is exploited indiscriminately for short-term gains and for territorial advantage. The choice we all have to make is very simple.

Index

Opposite: a diver – in this case in the Red Sea – looks like she is taking a casual walk on the seabed. With the light, relatively comfortable gear of today, many have taken up diving as a sport.

Picture Credits

Endpapers   Mary Evans Picture Library
Half Title   Intersub Limited
Title/Index pages   Seaphot
8   Zefa
10(T)   Courtesy ALCOA
10(B)   U.S. Geological Survey Photo Library
11(R)   Peter Sullivan © Aldus Books
12(T)   Dr. J. D. Taylor
12(B)   Solarfilma
13(L)   Woods Hole Oceanographic Institution
13(R)   Photomicrographs Institute of Oceanographic Science, Wormley, Surrey
14–15   © Geographical Projects Limited
16(T)   De Beers Consolidated Mines Ltd.
16(B)   British Museum/Michael Holford Library photo
17(T)   Leslie Salt Co.
17(B)   Photri
18   © Geographical Projects Limited
19(R)   © Aldus Books
20, 22(T)   Zefa
22(B)   Space Frontiers Ltd.
24(T)   Verner E. Suomi and Robert J. Parent, University of Wisconsin
24(B)   J. Allan Cash
25(L)   Sidney W. Woods © Aldus Books
25(R)   E. Silvester/Zefa
26(T)   Photri
26(B)   David Nash © Aldus Books
27(T)   Ostman Agency
27(B)   Mary Fisher/Colorific
28, 29(T)   BBC Hulton Picture Library
29(B)   Zefa
30(T)   Photri
30(B)   Colin Rivers/Seaphot
31(L)   Jon Kenfield/Seaphot
31(R)   Ralph Crane for *Life* © Time Inc.
32(T)   KLM Aerocarto b.v.
32(C)   Aerofilms Ltd.
32(B)   BBC Hulton Picture Library
33(L)   E. Winter/Zefa
33(R)   KLM Aerocarto b.v.
34(T)   Aldus Archives
34(B)   Victoria & Albert Museum/Michael Holford Library photo
35   International Tsunami Information Center
36(T)   © Aldus Books
36(B)   U.S. Navy photo
37   Compix, New York
38   Colour Library International
40(T)   British Museum/Michael Holford Library photo
40(B)   © Aldus Books
41   Reproduced by permission of the Trustees of the British Museum
42(T)   © Aldus Books
43(T)   Aschehoug & Co., Oslo
43(B)   Philippa Scott/Natural History Photographic Agency
44–45   © Geographical Projects Limited
46   Photri

47(T)   Offshore Environmental Systems Ltd.
47(B)   John D. Woods, Scientific Research/Seaphot
48(L)   Peter Sullivan © Aldus Books
48(R), 49(T)   Kevin Carver © Aldus Books after James E. McDonald, *The Coriolis Effect*, © May, 1952 by Scientific American, Inc. All rights reserved
50(T)   Aldus Archives
50(B)   Cooper-Bridgeman Library
51(T)   Photo M. Woodbridge Williams, Dickerson, Maryland
51(B)   © Geographical Projects Limited
52–53(T)   Kevin Carver © Aldus Books
52(B)   Popperfoto
53(C)   NASA
53(B)   Peter Sullivan © Aldus Books
54(L)   Michael Holford Library photo
54(R)   Spectrum Colour Library
55(T)   Mary Evans Picture Library
55(B)   Douglas Sneddon © Aldus Books
56(T)   Spectrum Colour Library
56(B)   Aerofilms Ltd.
57(TL)   W. F. Davidson/Zefa
57(TR)   N.B.A. (Controls) Limited
57(B)   National Film Board of Canada
58(T)   Kevin Carver © Aldus Books, after Morton and Miller, *The New Zealand Sea Shore*, Collins Publishers, London, 1968
58(B)   Aerofilms Limited
59   Starfoto/Zefa
60(T)   © Aldus Books
60(B)   Dr. D. James/Zefa
61   V. Stapelberg/Zefa
62   Aerofilms Limited
63(TL)   J. Allan Cash
63(TR)   Spectrum Colour Library
63(B)   Raymond Irons
64–65(B)   Kevin Carver © Aldus Books
65(T)   Gerolf Kalt/Zefa
66   Spectrum Colour Library
67(T)   Peter Sullivan © Aldus Books, from material supplied by Institute of Oceanographic Science, Wormley, Surrey
67(BL)   R. Mickleburgh/Ardea, London
67(BR)   Council of the Marine Biological Association of the United Kingdom
68   Heather Angel/Biophotos
69(T)   The Mansell Collection, London
69(B)   Photo John Webb © Aldus Books
71   Photo Michael Holford © Aldus Books
72, 73(T)   © Aldus Books
73(B)   Ann Ronan Picture Library
74(T)   Aldus Archives
74(B)   BBC Hulton Picture Library
75   National Maritime Museum, Greenwich/Michael Holford Library photo
76   Aldus Archives
77(TL)   Michael Holford Library photo
77(TC)   Aldus Archives
77(TR)   Mary Evans Picture Library
77(B)   Aldus Archives
78(TL)(B)   Michael Holford Library photos
79(T)   © Geographical Projects Limited
79(B)   Michael Holford Library photo
80(T)   Aldus Archives
80(B)   Courtesy Professor Sir Alister

Hardy, author of *Great Waters*
81(T)   Ann Ronan Picture Library
81(BL)   Musée Oceanographique de Monaco, Monte Carlo
81(BR)   Aldus Archives
83(TL)   Flip Schulke/Seaphot
83(TR)(B)   Decca Survey Ltd.
83(CR)   © Aldus Books
84   A. Defant, *Physical Oceanography*, Pergamon Press Limited
85(T)   Seaphot
85(B), 86   Offshore Environmental Systems Ltd.
87(T)   Space Frontiers Ltd.
87(BR)   E.M.I. Offshore Systems
88(L)   Peter Scoones/Seaphot
88(R)   Underwater & Marine Equipment Ltd., Farnborough, Hants.
89(T)   Sonarmarine Ltd./Photos John Cadd
89(B)   E.M.I. Electronics Ltd.
90(T)   Aldus Archives
90(B)   Kelvin Hughes, Ilford, Essex
91(TL)   © Aldus Books
91(TR)   Klein Associates, Inc.
91(B)   Offshore Environmental Systems Ltd.
92–93(T)   Klein Associates, Inc.
92(B)   Horizon Exploration Ltd.
93(BL)   © Aldus Books
93(BR)   Decca Survey Ltd.
94   NASA
95   © Geographical Projects Limited
96(T)   Courtesy Dr. R. W. Girdler and The Royal Society
96(B)   Edward Poulton © Aldus Books
97(T)   Mats Wibe Lund, Jr.
97(B)   © Aldus Books
98(T)   after *Scientific American*, September, 1969, © Scientific American, Inc. All rights reserved
98(C)   Lamont-Doherty Geological Observatory
98(B), 99   © Aldus Books
100(T)   © Geographical Projects Limited
100(B)   NASA
101(T)   Sidney W. Woods © Aldus Books
102(TL)   after William Bascom, *Technology and the Ocean*, © September, 1969 by Scientific American, Inc. All rights reserved
102(TR)   Global Marine Inc.
102(B)   Institute of Oceanographic Science, Wormley, Surrey
103(T)   Lamont-Doherty Geological Observatory
103(B)   DSDP photo by Scripps Institution of Oceanography
104(T)   Alan Hollingbery © Aldus Books
104(BL)   Aldus Archives
104(BR)   Minnesota Historical Society
105(T)   Photo John Steele
105(B)   Icelandic Photo & Press Service
106   Hervé Chaumeton/Photos Scientifiques
109   after *Scientific American* © Scientific American, Inc., September, 1969. All rights reserved
110(T)   John Lythgoe/Seaphot
110(B)   Dr. G. F. Leedale/Biophoto Associates

111(TR)   J. D. Guiterman/Seaphot
111(BR)   G. T. Boalch/Seaphot
112(T)   Oxford Scientific Films
112(B), 113(T)   Douglas P. Wilson
113(B)   Oxford Scientific Films
114(T)   Douglas P. Wilson
114(B), 115   Hervé Chaumeton/Photos
   Scientifiques
116(T)   Heather Angel/Biophotos
116(B)   Dick Clarke/Seaphot
117(T)   Peter David/Seaphot
117(B)   Rudolph Britto © Aldus Books,
   after F. S. Russell and C. M. Yonge,
   The Seas, Frederick Warne & Co., Ltd.,
   London
118   Heather Angel/Biofotos
119(T)   Spectrum Colour Library
119(B)   Heather Angel/Biofotos/Ian Took
120–122(T)   Hervé Chaumeton/Photos
   Scientifiques
122(B)   Heather Angel/Biofotos
123(T)(C)   Dr. G. F. Leedale/Biophoto
   Associates
123(B)   Mary Evans Picture Library
124   Peter David/Seaphot
125(T)   Ann Ronan Picture Library
125(C)(B)   Hervé Chaumeton/Photos
   Scientifiques
126(T)   Dr. G. F. Leedale/Biophoto
   Associates
126(B), 127(T)   Hervé Chaumeton/Photos
   Scientifiques
127(B)   P. M. David
128(T)   Jane Burton/Bruce Coleman Ltd.
128(B)   Hervé Chaumeton/Photos
   Scientifiques
129(T)   Dr. G. F. Leedale/Biophoto
   Associates
129(CR)   Hervé Chaumeton/Photos
   Scientifiques
129(B)   Heather Angel/Biofotos
130   Ardea, London
132(T)   Hervé Chaumeton/Photos
   Scientifiques
132(B), 133   Heather Angel/Biofotos
134, 135(T)   Ron and Valerie Taylor/
   Ardea, London
135(B)   Pix, Sydney/Camera Press,
   London
136(T)   Geoff Harwood/Seaphot
136(B)   Hervé Chaumeton/Photos
   Scientifiques
137(TL)   Tom McHugh/Photo
   Researchers, Inc.
137(TR)   Dr. J. David George, A.R.P.S.
137(B)   Douglas P. Wilson
138–139(T)   Hervé Chaumeton/Photos
   Scientifiques
138(B)   Douglas Sneddon © Aldus Books
139(C)   Jesus Navarro Perez/Seaphot
139(B)   Heather Angel/Biofotos
140   Hervé Chaumeton/Photos
   Scientifiques
141(T)(B)   Heather Angel/Biofotos
141(C)   Hervé Chaumeton/Photos
Scientifiques
142(T)   Oxford Scientific Films
142(B)   Peter Scoones/Seaphot
143(T)   Allan Power/Bruce Coleman Ltd.
143(B)   Jane Burton/Bruce Coleman Ltd.
144(T)   © Aldus Books

144(B)   Michael Holford Library photo
145(T)   A. Husmo/Mittet Foto
145(B)   Popperfoto
146–147(T)   Allan Power/Bruce Coleman
   Inc.
146(B)   BBC Hulton Picture Library
147(BL)   Seaphot
147(BR)   Dick Clarke/Seaphot
148–149(T)   Jeff Foott/Bruce Coleman
   Ltd.
148(B)   Heather Angel/Biofotos
149(B)   Solarfilma
150(T)   Heather Angel/Biofotos
150(BL)   © Aldus Books
150(BR)   Keith Gunnar/Bruce Coleman
   Ltd.
151(TL)   Jane Burton/Bruce Coleman
   Ltd.
151(TR)   © Aldus Books
151(B)   Heather Angel/Biofotos
152–153   © Geographical Projects Limited
154(T)   Peter David/Seaphot
154(B)   Popperfoto
155(T)   David Nash © Aldus Books
155(B)   Hervé Chaumeton/Photos
   Scientifiques
156(T)(C), 157(T)   Peter David/Seaphot
156–157(B)   Oxford Scientific Films
158(T)   Hervé Chaumeton/Photos
   Scientifiques
158(B)   Heather Angel/Biofotos
159(T)   Naval Ocean Systems Center,
   San Diego, California
159(B)   U.S. Navy Photo
160(T)   Kevin Carver © Aldus Books,
   after N. J. Berrill, *The Life of the Ocean*
   © 1966 by McGraw-Hill Inc. Used with
   permission of McGraw-Hill Book
   Company, New York
160(B)   Kenneth W. Fink/Ardea, London
161, 162(T)   Jen and Des Bartlett/Bruce
   Coleman Ltd.
162(B)   Heather Angel/Biofotos
163(T)   Dr. G. F. Leedale/Biophoto
   Associates
163(B)   Sidney W. Woods © Aldus Books
164   Erwin Christian/Zefa
166(T)   Photri
166(B)   © Geographical Projects Limited
167   Picturepoint, London
168(T)   FAO photo
168(B), 169(T)   Biophoto Associates
169(B)   © Aldus Books
170   Cooper-Bridgeman Library
171(T)   Michael Francis Wood
171(B)   Raymond Irons
172(T)   Picturepoint, London
172(B), 173(L)   Michael Francis Wood
173(R)   Raymond Irons
174–175(T)   Arthur B. Bowbeer ©
   Aldus Books
174–175(B)   W. Nygaard/O. Mustad &
   Son A/S
176   Michael Francis Wood
177   © Aldus Books
178   A. Husmo/Mittet Foto
179   Novosti Press Agency
180   A. Husmo/Mittet Fdto
181(L)   Bob Croxford/Zefa
181(R)   British Crown Copyright
   reserved. Reproduced by permission of

the Controller of Her Britannic Majesty's
   Stationery Office
182–183   © Geographical Projects Limited
184(T)   Zefa
184(B)   Heather Angel/Biofotos
185   Associated Press
186(L)   Janoud/Zefa
186(R)   John Tyler © Aldus Books, after
   ed. Rhodes Fairbridge, *Encyclopaedia
   of Oceanography*, University of
   Wisconsin Press, Madison, 1966
187   FAO photo
188(T)   Michael Francis Wood
188(B)   Photri
189(L)   Picturepoint, London
189(R)   Photo Michael Wood
190   Colour Library International
191(T)   Douglas P. Wilson
191(CR)   Heather Angel/Biofotos
192(T)   © Aldus Books
192(B)   Tom Hanley
193(L)   Ray Dean
193(R)   Photri
194(L)   Kennecott Exploration, Inc.
194(R)   Deepsea Ventures Inc.
195, 196(T)   © Aldus Books
196(B)   Ocean Minerals Company
197(L)   Royal Bos Kalis Westminster
   Group NV
197(R)   © Aldus Books
198–199   Royal Bos Kalis Westminster
   Group NV
200, 202   Colour Library International
203   Lockheed California Company
204(TL)   Harwell Design Studio/*Daily
   Telegraph* Colour Library
204(TR)(B)   © Aldus Books
205(T)   National Engineering Laboratory
205(B), 206(T)   © Aldus Books
206(B)   Colour Library International
207(R)   Novosti Press Agency
208   J. Pfaff/Zefa
209(L)   Camera Press, London
209(R)   J. Allan Cash
210(T)   © Aldus Books
210(B), 211(T)   Lockheed Missiles and
   Space Co. Inc.
211(B), 212(T)   © Aldus Books
212(B)   Zefa
213   Lockheed Corporation
214(T)   U.S. Department of Energy
214(B)   G. Kalt/Zefa
215(T)   Courtesy Taylor Woodrow/
   C.E.G.B./Department of Energy
215(B)   Mike Willoughby
216   Brian Coope
218   Zefa
219(T)   © Aldus Books
219(B)   Middle East Photographic Archive
220–221   © Geographical Projects
   Limited
222   © Aldus Books
223   © Geographical Projects Limited
224–225   Western Geophysical Company
226–227   Gulf Oil Corporation
228   © Aldus Books
229   Photri
230(L)   Colour Library International
230(R)   Fay Godwin
231   Brian Coope
232–233   Shell photos

234 Colour Library International
235 British National Oil Corporation
236(T) A Shell photo
236(B) British National Oil Corporation
237(T) A Shell photo
237(B) British National Oil Corporation
238 Heerema Engineering Services (U.K.) Ltd./Photo Niki
239(T) Mobil Oil Corporation
239(B) © Aldus Books
240, 241(T) Canocean Resources Ltd. (formerly Lockheed Petroleum Services)
241(B) © 1979 Society of Petroleum Engineers. Presented at Offshore Europe 1979 Conference in Aberdeen, U.K., September, 1979
243(T) BOC SubOcean Services/Photo John Cadd
243(B) Brian Coope
244(L) Seaphot
244(R) Brian Coope
245 Popperfoto
246–247 © Geographical Projects Limited
248 Seaforth Maritime
249 Fay Godwin
250(T) Brian Coope
250(B) Peter Scoones/Seaphot
251, 252(T) Fay Godwin
252(B) Brian Coope
253 Hereema Engineering Service
254 Peter Scoones/Seaphot
256, 257(L) Warren Williams/Seaphot
257(T) Michael Holford Library photo
257(BR), 258 © Aldus Books
259 Seaphot
260 Warren Williams/Seaphot
261(T) after information supplied by Westinghouse Underseas Division, East Pittsburgh, Pennsylvania
261(B) Seaphot
262(T) Comex
262(B) © Aldus Books
263 Oceaneering
264(T) Subsea International/Ocean

Inchcape Ltd. Photo John Cadd
264(B) Oceaneering
265 Brian Coope
266(L) Comex
266(R) Brian Coope/Seaphot
267(T) Subsea International/Ocean Inchcape Ltd. Photo John Cadd
267(B) Comex
268–269(T) J. David George/Seaphot
268(B), 269(CL) Seaphot
269(BR) U.S. Navy photo
270 Jean Deas/Seaphot
271 Flip Schulke/Seaphot
272(T) D. Clarke/Seaphot
272(B) Jon Kenfield/Seaphot
273(T) Warren Williams/Seaphot
273(B) Seaphot
274 J. W. Automarine
275(T) Seaphot
275(B) Subsea International/Ocean Inchcape Ltd. Photo John Cadd
276 Warren Williams/Seaphot
277(T) U.S. Navy photo
277(B), 278 Seaphot
280(T) The Mansell Collection, London
280(BL) Bibliotheque Royale de Belgique, Bruxelles
280(BR) © Aldus Books
281(L) Maurice Jarnoux/Stern
281(R) Brian Coope
282(L) Courtesy of Mike Borrow
282(R), 283(T) Admiralty Experimental Diving Unit, H.M.S. Vernon
283(B) Courtesy of Mike Borrow
284 Brian Coope
285, 286 Oceaneering
287(T) Courtesy Mike Humphrey, Underwater & Marine Equipment Ltd.
287(B) Admiralty Experimental Diving Unit, H.M.S. Vernon
288 Oceaneering
289 Warton Williams Ltd.
290 © Aldus Books
291(L) Brown Brothers

291(R) © Aldus Books
292, 293 U.S. Navy photos
294 Flip Schulke/Seaphot
295 U.S. Navy photos
296 Seaphot
297(T) Dick Clarke/Seaphot
297(B) Flip Schulke/Seaphot
298(T) Perry Oceanographics
298(B), 299(T) Intersub Limited
299(B) Brian Coope
300(T) Flip Schulke/Seaphot
300(B) Oceaneering
301(T) U.D.I. Group Ltd./Photo John Cadd
301(B) Seaphot
302 Zefa
304 Kennecott Exploration, Inc.
305(T) National Maritime Museum, Greenwich
305(B) Mittet Foto A/S
306–307(T) Robert Harding Associates
306(B) Reproduced from a Bow Group pamphlet, Ocean Space: Europe's New Frontier
307(BL) © Aldus Books
307(BR) Freeport Sulphur Company
308(T) British Crown Copyright
308(B) © Aldus Books
309 Popperfoto
310(L) Robert Harding Associates
311 © Aldus Books
312, 313 Martech International Inc.
314(T) Jeff Foott/Bruce Coleman Inc.
314(B) J. Allan Cash
315(C)(R) Photri
316 Ajax News and Features Service
317(T) Flip Schulke/Seaphot
317(B) Dick Clarke/Seaphot
318 General Electric Research and Development Center, Schenectady
319(T) Flip Schulke/Seaphot
319(B) Bos Kalis
320 Seaphot